Chemistry
for
University Students

Volume 2

H. C. Ananda Murthy M.Sc., M.Phil., Ph.D.
Assistant Professor
Department of Chemistry,
School of Physical Sciences,
College of Natural & Mathematical Sciences,
The University of Dodoma.

TANZANIA EDUCATIONAL PUBLISHERS LTD

Tanzania Educational Publishers Ltd,
TEPU House,
Uganda road, Plot No. 45 Block MDA,
Mob:0685 997583/0758 147871
Email: tepultd@yahoo.com
Website: www.tepu.co.tz
P.O. Box 1222,
Bukoba, Tanzania.

ISBN 978 9987 07 054 1

I dedicate this book to my wife and children

CONTENTS

ABOUT THE AUTHOR

Dr. H. C. Ananda Murthy

H. C. Ananda Murthy has been a sincere, committed and dedicated faculty member at various prestigious universities in India and Tanzania for the last 19 years. He is currently working as Assistant Professor, Department of Chemistry, The University of Dodoma, Tanzania, East Africa.

He completed his B.Sc. from Bangalore University, India in 1995. Later He completed his master's degree in Chemistry from Bangalore University, India in 1997. He also obtained M.Phil. degree in chemistry from Bharathidasan University, India in the year 2006. Then he went on to complete his Ph.D., for his work on corrosion behavior of aluminium composite materials from Visvesvaraya Technological University, Belgaum, India in the year 2012.

He has authored few books, compendia and published several research articles in the journals of international repute. He has taught General chemistry, Coordination chemistry and Organometallic Chemistry, Transition Elements Chemistry, Organometallic Chemistryand Electrochemistry & Corrosion protection, Advanced Coordination Chemistry, Surface Chemistry and Corrosion Science, Environmental Chemistry, Engineering chemistry and Advanced spectroscopyto the university students in India and Tanzania.

His research interest mainly includes Corrosion Science, Composite materials and Materials chemistry. He also presented many papers in national as well as international conferences. He has successfully completed a project titled "Amphibious bicycle" sanctioned by National Innovation Foundation of India (NIFI).

ACKNOWLEDGEMENTS

I express my profound gratitude and regards to the management of the University of Dodoma for providing me this opportunity to serve at CNMS, UDOM.

It is a great pleasure to express my gratitude to the Head, Department of Chemistry, the Dean, School of Physical Sciences and The Principal, CNMS for their constant support and encouragement in bringing out this book. I also place on record my appreciation to fellow colleagues at CNMS, UDOM for their cooperation. Finally, I thank my family members and friends for their constant support towards this work.

INTRODUCTION

Chemistry for University Students is one book published in three volumes made up of chapters as follows:

Volume 1

Chapter 1: Electronic Structure of Atoms

Chapter 2: Periodic Table

Chapter 3: Chemical Bonding

Chapter 4: Transition Elements Chemistry

Chapter 5: Inner Transition Elements Chemistry

Volume 2

Chapter 6: Coordination Chemistry

Chapter 7: Electronic Spectra of Coordination Compounds

Chapter 8: Organometalic Chemistry

Volume 3

Chapter 9: Electrochemistry

Chapter 10: Environmental Chemitry

Chemistry has great contribution to the world. Chemistry provides thousands of raw materials to get finished products that help keep you safe, warm, cool, on time, in motion, and connected. From heart monitors to satellites to cell phones to microwave ovens to synthetic fibers, packed food items to medicine chemistry makes the things that make modern life possible.

Today's world would be inconceivable without the developments that chemistry and its applications have brought to science, engineering and technology. The achievements of chemistry are omnipresent.

Students at universities definitely require chemical knowledge for their better understanding and academic performance. It is understood that a chemistry book comprising many aspects of chemistry is greatly needed by the students of the universities.

This book has been written keeping in mind the requirements of chemistry students at University level comprising of General chemistry, Periodic Table of elements, *d* and *f* block elements, Coordination chemistry, organometallic chemistry, Electrochemistry and Environmental Chemistry (All volumes).

I wish my book will be helpful to the students, as it is prepared to cater to the needs of the students, keeping in mind the relevance of each contents of the U G courses toward better understanding of the subject and with illustration including suitable figures, tables, and examples of structures and configurations of the complexes with simple presentation formats.

Chapter 1

COORDINATION CHEMISTRY

1.1 INTRODUCTION

Coordination compounds play vital role in biological systems as well as in Industrial processes. Photosynthesis in plants involves chlorophyll, which is a coordination complex of magnesium. Haem group of Hemoglobin that acts as oxygen carrier contains iron. A very large number of coordination compounds play the role of catalysts in synthetic chemistry. A few complexes of transition metals are used in medicine. In general coordination compounds find use in organic compound synthesis, polymerization, petrochemicals and catalysis.

The transition metals and their ions have much higher tendency to from coordination compounds as compared to the s- and p- block elements. It is because of their relatively smaller sizes, higher ionic charges and the availability of d orbitals for bond formation. Coordination compounds, unlike normal compounds, retain their identity even when dissolved in water or any other suitable solvent. The properties of these compounds are totally different from those of their constituents.

Alfred Werner can be considered to be the pioneer in the field of coordination chemistry and he received Nobel Prize in 1913 in recognition of his efforts. He showed that neutral molecules were bound directly to the metal so that complex salt such as $CoCl_3.6NH_3$ is correctly formulated as $[Co(NH_3)_6]^{3+}(Cl^-)_3$. G.N. Lewis and N.V. Sidgwick proposed that a chemical bond required the sharing of an electron pair. This led to the idea that a neutral molecule with an electron pair can donate these electrons to a metal ion or other electron acceptor.

Coordination Compound [Complex Compound]

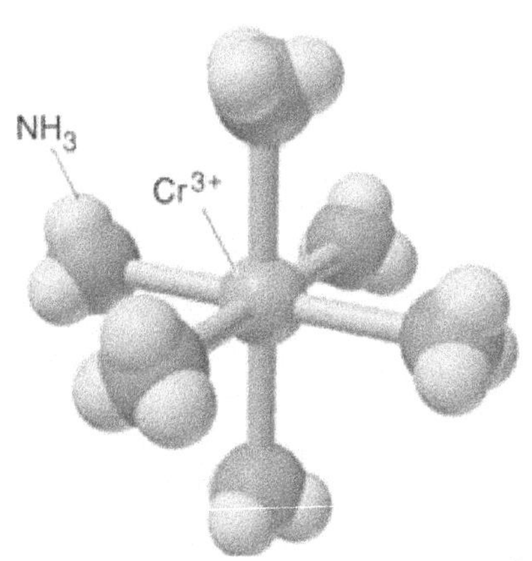

Fig. 49. Structure of $[Cr(NH_3)_6]^{3+}$

Coordinate compound is a compound in which a metal atom or ion is coordinated to two or more neutral molecules or ions. Coordinate compound retains its identity in the solid as well as in solution. A coordinate compound may be molecular or ionic. For example, hexamminechromium(III) chloride $[Cr(NH_3)_6]^{3+}$ as shown in Fig. 49.

$$\underset{\text{Neutral Coordination Compound}}{[Cr(NH_3)_6]Cl_3} \xrightarrow{H_2O} \underset{\text{Complex ion}}{[Cr(NH_3)_6]^{3+}} + \underset{\text{Simple ion}}{3Cl^-}$$

Other examples of coordination compounds include $[Cu(NH_3)_4]SO_4$, $K_4[Fe(CN)_6]$, $Ni(CO)_4$, $Fe(CO)_5$, $[Co(NH_3)_6]Cl_3$ etc.,

Inorganic salts are of three types:

1. Simple Salts: Simple salts contain simple cations and anions.

Example: Sodium chloride $[Na^+Cl^-]$, $AlCl_3$, KCl, $BeCl_2$, $CaSO_4$, $CuSO_4$ etc.

2. Double salts: [Addition compounds]

Double salts are molecular compounds made up of two simple salts. They exist only in crystal lattices but breakdown into their constituent compounds when dissolved in water.

Example: Mohr's salt - $FeSO_4\ (NH_4)_2SO_4.\ 6H_2O$

Mohr's salt, when dissolved in water gives Fe^{2+} NH_4^+ and SO_4^{2-} ions in solution. These ions can be tested positively in solution. Double salts lose their identity in aqueous solutions.

3. Complex salts: [Co-ordination compounds]

Complex salts contain complex ion and simple ions. Complex salts retain their identity even when dissolved in water. Complex ion does not give any test of Cu^{2+} ion.

$$[Cu(NH_3)_4]SO_4 \longrightarrow [Cu(NH_3)_4]^{2+} + SO_4^{2-}$$

Thus, in a coordination compound, the metal species acts as electron acceptor (**Lewis** acid) and neutral motecule with lone pair of electrons or anion as electron donor (**Lewis** bases). A metal atom or ion when bonded directly to a fixed number of anions or molecules constitutes a **coordination entity**. For example, $[Pt(NH_3)_2Cl_2]$ is a coordination entity in which platinum(II) is surrounded by two ammonia molecules and two chloride ions. Other examples are $[Fe(CN)_6]^{4-}$, $[Co(NH_3)_5Cl]^{2+}$, $[Ni(CO)_4]$ etc. In a coordination entity, the atom/ion to which a fixed number of ions/ groups are bound in a definite geometrical arrangement around it, is called the **central atom or ion.** For example, the central atom in $[Ni(NH_3)_6]^{2+}$ is Ni(II), in $[Mo(CN)_8]^{3-}$, Mo(V), and in $[Co(PR_3)_3Cl]$, Co(I). The charged ions or neutral motecules bound to the central atom in the coordination entitiy are called **ligands**.

For example, PR_3 and Cl^- are ligands in $[CoCl(PR_3)_3]$. The number of ligand donor atoms directly bonded to the central atom is defined as the **coordination number.** For example, in the complex species $[Mo(CN)_8]^{3-}$ and $[CoCl\ (PR_3)_3]$, the coordination number of Mo and Co are 8 and 4, respectively. The central atom and the ligands bonded to it are enclosed in square bracket and is collectively termed as the **coordination sphere.** For example in the complex $[Ni(NH_3)_6]Cl_2$, the coordination sphere is

$[Ni(NH_3)_6]^{2+}$. The spatial arrangement of the ligands around the central atom is termed as **coordination polyhedron**. The most common coordination polyhedra are octahedral, tetrahedral and square planar.

Co-ordination Compounds: Important Terms

1. Complex ion:

An electrically charged species in which a metal atom or a simple cation is coordinated to two or more anions or neutral molecules and which retain its identity in solution is called a complex ion. For example, $[Fe(CN)_6]^{4-}$ (ferrocyanide) is a complex ion in which the Fe^{2+} (ferrous) ion is surrounded by six CN^- (cyanide ions). Similarly, $[Cu(NH_3)_4]^{2+}$ is a complex ion in which Cu^{2+} ion is surrounded by four NH_3 molecules.

A complex ion may have positive or negative charge on it. A positively charged complex ion is called cationic complex ion, while the negatively charged complex ion is called anionic complex ion. For example,

Cationic complex ions: $[Cu(NH_3)_4]^{2+}$, $[Ag(NH_3)_2]^+$, $[Ni(NH_3)_6]^{2+}$ etc.

Anionic complex ions: $[Fe(CN)_6]^{3-}$, $[Fe(CN)_6]^{4-}$, $[CuCl_4]^{2-}$etc.

2. Central atom or ion

The metal atoms or ion to which two or more anions or neutral molecules are attached is called central atom or central ion. For example, in the complex ion $[Fe(CN)_6]^{4-}$, Fe^{2+} is the central ion.

3. Ligands

Ligands are atoms or ions or molecules which donate one or more pair of electrons to central metal atom or ion to form co-ordinate bonds.

For example, CN^-, OH^-, H_2O, NH_3, F^-, Cl^- etc.

Six CN^- ligands are linked to Fe^{3+} ion in ferrocyanide $[Fe(CN)_6]^{4-}$ complex.

Ligands are of two kinds:

(i) **σ-donar ligands**: These ligands donate electrons, via σ-bond, to the metal during complex formation.

Example, Cl^-, CN^-, CO, NH_3 etc:

(ii) **π-donar ligands**: These ligands donate the delocalized electrons, via π-bond, to the metal during complex formation.

Example, Alkene, Cp-C_5H_5, C_6H_6, C_7H_7

One way of classifying is structurally, i.e., by the number of connections they make to the central atom where only one atom becomes closely bonded, the ligand is said to be unidentate. When a ligand becomes attached by two or more atoms, it is a bidentate, tridentate and so on generally multidentate.
Bidentate ligands when bound entirely to one atom are called chelate.

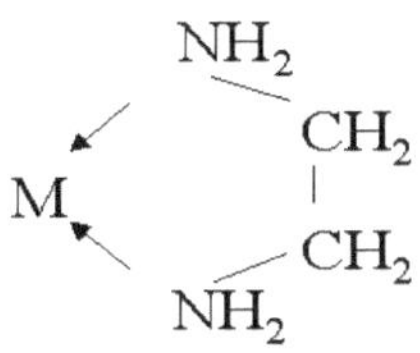

ethylenediamine

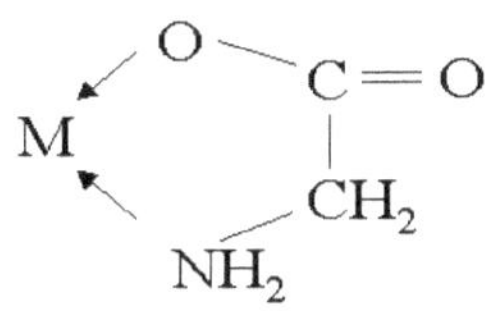

glycinato

In many cases, ligands serve as unidentate bridging ligand i.e., there is only one ligand atom that forms two or three bonds at two different metal atoms.

Example; NH_2, CO, O_2^{2-} etc.

4. Coordination Number

The total number of ligands attached directly to the central atom or ion in a complex is called the co-ordination number of that metal atom or ion.

For example in $[Fe(CN)_6]^{4-}$, the coordination number is 6. Each bidentate ligand contributes two to the coordination number. Similarly tridentate ligand contributes three and so on.

Coordination number is ***specific*** for a given metal ion in a particular oxidation state and compound. The most common coordination number in complex ions is 6, but 2 and 4 are often seen.

Coordination Number	Shape		Examples
2	Linear		$[CuCl_2]^-$, $[Ag(NH_3)_2]^+$, $[AuCl_2]^-$
4	Square planar	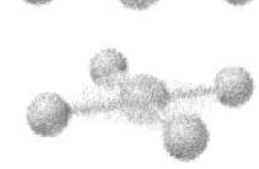	$[Ni(CN)_4]^{2-}$, $[PdCl_4]^{2-}$, $[Pt(NH_3)_4]^{2+}$, $[Cu(NH_3)_4]^{2+}$
4	Tetrahedral		$[Cu(CN)_4]^{3-}$, $[Zn(NH_3)_4]^{2+}$, $[CdCl_4]^{2-}$. $[MnCl_4]^{2-}$
6	Octahedral	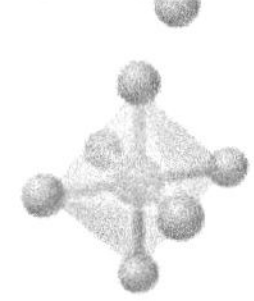	$[Ti(H_2O)_6]^{3+}$, $[V(CN)_6]^{4-}$, $[Cr(NH_3)_4Cl_2]^+$, $[Mn((H_2O_6]^{2+}$, $[FeCl_6]^{3-}$, $[Co(en)_3]^{3+}$

5. Coordination Sphere

The molecular or ionic species attached directly to the central ion constitute the coordination sphere of the central atom or ion.

$$\underbrace{[Co(NH_3)_6]}_{\text{Co-ordination sphere}}Cl_3$$

6. Oxidation state of the metal ion

It is determined by using following relation

$$\text{Oxidation state of the metal ion} = \text{Charge on the complex ion} - \text{Number of charges denoted by ligands}$$

For example, in case of $[Fe(CN)_6]^{4-}$, Oxidation state of Iron = -4-(-6) = +2

1.2 WERNER'S THEORY OF COORDINATION COMPOUNDS

Alfred Werner was awarded Nobel prize in chemistry in the year 1913 in recognition of his work on co-ordination chemistry. Alfred Werner is considered as father of coordination chemistry.

Nobel Prize in Chemistry- 1913 "in recognition of his work on the linkage of atoms in molecules by which he has thrown new light on earlier investigations and opened up new fields of research especially in inorganic chemistry."

Alfred Werner

In 1893, Alfred Werner proposed a theory to explain the existence of complexes such as $CoCl_3{\bullet}6NH_3$, $CoCl_3{\bullet}4NH_3$, $PtCl_2{\bullet}2NH_3$ and $Fe(CN)_3{\bullet}3KCN$. He began by assuming that a metal ion has two kinds of valence. The first, the *primary* valence, is satisfied by negative groups that balance the charge on the metal. For example, in $CoCl_3{\bullet}6NH_3$ the 3 valence of cobalt is satisfied by three Cl^- ions. A *secondary* valence is used to bind other groups that are usually specific in number.

Postulates of Werner's Theory

1. Metal shows two different types of valency in complexes. These are

a) Primary valency or Ionisable valency

b) Secondary valency or Non-ionisable valency

2. Primary Valency [Oxidation number]

The primary valency is the number of charges on the complex ion. Primary valency is variable. These are non-directional.

Ex: The primary valencies of Iron in $K_3[Fe(CN)_6]$ and $K_4[Fe(CN)_6]$ are 3+ and 2+.
The primary valencies are satisfied only by Negative ions.

3. Secondary Valency [Coordination number]

The number of secondary valencies is equal to the number of ligands coordinated to the metal. Secondary valency has a fixed value for a given metal. These are directional.

Ex: Secondary valency of Fe in $K_3[Fe(CN)_6]$ and $K_4[Fe(CN)_6]$ is 6.
Secondary valencies are satisfied by either Neutral ligands or Negative ligands.

4. Ligands

Ligands are bound to the metal by donating a pair of electrons to the metal ion. These ligands are directed in space in particular directions with the metal ion at the centre. This leads to specific structures for complexes depending on coordination number. Werner deduced that in $CoCl_3$. $6NH_3$, the three chlorines acted as primary valencies and the six ammonias as secondary valencies. In modern terms the complex is written as $[Co(NH_3)_6]Cl_3$.

The three chlorines are ionic and the six ammonia ligands form co-ordinate bonds to Co^{3+} forming a complex ion $[Co(NH_3)_6]^{3+}$. Structures of (a) $[Co(NH_3)_6]Cl_3$ and (b) $[Co(NH_3)_3]Cl_3$

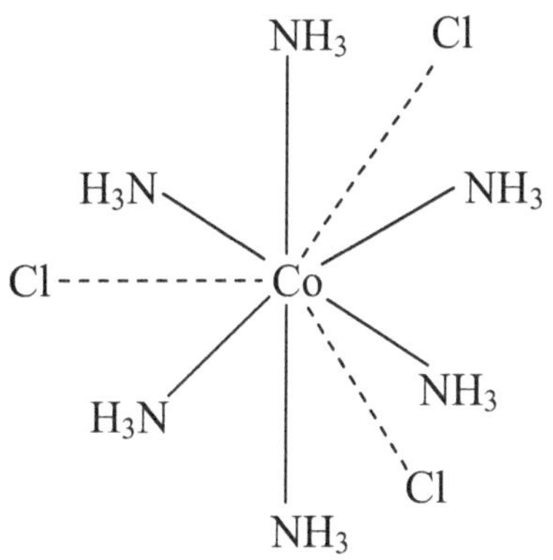

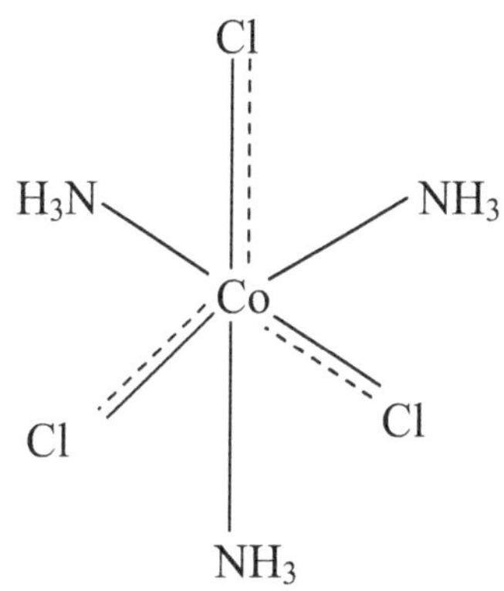

Alfred Werner was first to explain the existence and properties of a variety of cobalt (III) chloride compounds with ammonia:

Compound	Formula	Color
1	$CoCl_3 \bullet 6NH_3$	orange-yellow
2	$CoCl_3 \bullet 5NH_3$	purple
3	$CoCl_3 \bullet 4NH_3$	green
4	$CoCl_3 \bullet 3NH_3$	green

In addition:

When an aqueous solution of HCl is added to either compounds 1, 2, 3 or 4, NH_3 is not removed. Compound 1, when treated with an aqueous solution of $Ag(NO_3)$ precipitates all the chloride as AgCl. Compound 2, when treated with an aqueous solution of $Ag(NO_3)$ precipitates only 2/3 of the Cl. Compound 3, when treated with an aqueous solution of $Ag(NO_3)$ precipitates only 1/3 of the Cl. Compound 4, when treated with an aqueous solution of $Ag(NO_3)$ precipitates no AgCl.

1.3 TYPES OF LIGANDS

Types of ligands: The ligands can be classified structurally as follows:

1. **Unidentate ligands**-These are ligands which bonds to the central metal atom through a single atom, e.g. ammonia (NH_3) which bonds to the central metal atom through nitrogen atom. Other examples are Cl^-, CN^-, H_2O, etc.

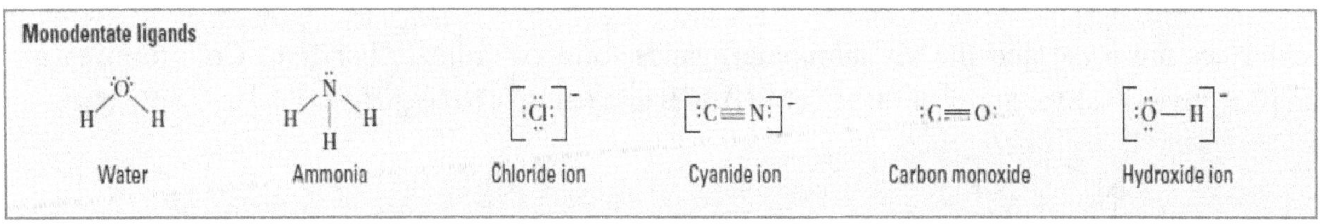

2. **Bidentate ligands**- These are ligands which link to the central metal atom through two atoms, e.g. carbonate ion (CO_3^{2-}), oxalate ($C_2O_4^{2-}$), ethylenediamine (en), acetylacetonato (acac-), ortho-phenanthroline etc.

3. Polydentate ligands- These include all types of ligands with denticity higher than two. The various types of ligands included are:

   i) tridentate (denticity of 3), e.g. diethylenetriamine (dien), terpyridine (terpy) etc.

   $H_2N\text{-}CH_2\text{-}CH_2\text{-}NH\text{-}CH_2\text{-}CH_2\text{-}NH_2$

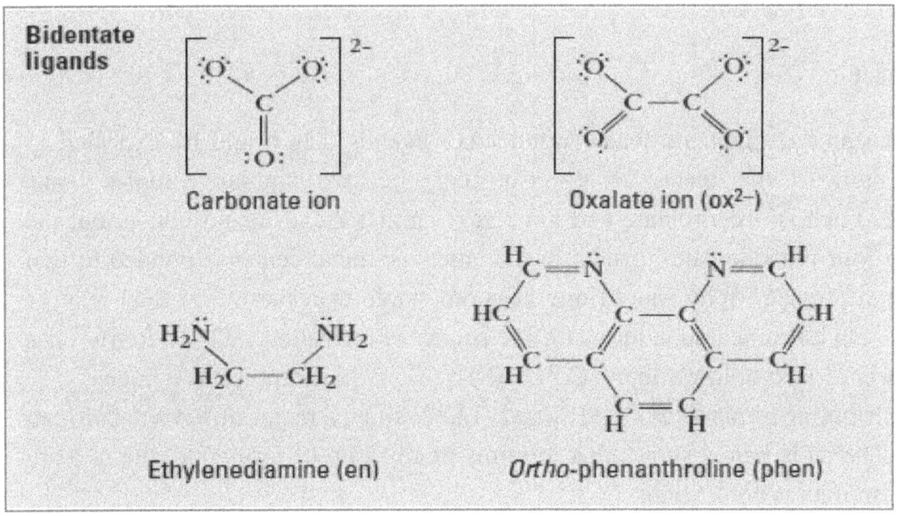

ii) tetradentate (denticity of 4), e.g. triethylenetetramine (triene),

H_2N-CH_2- CH_2-NH- CH_2- CH_2-NH-CH_2- CH_2-NH_2 (trien)

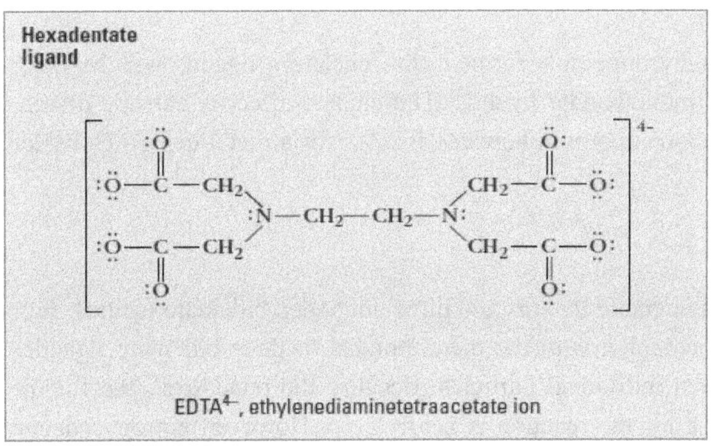

Here coordination to the metal atom can take place through the four nitrogen atoms.

iii) pentadentate ligand (denticity of 5), e.g. ethylenediaminetriacetato,

iv) hexadentate ligand (denticity of 6) , e.g. ethylenediaminetetraacetato ($EDTA^{4-}$). Here coordination to the metal atom can take place through the two nitrogen atoms and four oxygen atoms of the acetate group.

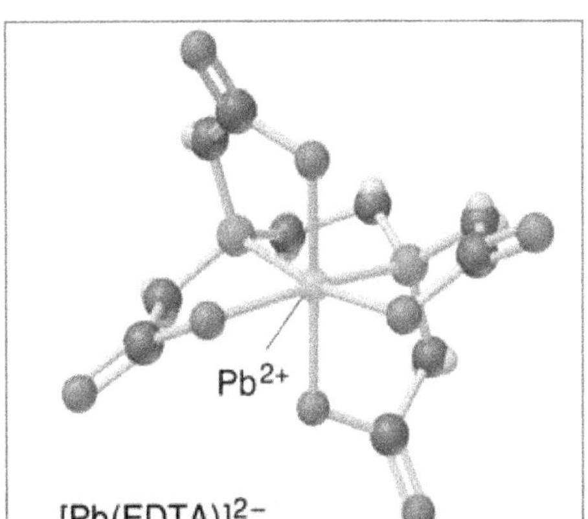

In many bidentate to polydentate ligands, the ligand is bonded to the same central metal atom at two or more places due to which a ring structure is formed. Such a ring structure formed by a bidentate (or polydentate) ligand is known as a chelate. The term chelate, derived from the Greek word **chela** meaning great claw of the lobster or other crustaceans, is suggested for the cliperlike groups whiuch function as two associating units. Chelated complexes are known to have a higher stability as compared to ordinary complexes. Ethylenediaminetetraacetic acid (H_4EDTA) is an important chelating ligand. It binds the metal atom through two nitrogen and four oxygen atoms and hence acts as a hexadentate ligands and forms five rings.

1.4 STABILITY AND CHELATE EFFECT

Stability and the Chelate Effect

Chelating ligands make more stable compounds than monodentate ligands that would be expected to form bonds of similar strength to the metal center. Consider the two similar complex ions, $Cd(CH_3NH_2)_4^{2+}$ and $Cd(en)_2^{2+}$. Each is 4-coordinate and have very similar metal-ligand bonds, but the former the metal is bonded to four monodentate ligands; in the latter the metal center is bonded to two bidentate ligands. What's the difference? Why should one be more stable than the other? The relative stability of the two compounds can be compared in terms of free energy of formation, ΔG^o_f. $Cd(en)_2^{2+}$ has a free energy of formation that is 23 kJ/mol lower than $Cd(CH_3NH_2)_4^{2+}$, which represents a huge difference in stability. The contributing terms to ΔG^o_f, ΔH^o_f and $-T\Delta S^o_f$, show a major difference between enthalpy and entropy factors. The table below shows that in terms of enthalpy of formation, the two are indeed nearly identical. Bond strength is not a factor.

Thermodynamics of Formation

Ligands	ΔH_o(kJ/mol)	ΔS_o(J/K-mol)	$-T\Delta S_o$(J/mol)	ΔG_o(kJ/mol)
4 CH_3NH_2	-57.3	-67.3	20.1	-37.2
2 en	-56.5	+14.1	-4.2	-60.7

This is called the chelate effect: coordination compounds formed with chelating ligands will be more stable than compounds formed with similar monodentate ligands. The chelate effect is entropy driven. The chelate effect can also be considered as a competition between ligands for a metal center. Consider the equilibrium below,

$$[Ni(NH_3)_6]^{2+} + 3\ en \leftrightarrow [Ni(en)_3]^{2+} + 6\ NH_3 \qquad K = 10^{9.67}$$

The reactant side has a complex with six monodentate ligands and three unbonded bidentate ligands; four individual particles. The product side has a complex with the metal bonded to three bidentate ligandes and six unbonded monodentate ligands; seven individual particles. Because the bond strengths for the different ligands are similar, enthalpy change for the reaction is nearly zero. However, entropy change strongly favors formation of the products due to the greater number of independent species.

Ligands which ligate to metal ions through two atoms of different elements present in it are called ambidentate ligands. Example of such ligands is NO_2^- and SCN^- ions. NO_2^- ion can coordinate either through nitrogen or through oxygen to a central metal. Similarly, SCN^- ion can coordinate through sulphur or nitrogen atom. **The same groups can behave as ambidentate ligands using either end of the ion in binding to the metal ion.**

$[Co(NH_3)_5NO_2]^{++}$ — *Nitropentammineecobalt (III) ion*

$[Co(NH_3)_5(ONO)]^{++}$ — Nitritopentammineecobalt (III) ion

It should be noted that, by convention, the ambidentate ligand is always written with the donor atom first, i.e., NO_2 for the nitro, ONO for the nitrito and NCS for the N-thiocyanato and SCN for S-thiocyanato complex.

In addition there are macroacyclic and macrocyclic ligands of importance. Macroacyclic ligands derived from β – diketones and alkylamines are commonly known as Schiff bases, e.g. acacen. Macrocyclic ligands, such as porphyrin and its derivatives, phthalocyanine and similar others have important role in biological systems.

Macrocyclic Effect –

Macrocyclic ligand complexes are more stable than open chain analogues. Depending on the system, can have favourable ΔS and/or ΔH, although enthalpy generally dominates.

Major contributing factors to favourable ΔH are:

a) greater basicity of donor atoms due to increased alkylation.

b) Macrocyclic ligands are frequently less strongly solvated than acyclic analogues.

(fewer ligand-solvent bonds to break on complexation). Overcomes repulsive forces between polar donor groups. Favourable ΔS results from macrocycles which are less conformationally flexible, i.e. lose fewer degrees of freedom on complexation.

Sidgwick's Electronic Theory

In 1927, Sidgwick extended the electronic theory of Lewis for the formation of coordination compounds. He introduced a new concept of a 'coordination bond'. According to Sidgwick's concept, the ligands donate the electron pair(s) to the central metal ion and thus form a number of coordinate bonds.

Example: Cuprammonium ion is formed by the donation of four electron pairs by four ammonia molecules towards cupric ion.

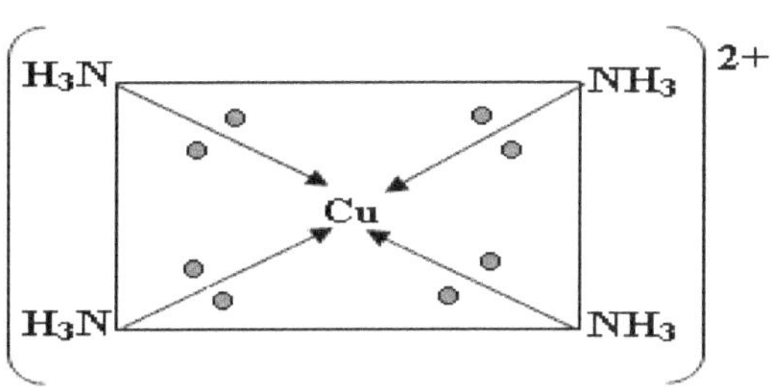

1.5 EFFECTIVE ATOMIC NUMBER

It is the total number of electrons around the central metal ion present in a complex and calculated as the sum of the electrons on the metal ion and the number of electrons donated by the ligands. It can be calculated using the formula.

$$EAN = Z - X + Y$$

where Z = Atomic number of the metal, X = number of electrons lost during the formation of the metal ion from its atom, Y = number of electrons donated by the ligands.

Example, In ferrocyanide ion $[Fe(CN)_6]^{4-}$,

$$Fe \xrightarrow{-2e^-} Fe^{2+} \xrightarrow[+6 \times 2e^-]{+6CN^-} [Fe(CN)_6]^{4-} \quad 24 + 12 = 36e^-$$

Z, Atomic number of iron = 26.

X, Number of electrons lost due to the formation of ferrous ion from an iron atom = 2.

Y, Number of electrons donated by $6CN^-$ ions = 6 x $2e^-$ or $12e^-$. ∴ EAN = 26 - 2 + 12 = 36.

Similarly, EAN of cobalt in $[Co(NH_3)_6]Cl_3$ and zinc in $[Zn(NH_3)_4]SO_4$ are also 36 each.

EAN Rule

A metal ion continues to accept electron pairs from the ligands till the total number of electrons present around the metal ion in the complex becomes equal to the atomic number of the next noble gas atom. In $[Fe(CN)_6]^{4-}$, $[Zn(NH_3)_4]^{2+}$ and $[Co(NH_3)_6]^{3+}$ ions, the EAN of the metal ions are 36 each, which is the atomic number of next noble gas, Krypton. Similarly, EAN of platinum in $[Pt(NH_3)_6]Cl_4$ is 86 which is the atomic number of next noble gas, radon.

Table 17. EAN of few metal complexes

Metal	Complex	Z	X	Y	EAN
Ni	$Ni(CO)_4$	28	0	8	36
Fe	$Fe(CO)_5$	26	0	10	36
Cr	$Cr(CO)_6$	24	0	12	36
Fe	$[Fe(CN)_6]^{4-}$	26	2	12	36
Co	$[Co(NH_3)_6]^{3+}$	27	3	12	36
Zn	$[Zn(NH_3)_4]^{2+}$	30	2	8	36
Pt	$[Pt(NH_3)_6]^{4+}$	78	4	12	86

Examples for the complexes deviating from the EAN

Metal	Complex	Z	X	Y	EAN
Fe	$K_3[Fe(CN)_6]$	26	3	12	35
Cu	$[Cu(NH_3)_4]^{2+}$	29	2	08	35

18 electron rule

Stable low oxidation state complexes are found to have a total of 18 bonding electrons.

Metal valence electrons + lone pairs from ligands = 18

$Ni(CO)_4$ - $4s^2 3d^8$ and 4 lone pairs

$Fe(CO)_5$ - $4s^2 3d^6$ and 5 lone pairs $Cr(CO)_6$ - $4s^2 3d^4$ and 6 lone pairs

The stability of these 18 electron species can be explained using MO theory. This Corresponds to filling all the molecular bonding orbitals and none of the antibonding orbitals. However, this rule only works for species with metals in a low oxidation state not for most complexes.

1.6 NOMENCLATURE OF COORDINATION COMPOUNDS

Nomenclature is important in Coordination Chemistry because of the need to have an unambiguous method of describing formulae and writing systematic names, particularly when dealing with isomers. This nomenclature system is in accordance with recommendation by the IUPAC (International Union of Pure and Applied Chemistry).

The names of co-ordination compounds are derived by following the principles of additive nomenclature. The groups that surround the central atom or structure must be identified in the name. In a coordination formula, the central atom is listed first. The formally anionic ligands appear next, listed in alphabetical order according to the first symbols of their formulae. The neutral ligands follow, also in alphabetical order, according to the same principle. The formula of the entire coordination entity, whether charged or not, is enclosed in square brackets. If the coordination entity is negatively charged, the formula is preceded by the cation formula. When ligands are polyatomic, their formulae are enclosed in parentheses. Ligand abbreviations are also enclosed in parentheses.

1. Order of naming ions: In ionic complexes, the cation is named first and then the anion irrespective whether the cation or the anion is the complex species (which is enclosed in square brackets). No spaces are left between parts of the name that refer to the same coordination entity. For example:

$K_4[Fe(CN)_6]$ is named as potassium hexacyanoferrate(II)

$[Ni(NH_3)_6]Cl_2$ is named as hexaamminenickel (II) chloride

$[Co(NH_3)_5 CO_3] Cl$ is named as pentaaminecarbonatocobalt(III) chloride

Non-ionic complexes are given a one word name, e.g. $[Pt(NH_3)_2Cl_4]$ is named as diammine tetrachloroplatinum(IV).

2. Names of ligands

In the complex ion, the name of the ligand(s) precedes that of the central metal atom. (This procedure is reversed from writing formulas). Neutral ligands are named as the molecules. However, water molecule is named as 'aqua', ammonia as 'ammine', carbon monoxide as 'carbonyl' and nitric oxide as 'nitrosyl'. Negative ligands generally end in **o** as illustrated below with a few examples: Cl^- - chloro, CN^- - cyano, H^- - hydrido.

Neutral		Anionic	
Name	**Formula**	**Name**	**Formula**
Aqua	H_2O	Fluoro	F^-
Ammine	NH_3	Chloro	Cl^-
Carbonyl	CO	Bromo	Br^-
Nitrosyl	NO	Iodo	I^-
		Hydroxo	OH^-
		Cyano	CN^-

(Note that the 2004 draft on naming coordination compounds has recommended that the anionic ligands ending in 'ide', 'ite' or 'ate', the final 'e' is replaced by 'ido', 'ito' and 'ato', respectively; thus, Cl- - chlorido, CN- - cyanido, H- - hydrido, CH_3COO- - acetato or ethanoato)

3. Order of naming ligands

If more than one ligand is present in the complex species, then the ligands are named in alphabetical order without any separation by hyphens. For example,$[Pt(NH_3)_4(Cl)(NO_2)]SO_4$ is named as tetramminechloronitroplatinum(IV)sulphate;

$[Cu(CH_3NH_2)_2Cl_2]$ is named as dichloro-bis(methylamine)copper(II). (This is in accordance with the 1971 rule. Some texts still say that ligands are named in the order: neutral then anionic).

4. Numerical prefixes to indicate number of ligands: Two kinds of numerical prefix are used for indicdating the number of each type of ligand within the name of the coordination entity.

i) If ligands have simple names such as chloro, iodo, nitro etc., their number is indicated by prefixes such as di, tri, tetra, penta etc.

ii) If the ligands have complex names, their number is given by prefixes such as bis, tris, tetrakis etc. for examples:

$[Co(CH_2NH_2)_2Cl_2]_2SO_4$ is named as dichlorobis(ethylenediamine)cobalt(III) sulphate.

$[Cr\{(CH_2NH_2)_2\}_3]$ Cl_3 is named as *d* or *l* tris(ethylenediamine)chromium(III)

5. Ending of names

The name of an anionic complex which ends in 'ate', e.g. $K_3[Al(C_2O_4)_3]$ is named as potassium trioxalatoaluminate(III). In case of cationic and neutral complexes, the name of the metal is written without any characteristic ending. For example, $[Fe(H_2O)_6]SO_4$, which contains the cationic complex $[Fe(H_2O)_6]^{2+}$, is named as hexaaquoiron(II)sulphate. Coordination compound $K[Pt(NH_3)Cl_5]$ having the anionic complex $[Pt(NH_3)Cl_5]^-$ is named as potassium amminepentachloroplatinate(IV).

Metal	Name in Anion
Iron	Ferrate
Copper	Cuprate
Lead	Plumbate
Silver	Argentate
Gold	Aurate
Tin	Stannate

6. Oxidation states of the central ion

The oxidation state of the central ion is designated by a roman numeral i.e., I, II, III, IV etc. in parantheses at the end of the name of the central metal atom. For zero oxidation state, the letter 0 is written in parantheses. A number of examples have already been given to illustrate this rule.

7. Bridging groups in bidentate ligands

Ligands which link two central metal atoms are usually separated from the rest of the complex by a hyphen and is/are denoted by the Greek letter **μ**. This is repeated before the name of each bridging group in the complex. The following example illustrates the rule.

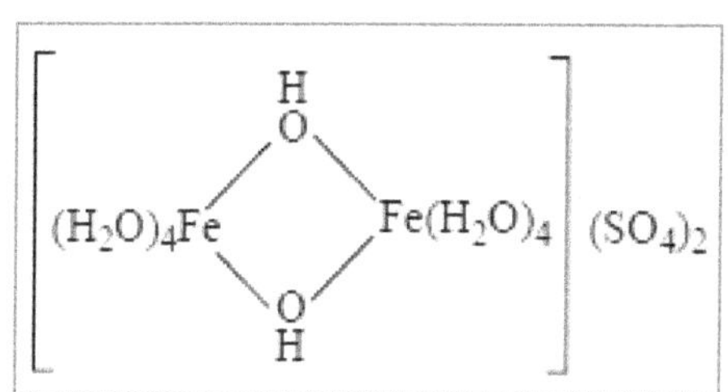

This coordination compound is named as -μ-dihydroxo octaaquodiiron(III) sulphate and written as $[(H_2O)_4Fe(\mu\text{-}OH)_2Fe(H_2O)_4]$ $(SO_4)_2$.

8. Abbreviations for complicated molecules

While writing the structure of complex for complicated molecules, generally abbreviations are used. For example, en for ethylenediamine, py for pyridine etc.,

$Cr(en)_3]Cl_3$ - tris(ethylenediamine)chromium(III) chloride

K[CoEDTA] - potassium ethylenediaminetetracetatocobalt(III)

Additional Illustrations of the Rules:

1. $[Co(NH_3)_5Cl]^{2+}$ - pentamminechlorocobalt(III) ion
2. K_2FeO_4 - potassium tetraoxoferrate(VI)
3. $[Cr(en)_3]Cl_3$ - tris(ethylenediamine)Chromium(III) chloride
4. $[(NH_3)_5Co\ NH_2\ Co(NH_3)_5](NO_3)_3$ - μ-amido bis(pentamminecobalt(II)) nitrate.
5. $[Pt(py)_4]\ [PtCl_4]$ - tetrapyridineplatinum(II) tetrachloroplatinate(II)
6. $K_2[Cr(CN)_2O_2(O_2)NH_3]$ - potassium amminedicyanodioxoperoxochromate(VI)
7. $[Co(NH_3)_3NO_2ClCN]$ - triamminechlorocyanonitro cobalt(III)
8. K[CoEDTA] - potassium ethylenediaminetetracetatocobalt(III)
9. 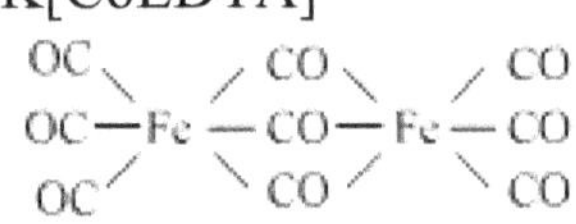 - tri μ-carbonyl bis[tricarbonyl iron(III)]
10. $Na_2\ [Fe(CN)_5NO]$ - sodium pentacyanonitrosylferrate(III)
11. $Ni(CO)_2(PPh_3)_2$ - dicarbonyl bis(triphenylphosphine) nickel(0)
12. $[Ag(NH_3)_2]Cl$ - diamminesilver(I) chloride
13. $Na_3[Ag(S_2O_3)_2]$ - sodium dithiosulphatoargentite(I)
14. $[Pd(H_2O)_2(ONO)_2I_2]$ - diaquodiiododinitritopalladium(IV)
15. $K[Co(NH_3)_2\ Cl_4]$ - potassium diammnetetrachlorocobaltate(III)
16. $Na_3[AlF_6]$ - sodium hexafluoroaluminate
17. $[Co(en)_2Cl_2]NO_3$ - dichlorobis(ethylenediamine)cobalt(III) nitrate
18. $[Pt(NH_3)_4BrCl]Cl_2$ - tetramminebromochloroplatinum(IV) chloride

19. $[Co(NH_3)_6][FeCl_4]$ - hexamminecobalt(II) tetrachloroferrate(II)

20. $[Fe(en)_2(NO_2)_2]_2SO_4$ - bis(ethylenediamine)dinitroiron(III) sulphate

21. $[Cr(H_2O)_5Cl]Cl_2$ - H_2O - pentaaquochlorochromium(III) chloride monohydrate

22. $[Cr(H_2O)_4Cl_2]Cl.2H_2O$ - tetraaquodichlorochromium(III) chloride dihydrate

23. $[Pt(NH_3)_4SO_4]\ (OH)_2$ - tetramminesulphatoplatinum(IV) hydroxide

24. $[Pd(PPh_3)_2(NCS)_2]$ - diisithiocyanatobis(triphenylphosphine)palladium(II)

25. $[Pd(PPh_3)_2\ (SCN)_2]$ - dithiocyanatobis(triphenylphosphine)palladium(II)

26. 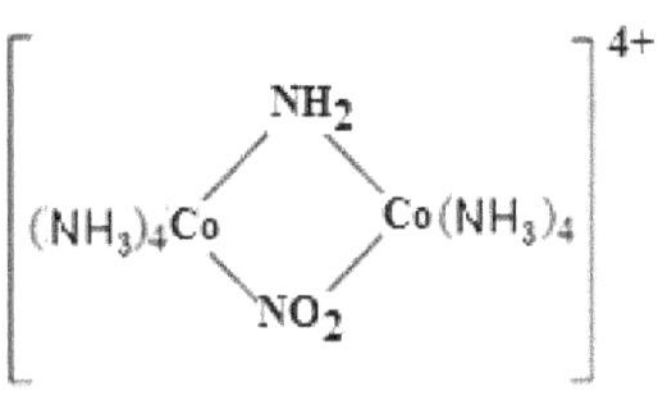

- μ-amido-μ-nitrobis(tetramminecobalt(III)) ion

27. 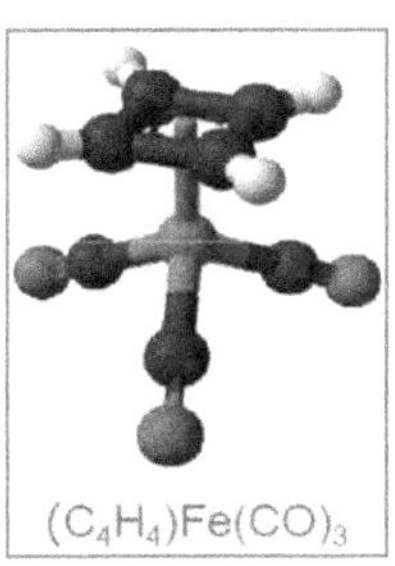

- tricarbonyl(η^4-cyclobutadienyl)iron(0)

28. $K_2[ReF_8]$ - potassium octafluororhenate(VI)

29. $[Ru(NH_3)_5(N_2)]^{2+}$ - pentamminedintrogenruthenium(II) ion

30. $(NH_4)_2[SnCl_6]$ - ammonium hexachlorostannate(IV)

1.7 ISOMERISM IN COORDINATION COMPOUNDS

Compounds having same chemical formula but different structures are called isomers. Many different types of isomerism occur because of complicated formulas of many coordination compounds, the variety of bond types and the number of shapes possible. These are described below:

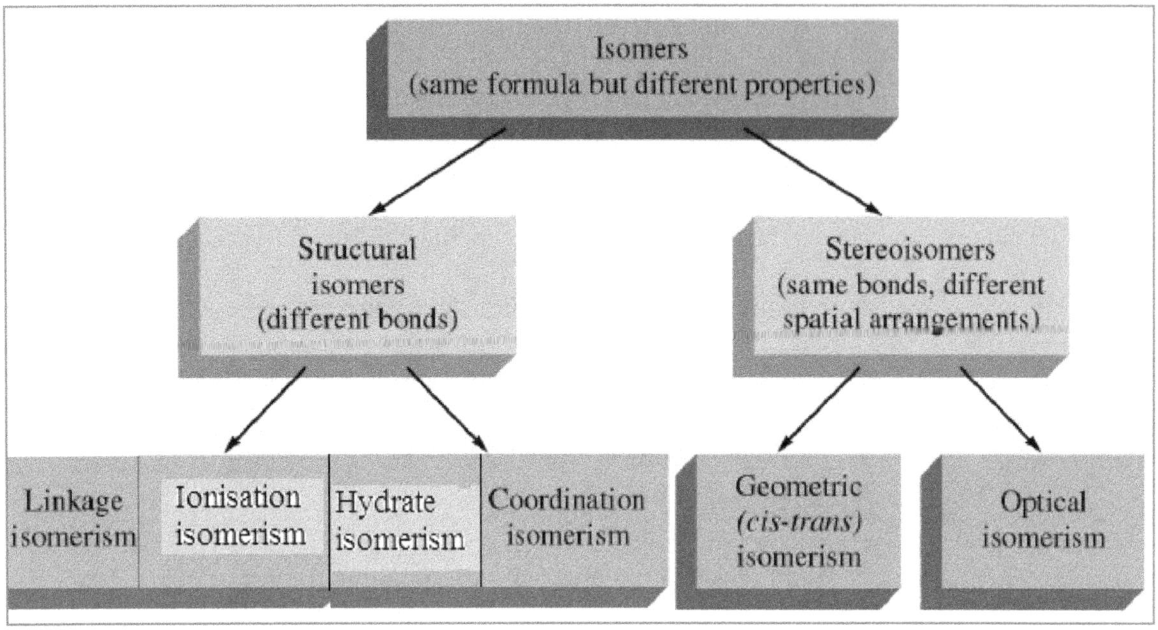

A. Structural Isomerism

1. **Linkage isomerism:** This isomerism arises, if the ligand has two different atoms, either of which can coordinate with a metal ion, i.e. the same groups can behave as ambidentate ligands using either end of the ion in binding to the metal ion.

$[Co(NH_3)_5NO_2]^{++}$
pentamminenitrocobalt(III) ion

$[Co(NH_3)_5(ONO)]^{++}$
pentamminenitritocobalt(III) ion

Fig. 50. Structures of Nitro and nitrito complexes of Cobalt

In the first complex coordination occurs through nitrogen of the NO_2 group while in the second complex coordination occurs through oxygen (Fig. 50). The nitro-compound (like organic nitrates R-NO_2) is

stable to acids while the nitrite compound (like organic nitrites R-ONO) is decomposed by acids to give nitrous acid.

Other ligands which may give rise to linkage isomerism are

-CN (cyano) and -NC (isocyano)　　　-SCN (thiocyanato) and -NCS (isothiocyanato)

Ex., $[Pd(PPh_3)_2(NCS)_2]$, $[Pd(PPh_3)_2(SCN)_2]$

It should be noted that, by convention, the ambidentate ligand is always written with the donor atom first, i.e., NO_2 for the nitro, ONO for the nitrito and NCS for the N-thiocyanato and SCN for S-thiocyanato complex.

2. **Ionisation isomerism**: This type of isomerism arises due to difference in the position of groups being inside or outside the coordination sphere of the complex. This is due to a ligand exchanging positions between the coordination sphere and outside. The isomers behave differently when dissolved in water because the ligands in the inner coordination sphere do not ionize. There is interexchange of Br^- and SO_4^{2-} in the isomers and since the two isomers on ionisation give different ions these are called ionisation isomers. Complex 2 on ionisation will give rise to $[Co(NH_3)_5Br]^{2+}$ and SO_4^{2-} ions while complex 1 on ionisation will give rise to $[Co(NH_3)_5SO_4]^+$ and Br^- ions.

(i)	$[Co(NH_3)_5SO_4]Br$	and	$[Co(NH_3)_5Br]SO_4$
	pentamminesulphatocobalt(III) bromide		pentamminebromocobalt(III) sulphate

The first compound answers tests for bromide and not for sulphate, while the second one answers for sulphate and not for bromide. Other examples are,

(ii)	$[Pt(NH_3)_4SO_4](OH)_2$	and	$[Pt(NH_3)_4(OH)_2]SO_4$
	tetramminesulphato platinum(IV) hydroxide		tetramminedihydroxo platinum(IV) sulphate
(iii)	$[Pt(NH_3)_4Cl_2]Br_2$	and	$[Pt(NH_3)_4Br_2]Cl_2$
	tetramminedichloro platinum(IV) bromide		tetramminedibromo platinum(IV) chloride
(iv)	$[Co(NH_3)_4Cl_2]NO_2$	and	$[Co(NH_3)_4ClNO_2]Cl$
	tetramminedichloro cobalt(III) nitrite		tetramminechloronitro cobalt(III) chloride

3. **Hydrate isomerism:** This isomerism arises due to the replacement of coordinated group by water of hydration. In this, the isomers differ in the number of water molecules in the co-ordination sphere (i.e. water may be inside or outside the coordination sphere). These isomers also differ largely from one another in their physical and chemical properties.

Examples:

	Formula and name of complex	Treatment with conc. H_2SO_4	Addition of $AgNO_3$
i)	$[Cr(H_2O)_6]Cl_3$ hexaaquochromium (III) chloride (violet)	No water is lost	3 Cl^- are precipitated
ii)	$[Cr(H_2O)_5Cl]Cl_2 - H_2O$ pentaaquochlorochromium(III) chloride monohydrate (pale green)	One water molecule is lost	2 Cl^- are precipitated
iii)	$[Cr(H_2O)_4Cl_2]Cl.2H_2O$ tetraaquodichlorochromium(III) chloride dihydrate (green)	Two water molecules are lost	1 Cl^- is precipitated

Other examples:

$[Co(en)_2(H_2O)Cl]Cl_2$

aquochlorobis(ethylene diamine)cobalt(III) chloride

$[Co(en)_2Cl_2]Cl.(H_2O)$

Dichlorobis(ethylenediamine) cobalt(III) chloride monohydrate

4. **Coordination isomerism:**
This type of isomerism is observed in case of compounds in which both, anionic and cationic parts of the coordination compound are complexes. An example is provided by $[Co(NH_3)_6][Cr(CN)_6]$, in which the NH_3 are bound to Co^{3+} and the CN^- to Cr^{3+} and $[Co(CN)_6][Cr(NH_3)_6]$ where the NH_3 are bound to Cr^{3+} and the CN- to Co^{3+}, both the anionic and cationic parts are complex ions. A special case of coordination isomerism is **coordination position isomerism** which involves different placements of ligands in a bridged complex as shown,

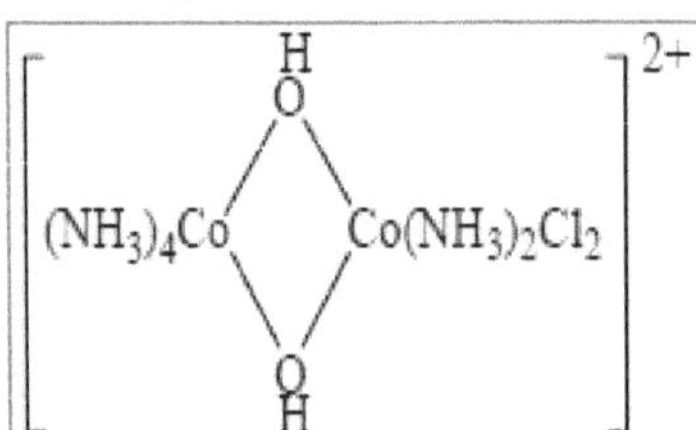

1

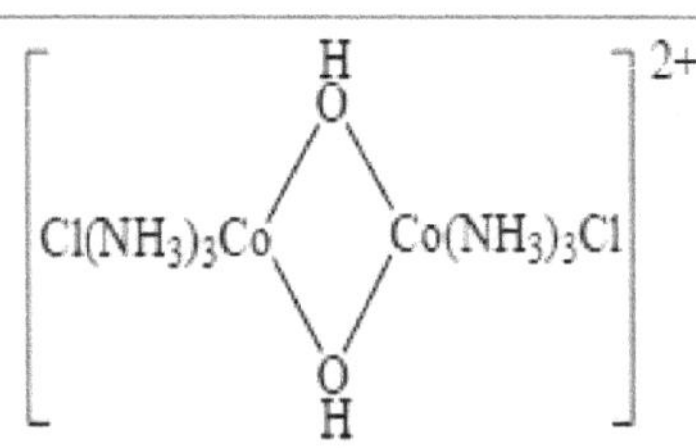

2

A special case of coordination isomerism has sometimes been given the name **polymerization isomerism** since the various isomers differ in formula weight from one another. The isomers are represented by compounds such as $[Co(NH_3)_4(NO_2)_2]$ $[Co(NH_3)_2(NO_2)_4]$, $[Co(NH_3)_6]$ $[Co(NO_2)_6]$, $[Co(NH_3)_5(NO_2)]$ $[Co(NH_3)_2(NO_2)_4]_2$, $[Co(NH_3)_6]$ $[Co(NH_3)_2(NO_2)_4]_3$, and $[Co(NH_3)_5(NO_2)]_3$ $[Co(NO_2)_6]_2$. All these compounds have the empirical formula $[Co(NH_3)_3(NO_2)_3]$, but they have formula weights that are 2,2,3,4 and 5 times this respectively.

B. **Stereoisomesism:** Stereoisomers are compounds that have the same atoms and the same interatomic connections, but differ the spatial arrangements of the atoms, or in particular here, the ligands. Stereoisomers have the same atoms, same sets of bonds, but differ in the relative orientation of these bonds.

❖ **Geometric isomerism**

Geometric isomers are possible for both square planar and octahedral complexes, but not tetrahedral. The two most important types of geometric isomers are cis-trans and fac-mer isomers.

Geometric or *cis-trans* isomers (Fig. 51) occur when atoms or groups can either be arranged on the ***same*** side or on ***opposite*** sides of the compound relative to the central metal ion.

a) In square planar Ma_2b_2, the two ligands 'a' may be arranged adjacent to each other, in a **cis** isomer, or opposite to each other, in a **trans** isomer as depicted below:

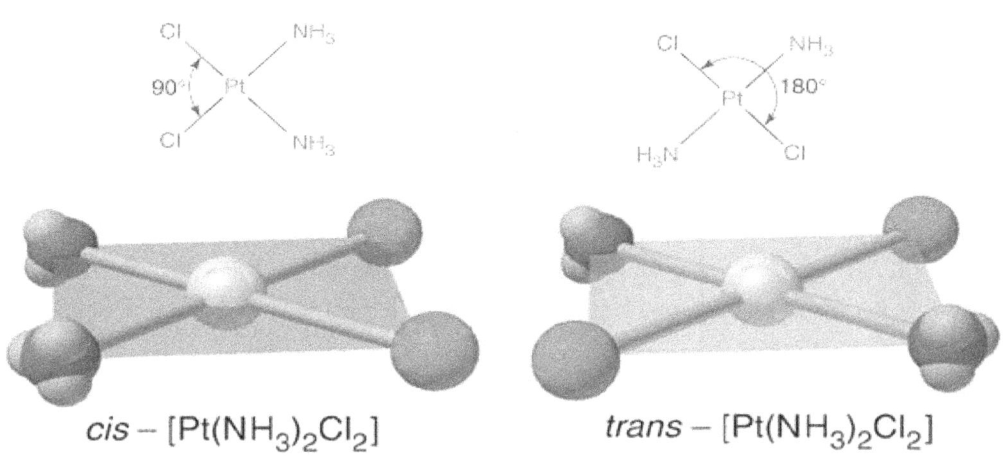

Fig. 51. cis and trans isomers of diamminodichloro platinum(II)

b) Among the octahedral complexes of the types Ma_6, Ma_5b, Ma_4b_2, Ma_3b_3, only the last two show two isomeric forms. In an octahedral (Ma_2b_4) complex (coordination number 6, a and b are unidentate ligands) there are two different positions that ligands can occupy relative to one another.

In the two structures below it can be seen that the two ammonia ligands could be 90° or 180° relative to one another. This gives a cis-isomer and trans-isomer (Fig. 52).

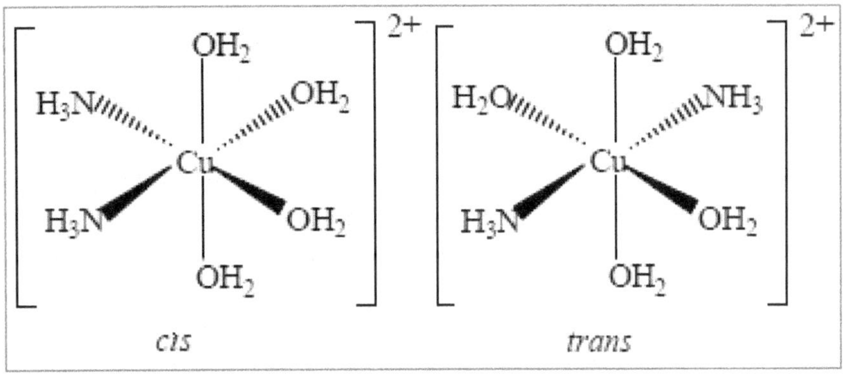

Cis-trans tetramminedichloro cobalt isomers

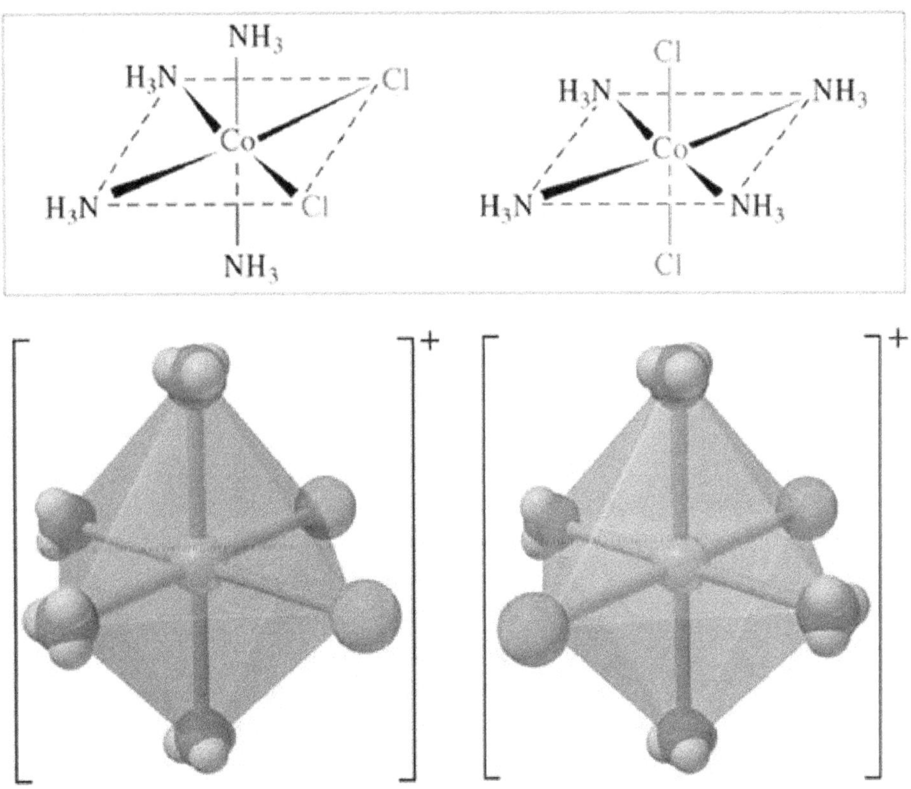

cis – $[Co(NH_3)_4Cl_2]^+$ *trans* – $[Co(NH_3)_4Cl_2]^+$

Fig. 52. cis and trans isomers of tetramminedichlorocobalt(III) ion

This type of cis-trans isomerism also arises when bidentate ligands such as $NH_2CH_2CH_2NH_2$ (en) are present in the complex, e.g. $[CoCl_2(en)_2]^+$ (Fig. 53).

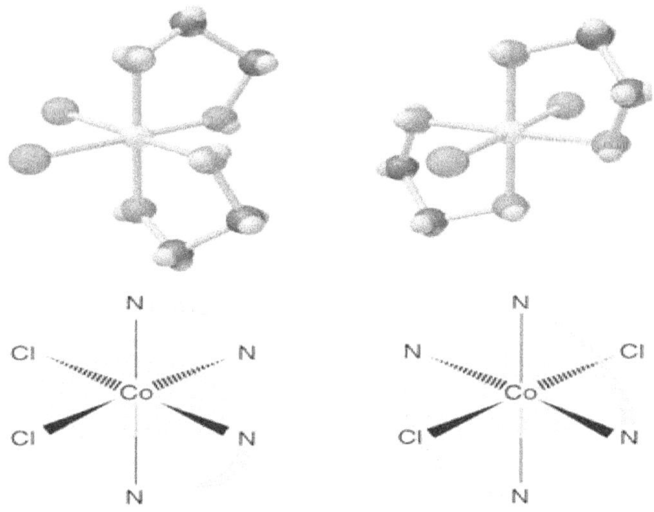

cis – $[Co(en)_2Cl_2]^+$ *trans* – $[Co(en)_2Cl_2]^+$

Fig. 53. cis and trans isomers of dichlorobis(ethylenediamine)cobalt(III) ion

c) Among the complexes of the type Ma_3b_3 in octahedral geometry, another type of isomerism is observed in which the isomers are known as **facial (fac)** and **meriodional (mer) isomers**. If the three ligands 'a', or 'b', all are in the same plane it gives a **meridional (mer)** isomer. On the other hand, if they are adjacent forming a triangular face of the octahedron it gives a **facial (fac)** isomer.

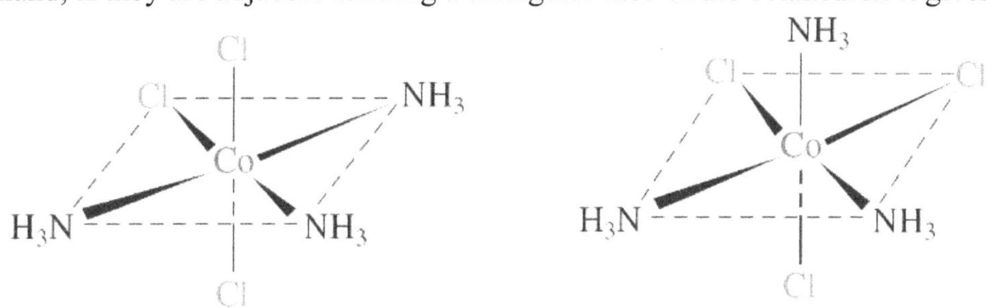

mer isomer fac isomer

Fig. 54. Structure of mer and eac isomers of cobalt complex

❖ **Optical isomerism**

Optical isomers are possible for both tetrahedral and octahedral complexes, but not square planar. ***Optical*** isomers (***enantiomers***) are non-superimposable mirror images of each other. With certain ligands an octahedral complex ion can exhibit optical isomerism. This has nothing to do with the idea of a central carbon atom surrounded by four different groups, as in organic chemistry. However, it does have to do with the idea of two molecules that are not superimposable onto each other. They are also known as *enantiomers*.

For structures that do not possess a plane of symmetry, the mirror images are not superimposable. Known as *chiral* structures, such molecules rotate a beam of polarized light. If the beam is rotated to the right

(when looking along the beam in the direction of propagation), the substance is said to be *dextrorotatory* (or simply *dextro*) and indicated by (+). Those substances that rotate the plane of polarized light to the left are said to be *levorotatory* or *levo* and indicated as (-). A mixture of equal amounts of the two forms is called a racemic mixture, and it produces no net rotation of the polarized light.

The dichlorobis(ethylenediamine) cobalt(II) ion can exist in two geometrical isomers. For the *trans* isomer, there is a plane of symmetry that bisects the cobalt ion and the ethylenediamine ligands, leaving one Cl on either side of the plane. However, the *cis* isomer has no plane of symmetry so two optical isomers exist. This is also the case for $[Co(en)_3]^{3+}$ as is illustrated in Figure 55. Light behaves as though it consists of waves vibrating in all directions around the direction of wave propagation. For polarized light, the propagation can be regarded as a vector, which can be resolved into two circular vectors. If there is no rotation of the plane, it is expected that motion along each vector is equivalent so that each vector traverses an equal distance around the circle as shown in Figure.

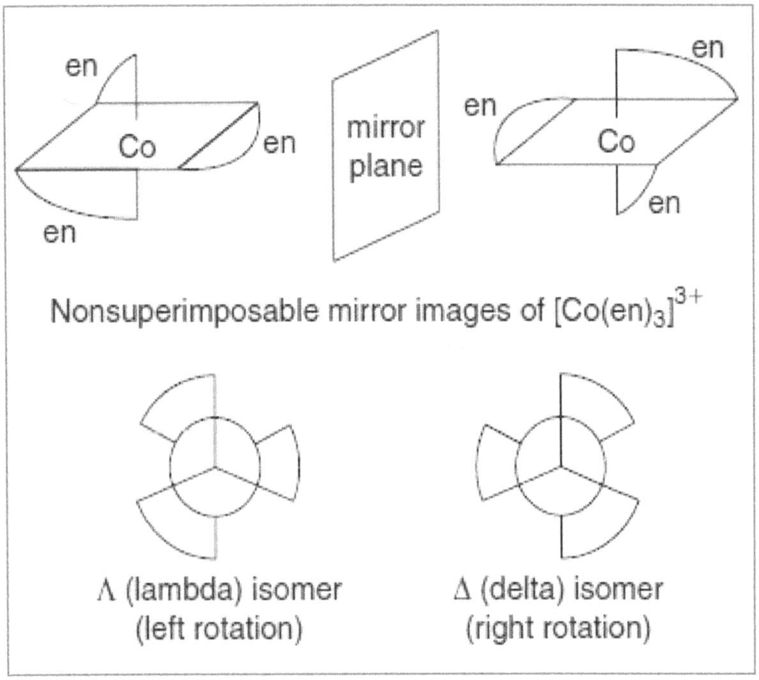

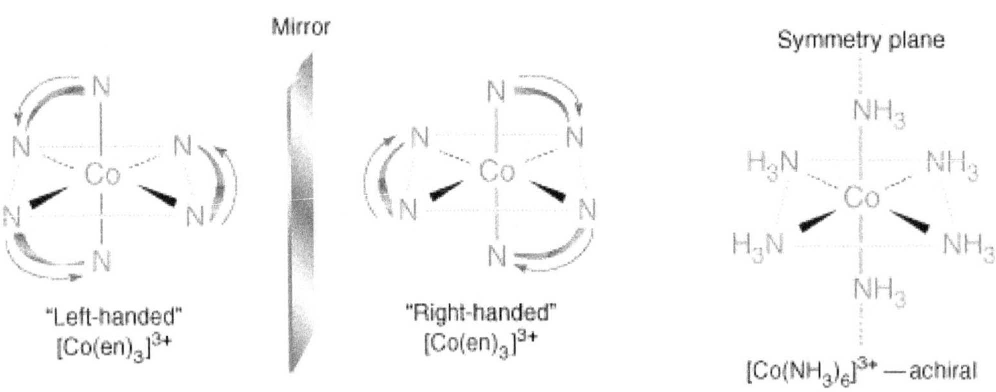

Fig. 55. Otical isomers of Cobalt complexes

Structure I and its mirror image, structure II, are optical isomers of *cis*-$[Co(en)_2Cl_2]^+$ (Fig. 56).

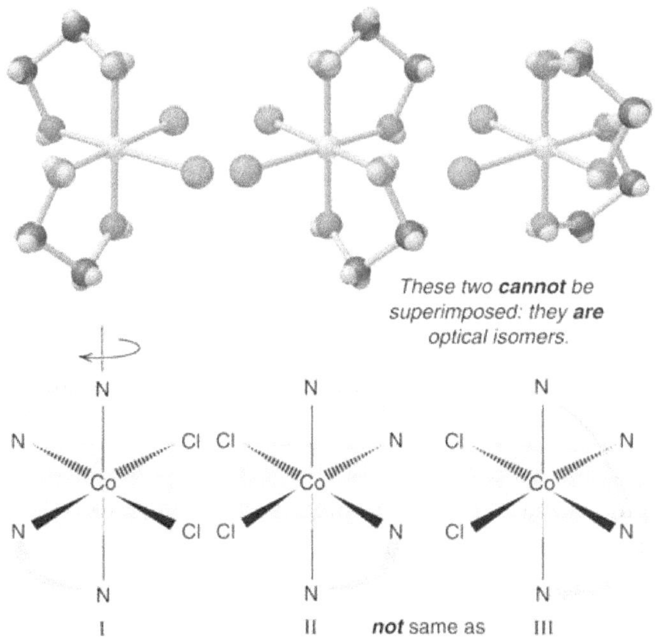

Fig. 56. Optical isomers of *cis* – $[Co(en)_2Cl_2]^+$

The *trans* isomer of $[Co(en)_2Cl_2]^+$ does not have optical isomers. Structure I can be superimposed on its mirror image, structure II (Fig. 57).

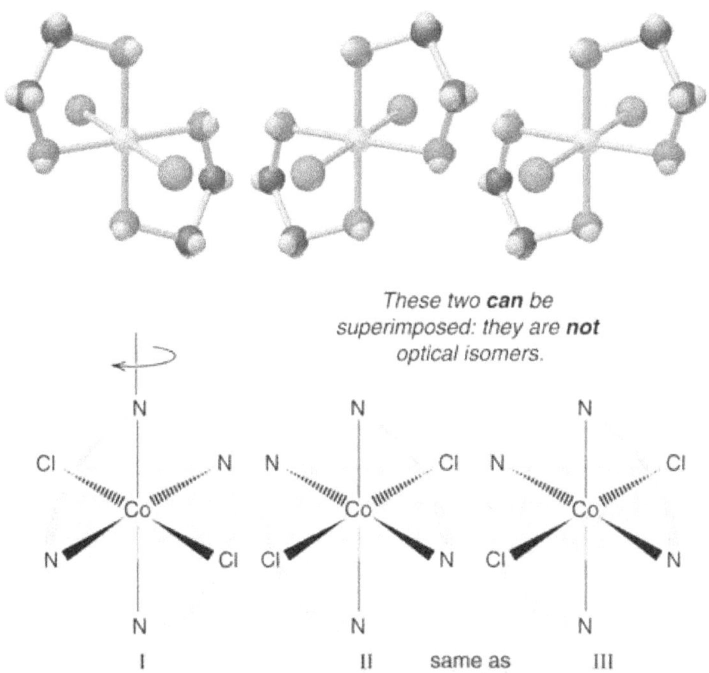

Fig. 57. Optically inactive isomers of *trans* – $[Co(en)_2Cl_2]^+$

Theories of bonding in coordination chemistry

Werner made the first attempt to explain bonding in the coordination compounds. With the development of modern theories of valence, it has now become possible to explain some properties and bonding in coordination compounds in terms of the following theories:

i) Valence Bond Theory

ii) Crystal Field Theory

iii) Molecular Orbital Theory

1.8 VALENCE BOND THEORY (VBT)

In the area of coordination chemistry, a simple valence-bond approach is successful in explaining some of the characteristics of the complexes. The valence bond theory of coordination compounds was mainly developed by Linus Pauling. According to this theory, the formation of a complex is considered to be a reaction between a Lewis base (ligand) and a Lewis acid (metal or metal ion) with the formation of a coordinate covalent bond between the ligand and the metal. The metal atom or ion under the influence of ligands can use (n-1)d, ns, np or ns, np, nd orbital for hybridization to yield a set of equivalent orbitals of definite geometry such as octahedral, tetrahedral, square planar and so on. In this approach, the empty orbitals on the metal ions are viewed as hybrid orbitals in sufficient number to accommodate the number of electron pairs donated by the ligands.

The important features of VBT are:

(1) The central metal ion in the complex makes available a number of empty orbitals for the formation of coordinate bonds with suitable ligands. The number of empty orbitals made available for bonding purposes is equal to the coordination number of the central ion.

(2) The available orbitals on the metal ion hybridise to give an equal number of new hybrid orbitals of the equivalent energy. The hybrid orbitals are described according to the geometry of the complex.

(3) Each ligand has at least one sigma orbital containing a lone-pair of electrons.

(4) The empty hybrid orbitals of the metal ion overlap with the filled sigma orbitals of the ligand forming ligand $\rightarrow$ metal coordinate bond.

(5) The d-orbitals involved in the hybridization may be inner d-orbitals viz., (n-1) d-orbitals, or the outer d-orbitals. The complexes formed in these two ways are referred to as low spin and high spin complexes or, inner and outer orbital complexes respectively.

(6) The geometrical shape of the complex ion depends on the hybridization of the metal orbitals.

Geometry		No. of ligands	hybridization
Octahedral	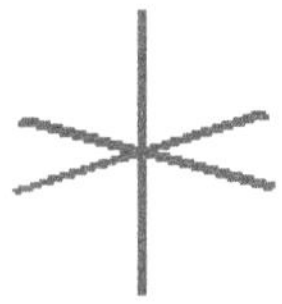	**6**	**d^2sp^3**
Trigonal bipyramid	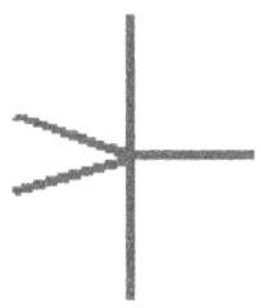	**5**	**dsp^3**
Tetrahedral		**4**	**sp^3**
Square planar		**4**	**dsp^2**
Linear		**2**	**sp**

Structure and bonding in $[Cr(NH_3)_6]^{3+}$ complex ion

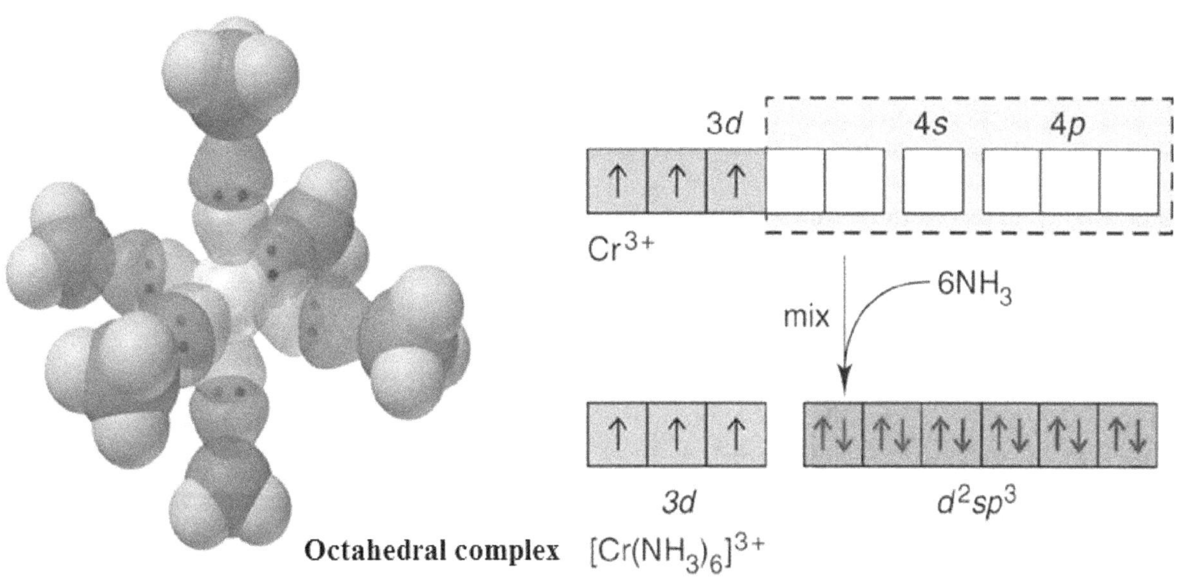

Structure and bonding in $[Ni(CN)_4]^{2-}$ complex ion

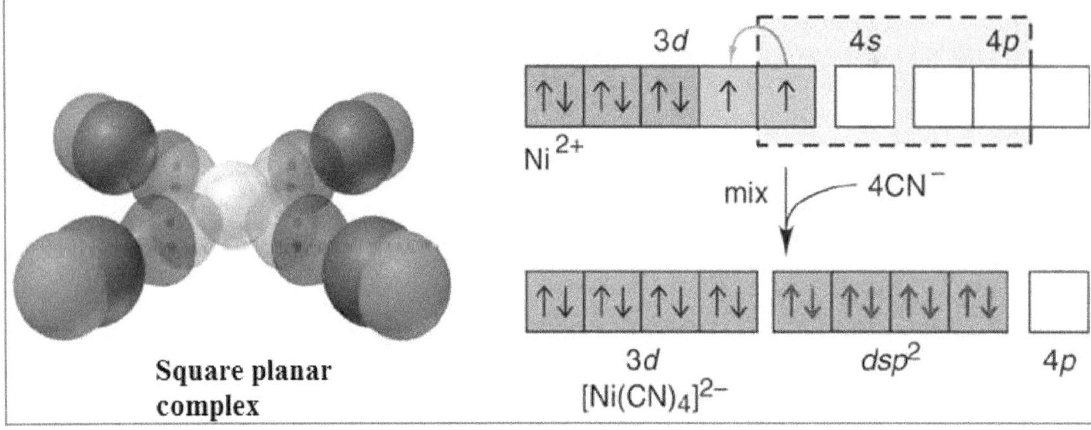

Structure and bonding in $[Zn(OH)_4]^{2-}$ complex ion

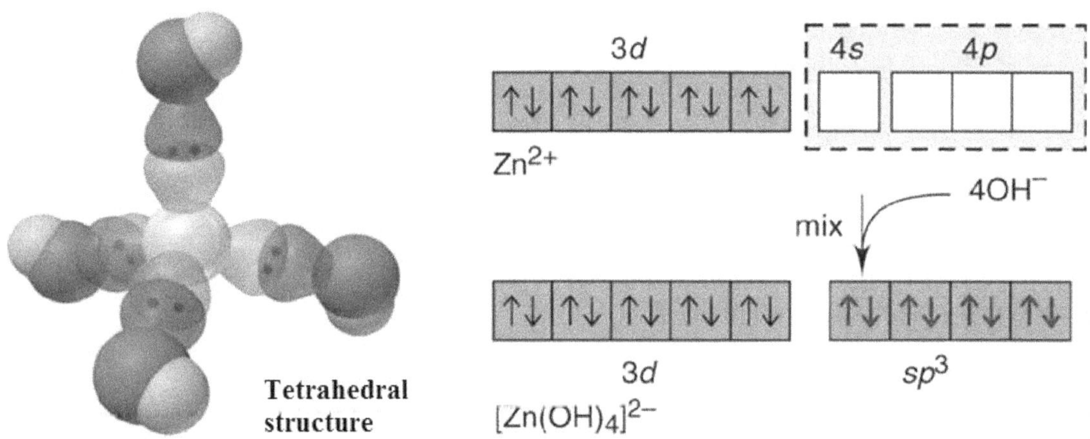

In the octahedral complex, $[Co(NH_3)_6]^{3+}$. The cobalt (III) has the electronic configuration $3d^6$. The complex is diamagnetic. The hybridization scheme can be given as:

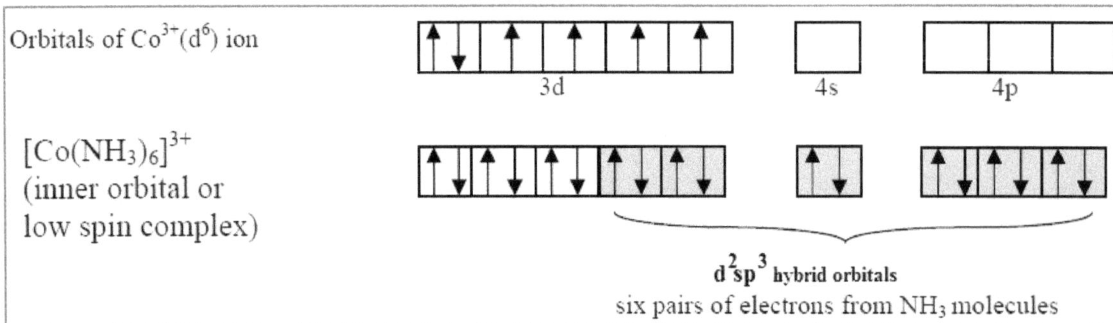

Six pairs of electrons, one from each NH_3 molecule occupy the six d^2sp^3 hybrid orbitals. Thus, the complex has octahedral geometry and its diamagnetic character is justified because of the absence of unpaired electron. Since in the formation of the complex, the inner d-orbitals (3d) are used in hybridization, the complex, $[Co(NH_3)_6]^{3+}$ is called an **inner orbital or low spin** or **spin paired complex.**

Similar octahedral complex, $[CoF_6]^{3-}$ is, however, paramagnetic. In such a situation Co^{+3} can be considered to use outer d orbitals (4d) in hybridization (sp^3d^2) and leaving four unpaired electrons to satisfy its paramagnetic nature. It is thus called **outer orbital** or **high spin** or **spin free complex.**

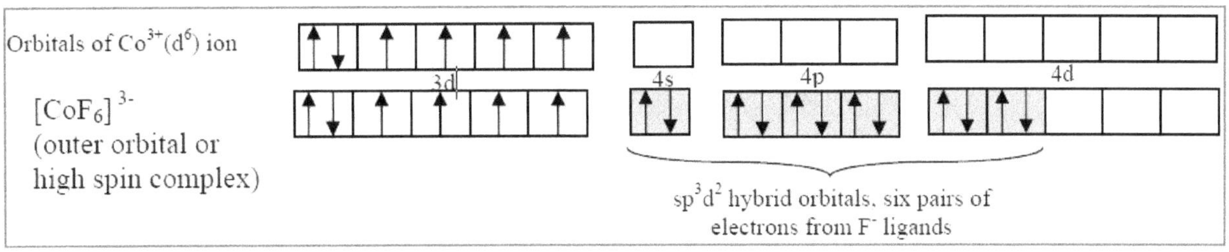

In tetrahedral complexes one s and three p-orbitals are hybridized to form four equivalent orbitals oriented tetrahedrally. This is illustrated below for a paramagnetic $[NiCl_4]^{2-}$. Here nickel(II) has the electronic configuration $3d^8$. The hybridization scheme is as follows:

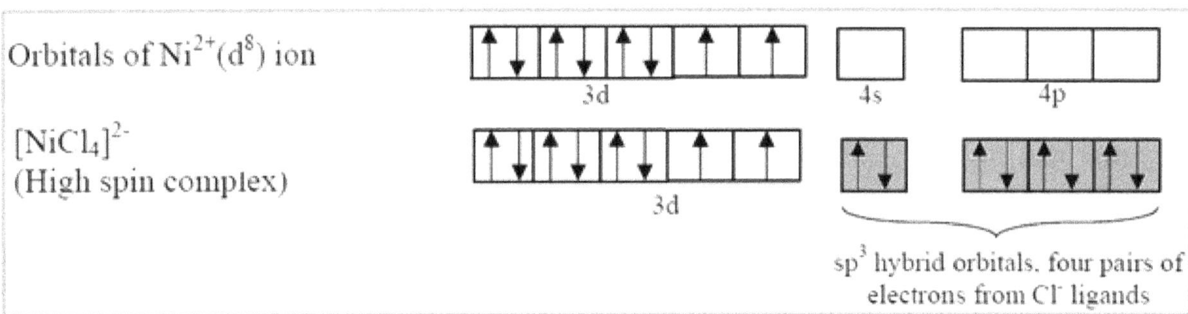

Each Cl- ion donates a pair of electrons. The compound is paramagnetic since it contains two unpaired electrons. The $[Ni(CO)_4]$ also has a tetrahedral geometry but is diamagnetic since it contains no unpaired electron. The two electrons of 4s orbital under the influence of CO ligand molecules pair up with the two unpaired electrons of the 3d orbitals leaving the 4s orbital vacant.

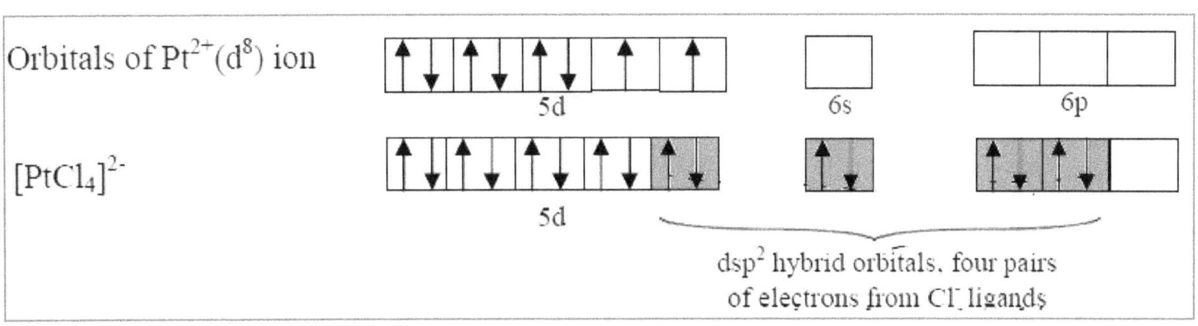

The four pairs of electrons from 4 CO molecules then occupy the sp^3 hybrid orbitals. Complexes of Pd^{2+} and Pt^{2+} having d^8 configuration are 4-coordinate, square planar and diamagnetic and this arrangement is found in Ni^{2+} (e.g. $[Ni(CN)_4]^{2-}$) as well. Since the complexes are diamagnetic, the ligands in these complexes must cause pairing up the two unpaired electrons. Consider the example of $[PtCl_4]^{2-}$, where the hybridization scheme can be given as follows:

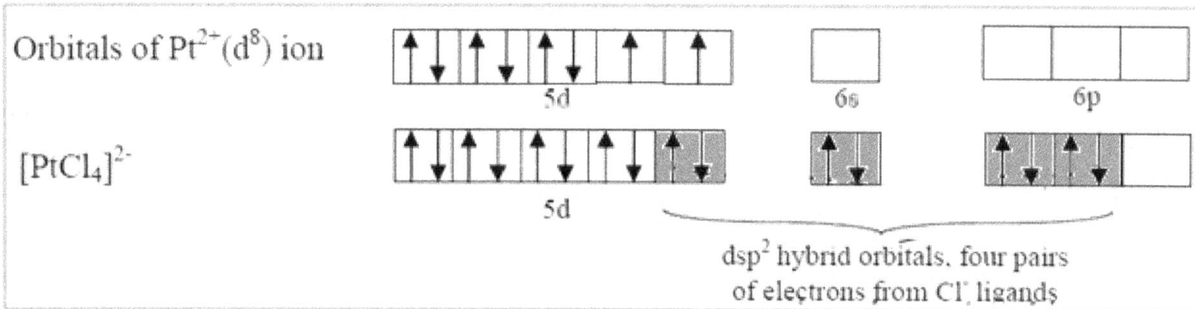

Since the 4 donated electron pairs occupy one 5d, one 6s and two 6p orbitals, the complex $[PtCl_4]^{2-}$ involves dsp^2 hybridization. As a consequence the geometry of the complex is square planar with platinum at the centre and the chloride ions at the four corners. The coordination entities like $[PtCl_4]^{2-}$ and $[Ni(CN)_4]^{2-}$ on one side and $[NiCl_4]^{2-}$ and $[Ni(CO)_4]^0$ on the other side illustrate the useful generalization called, the **magnetic criteria of the bond type**, this is, the geometry of the coordination entity can be predicted if its magnetic behavior is known. In the above cases, we find that diamagnetic species are square planar while the paramagnetic species are tetrahedral.

It is generally true that first-row transition metals form complexes that in many cases contain six coordinate bonds. This is especially true if the charge on the metal ion is +3. The reason for this is that a +3 metal ion has a high charge-to-size ratio that exerts a strong attraction for pairs of electrons. If the metal ion has a +2 charge, it is sometimes found that only four ligands will bind to the metal ion, but it also depends on the availability of empty orbitals on the metal ion. In this case, the charge-to-size ratio is considerably smaller (the charge is only 67% as high and the size is usually larger), so the metal ion has a much smaller affinity for electron pairs. This is the case for +2 ions such as Cu^{2+} and Zn^{2+}, which form many complexes in which the coordination number is 4, although as we shall see there are other factors involved. When Zn^{2+} forms a complex such as $[Zn(NH_3)_4]^{2+}$, it is easy to rationalize the bonding in terms of hybrid orbitals. Zn^{2+} has a d^{10} configuration, but the 4*s* and 4*p* orbitals are empty. Therefore, a set of four empty sp^3 hybrids can result. As shown next, four orbitals can accept four pairs of electrons that are donated by the ligands.

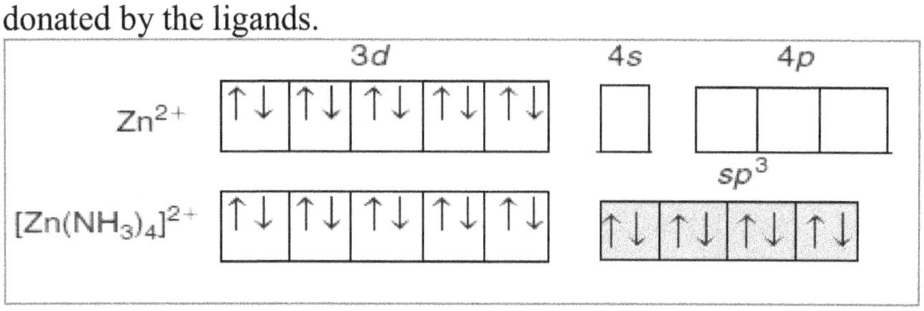

Likewise, it is easy to rationalize how Ag^+ (a d^{10} ion) can form a linear complex with two ligands like NH_3.

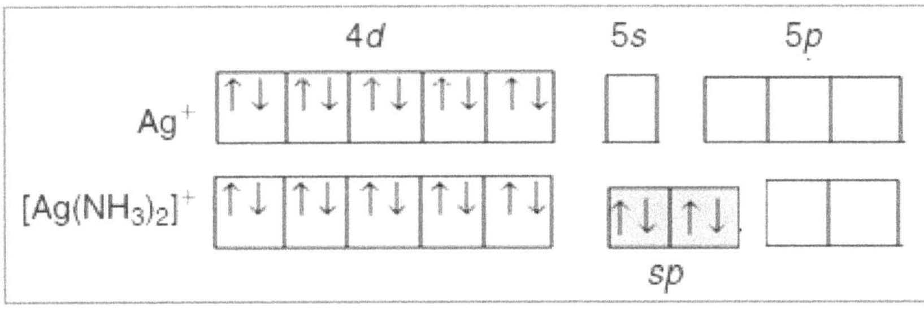

When the scheme above is examined, it is easy to see that if two of the 5 p orbitals are used, a set of sp^2 hybrid orbitals could be produced that would accommodate pairs of electrons from three ligands. It is not surprising that only a few complexes are known in which Ag^+ exhibits a coordination number of 3. If all of the 5 p orbitals were used in the hybrids, the resulting sp^3 hybrids could accommodate four pairs of electrons, and a tetrahedral complex could result. Although complexes like $[Ag(CN)_3]^{2+}$ and $[Ag(CN)_4]^{3+}$ have been identified in solutions where the concentration of CN^- is high, such complexes are unusual and of low stability. The vast majority of complexes of Ag^+ have a coordination number of 2 and linear structures. Part of the reason for this is that the silver ion is relatively large and has a low charge, which results in a low charge density. Such an ion does not assimilate the electron density from added ligands as readily as do those for which the charge density is high.

For metal ions having configurations d^0, d^1, d^2, or d^3, there will always be two of the d orbitals empty to form a set of d^2sp^3 hybrids. Therefore, we expect complexes of these metal ions to be octahedral in which the hybrid orbital type is d^2sp^3. If we consider Cr^{3+} as an example, the formation of a complex can be shown as follows:

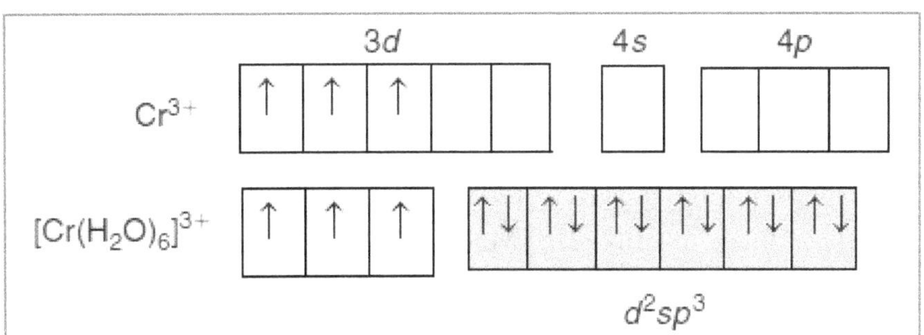

In an octahedral complex, the coordinate system is set up so that the ligands lie on the axes. The d orbitals that have lobes lying along the axes are the d_{z2} and the d_{x2-y2}. The d_{xy}, d_{yz}, and d_{xz} orbitals directed between the axes are considered to be nonbonding. It should not be inferred that only octahedral complexes are produced by Cr^{3+}. Although tetrahedral complexes would not be expected in solutions that contain Cr^{3+} because of its high charge density, solid $[PCl_4][CrCl_4]$ contains the tetrahedral $CrCl_4^-$ ion. With the 4s and 4p orbitals being empty, it is easy to see how sp^3 hybrids are obtained. When the number of electrons in the *d* orbitals is four, as in the case of Mn^{3+}, there exists more than one possible type of hybrid orbital. For example, if the electrons remain unpaired in the *d* orbitals, there is only one orbital in the set that is empty. As a result, if an octahedral complex is formed, making use of two *d* orbitals requires that the 4*d* orbitals be used so that sp^3d^2 hybrid orbitals would be used by the metal. This case can be shown as follows:

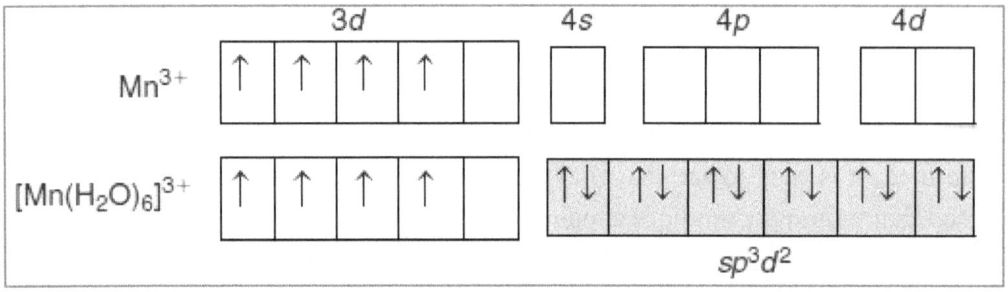

This bonding arrangement results in four unpaired electrons in the 3*d* orbitals. But with some ligands, the situation is different; In those cases, two of the 3*d* orbitals are made available to form a set of hybrid orbitals by means of electron pairing. When this occurs, the bonding arrangement can be shown as follows:

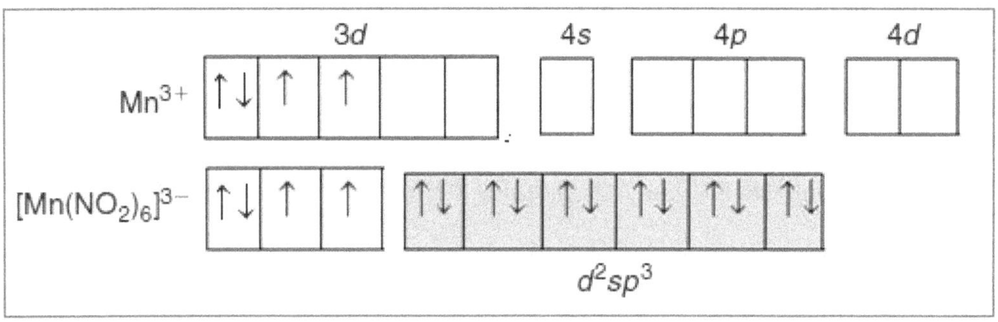

In $[Mn(H_2O)_6]^{3+}$, the 4 *d* orbitals used to form hybrids are those outside the usual valence shell that consists of 3*d*, 4*s*, and 4*p* orbitals. Consequently, such a complex is often referred to as an *outer orbital* complex. To identify the hybrid orbitals, the symbol sp^3d^2 is used to indicate that the *d* orbitals are part of the shell with $n = 4$ and they follow the *s* and *p* orbitals in filling. In $[Mn(NO_2)_6]^{3-}$, the *d* orbitals are those in the valence shell so the complex is called an *inner orbital* complex and the hybrid orbital type is designated d^2sp^3 to show that the principal quantum number of the *d* orbitals is lower than that of the *s* and *p* orbitals. Another way in which the two types of complexes are distinguished is by the terms *high spin* and *low spin*. In $[Mn(H_2O)_6]^{3+}$, there are four unpaired electron spins, whereas in $[Mn(NO_2)_6]^{3-}$, there are only two unpaired electrons. Therefore, the former is referred to as a high-spin complex, and the latter is designated a low-spin complex.

For a d^5 ion such as Fe^{3+}, the five electrons may be unpaired in the set of five *d* orbitals, or they may be present as two orbitals occupied by pairs of electrons and one orbital having single occupancy. A complex such as $[Fe(H_2O)_6]^{3+}$ is typical of the first bonding mode (five unpaired electrons, high spin,

outer orbital), whereas $[Fe(CN)_6]^{3-}$ is typical of the second (one unpaired electron, low spin, inner orbital). Three orbitals can accommodate six electrons, so a d^6 ion such as Co^{3+} should form two series of complexes in which the six electrons occupy all five orbitals with two in one orbital and one in each of the others.
This results in four unpaired electrons being found in the outer-orbital, high-spin complex $[CoF_6]^{3-}$. On the other hand, a low-spin, inner-orbital complex such as $[Co(NH_3)_6]^{3+}$ has no unpaired electrons because all six of the 3*d* electrons are paired in three of the orbitals.

When the d^7 and d^8 electron configurations are considered, there are additional possibilities for forming bonding orbitals. With seven or eight electrons in the five orbitals, there is no possibility of having two of the 3*d* orbitals vacant so if an octahedral complex is formed, it must be of the high-spin, outer-orbital type. Both Co^{2+} (d^7) and Ni^{2+} (d^8) form many complexes of this type. With all of the 3*d* orbitals being populated (not necessarily *filled)*, the possibility exists for using the 4*s* and 4*p* orbitals in forming tetrahedral complexes and $[CoCl_4]^{2-}$ and $[Ni(NH_3)_4]^{2-}$ have tetrahedral structures. Although it is not possible to have two of the 3*d* orbitals vacant, it is certainly possible to have one empty, so the possibility of the dsp^3 hybrid orbital type exists. With regard to the orbitals used, the Co^{2+} ion behaves as follows:

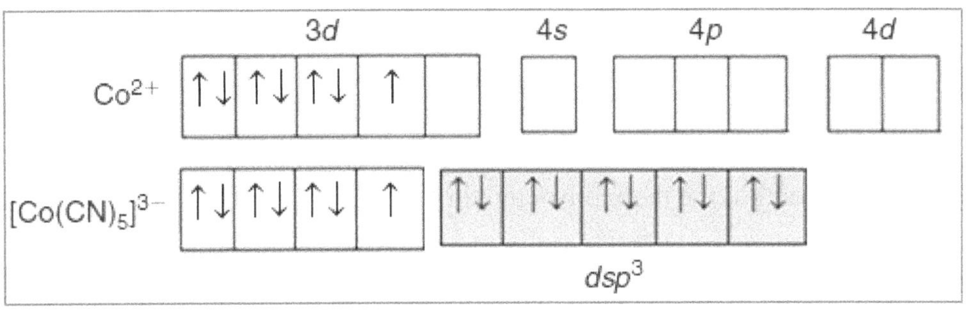

For the d^8 Ni^{2+} ion, the orbitals and bonding can be shown as follows:

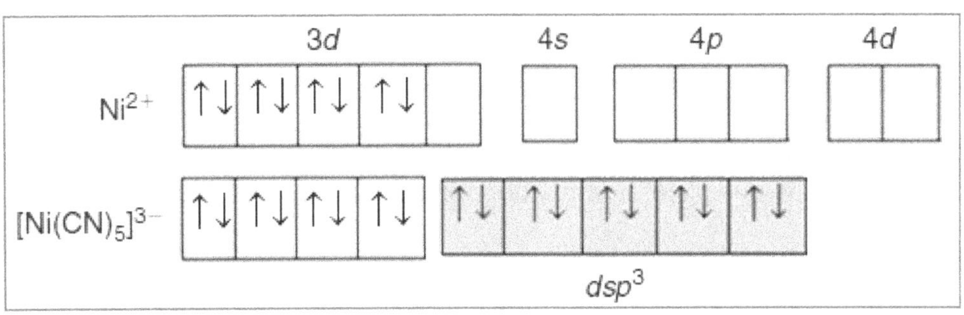

If a complex having a coordination number of 4 is produced, only two of the 4*p* orbitals are used and the hybrid orbital type is dsp^2, which is characteristic of a square planar complex. For Ni^{2+}, this is illustrated by the following scheme:

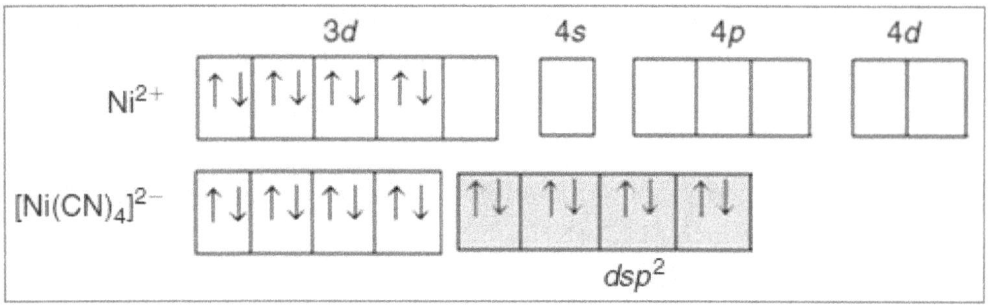

Both Co^{2+} and Ni^{2+} also have the ability to make use of 4*s*, 4*p* and 4*d* orbitals in another way by forming sp^2d^2 hybrids in the formation of complexes having a square-based pyramid structure. In fact, trigonal bipyramid and square-based pyramid structures are both observed for $[Ni(CN)_5]^{3-}$.

Limitations of Valence Bond Approach

While the VB approach, to a larger extent offers explanation of structure and magnetic behaviour of coordination compounds, it has following shortcomings:

i) It involves a number of assumptions.

ii) It does not give quantitative interpretation of magnetic data.

iii) It does not explain the colour exhibited by coordination compounds.

iv) It does not give a quantitative interpretation of the thermodynamic or kinetic

stabilities of coordination compounds.

v) It does not make exact predictions regarding the tetrahedral and square planar

structures of 4-coordinate complexes.

vi) It does not distinguish between weak and strong ligands.

Although the simple valence-bond approach to the bonding in coordination compounds has many deficiencies, it is still useful as a first attempt to explain the structure of many complexes.

SURVEY OF COMPLEXES OF FIRST-ROW METALS

Complexes form as the result of the attraction of a metal ion for electron pairs on the ligands (electron pair donors or nucleophiles). One of the most important considerations is what types of orbitals are available on the metal ion to accept electron pairs. If a metal ion has at least two empty *d* orbitals, it is possible to form a set of d^2sp^3 hybrid orbitals so that six pairs of electrons can be accommodated. Therefore, for ions that have the d^0, d^1, d^2 and d^3 configurations, complexes having a coordination number of 6 and octahedral geometry are the norm.

A second important consideration in the formation of complexes is the charge-to-size ratio of the metal ion. Ions of high charge and small size (high charge density) have the greatest affinity for electron pairs and can generally accommodate the greater buildup of negative charge that results from the acceptance of electron pairs. As a general rule, metal ions of high charge tend to form complexes in which the coordination numbers are higher, especially if they are larger ions from the second and third transition series.

A Summary of Types of Complexes Formed by First-Row Metal Ions.

Number of *d* Electrons	Most Common Ion	Usual Geometry[b]	Hybrid Orbital	Example
0	Sc^{3+}	Octahedral	d^2sp^3	$Sc(H_2O)_6^{3+}$
1	Ti^{3+}	Octahedral	d^2sp^3	$Ti(H_2O)_6^{3+}$
2	V^{3+}	Octahedral	d^2sp^3	VF_6^{3-}
3	Cr^{3+}	Octahedral	d^2sp^3	$Cr(NH_3)_6^{3+}$
4[a]	Mn^{3+}	Octahedral (h.s.)	sp^3d^2	$Mn(H_2O)_6^{3+}$
4[a]	Mn^{3+}	Octahedral (l.s.)	d^2sp^3	$Mn(CN)_6^{3-}$
5	Fe^{3+}	Octahedral (h.s.)	sp^3d^2	$Fe(H_2O)_6^{3+}$
5	Fe^{3+}	Octahedral (l.s.)	d^2sp^3	$Fe(CN)_6^{3-}$
5	Fe^{3+}	Tetrahedral	sp^3	$FeCl_4^-$
6	Co^{3+}	Octahedral (h.s.)	sp^3d^2	CoF_6^{3-}
6	Co^{3+}	Octahedral (l.s.)	d^2sp^3	$Co(H_2O)_6^{3+}$
7	Co^{2+}	Octahedral (h.s.)	sp^3d^2	$Co(H_2O)_6^{2+}$
7	Co^{2+}	Trigonal bipyramid	dsp^3	$Co(CN)_5^{3-}$
7	Co^{2+}	Tetrahedral	sp^3	$CoCl_4^{2-}$
7	Co^{2+}	Square planar	dsp^2	$Co(CN)_4^{2-}$
8	Ni^{2+}	Tetrahedral	sp^3	$Ni(NH_3)_4^{2+}$
8	Ni^{2+}	Octahedral	sp^3d^2	$Ni(NH_3)_6^{2+}$
8	Ni^{2+}	Trigonal bipyramid	dsp^3	$Ni(CN)_5^{3-}$
8	Ni^{2+}	Square base pyramid	sp^2d^2	$Ni(CN)_5^{3-}$
8	Ni^{2+}	Square planar	dsp^2	$Ni(CN)_4^{2-}$
9[a]	Cu^{2+}	Octahedral	sp^3d^2	$CuCl_6^{4-}$
9[a]	Cu^{2+}	Tetrahedral	sp^3	$Cu(NH_3)_4^{2+}$
10	Zn^{2+}	Octahedral	sp^3d^2	$Zn(H_2O)_6^{2+}$
10	Zn^{2+}	Tetrahedral	sp^3	$Zn(NH_3)_4^{2+}$
10	Ag^+	Linear	sp	$Ag(NH_3)_2^+$

[a] Irregular structure due to Jahn-Teller distortion.
[b] h.s. and l.s. indicate high and low spin, respectively.

COMPLEXES OF SECOND- AND THIRD-ROW METALS

Much of what has been said so far in applies equally well to complexes of second- and third-row transition metals. However, there are some general differences that result from the fact that atoms and ions of the second- and third-row metals are larger in size than those of first-row metals. For example, because of their larger size (when in the same oxidation state as a first-row ion), ions of metals in the second and third rows form many more complexes in which they have a coordination number greater than 6. Whereas chromium usually has a coordination number of 6, molybdenum forms $[Mo(CN)_8]^{4-}$ and other complexes in which the coordination number is 8. Other complexes of second- and third-row metals exhibit coordination numbers of 7 and 9.

Because of their having larger sizes and more filled shells of electrons between the outer shell and the nucleus, the ionization energies of second- and third-row metals are lower than those of first-row metals. Consequently, it is easier for the heavier metals to achieve higher oxidation states, which also favors higher coordination numbers.

In general, there is also a greater tendency of the heavier metals to form metal-metal bonds. The result is that there are numerous complexes that contain clusters of metal atoms or ions when the metals are from the second and third rows. This will be shown in more detail in later chapters. Another significant difference between complexes of first-row metals and those of the second and third rows involves the pairing of electrons. Earlier in this chapter, it was shown that for the d^4 ion Mn^{3+} there are two series of complexes. One series has all four of the electrons unpaired, and the bonding was interpreted as involving hybrid orbitals of the sp^3d^2 type. In the other series of compounds, there are only two unpaired electrons and the hybrid orbital type is considered to be d^2sp^3.

However, because the second- and third-row metals are larger and their orbitals are larger, it requires less energy to pair the electrons than it does for a first-row metal ion. As a result, ligands that do not force the pairing of electrons in complexes of a first-row metal may do so if the metal is from the second or third row. For example, the complex $[CoF_6]^{3-}$ is high spin, but $[RhF_6]^{3-}$ is a low-spin complex.

As a general rule, there are far fewer high-spin complexes of second- and third-row metals than there are of first-row metals. In the case of Ni^{2+} which is a d^8 ion, both tetrahedral (sp^3 hybrid orbitals) and square planar (dsp^2 hybrid orbitals) complexes are known. Complexes of the ions Pd^{2+} and Pt^{2+} (both of which are also d^8) are square planar because it is easier for the electrons to pair, which gives one empty *d* orbital, making it possible for the dsp^2 hybrid orbitals to be used. This difference in behavior will be interpreted in terms of the effects of the ligands on the *d* orbitals.

1.9 CRYSTAL FIELD THEORY (CFT)

There are several characteristics of coordination compounds that are not satisfactorily explained by a simple valence bond description of the bonding. For example, the magnetic moment of $[CoF_3]^{3-}$ indicates that there are four unpaired electrons in the complex, whereas that of $[Co(NH_3)_6]^{3+}$ indicates that this complex has no unpaired electrons, although in each case Co^{3+} is a d^6 ion. According to VBT, bonding types in these complexes involve sp^3d^2 and d^2sp^3 hybrid orbitals, respectively, but that does not provide an explanation as to why the two cases are different. Another area that is inadequately explained by a simple valence-bond approach is the number of absorption bands seen in the spectra of complexes. One of the most successful approaches to explaining these characteristics is known as crystal or ligand field theory.

When a metal ion is surrounded by anions in a crystal, there is an electrostatic field produced by the anions that alters the energies of the *d* orbitals of the metal ion. The field generated in this way is known as a *crystal field.* Crystal field theory was developed in 1929 by Hans Bethe in an attempt to explain the spectral characteristics of metal ions in crystals. It soon became obvious that anions surrounding a metal in a crystal gave a situation that is very similar to the ligands (many of which are also anions) surrounding a metal ion in a coordination compound. In cases where the ligands are not anions, they may be polar molecules, and the negative ends of the dipoles are directed toward the metal ion generating an electrostatic field. Strictly speaking, the *crystal field* approach is a purely electrostatic one based on the interactions between point charges, which is never exactly the case for complexes of transition metal ions. In view of the fact that coordinate bonds result from electron pair donation and have some covalency, the term *ligand field* is used to describe the effects of the field produced by the ligands in a complex.

In the 1930s, J. H. Van Vleck developed *ligand* field theory by adapting the crystal field approach to include some covalent nature of the interactions between the metal ion and the ligands. Before we can see the effects of the field around a metal ion produced by the ligands, it is essential to have a clear picture of the orientation of the *d* orbitals of the metal ion. The Figure 58 shows a set of five *d* orbitals, and for a gaseous ion, the five orbitals are degenerate.

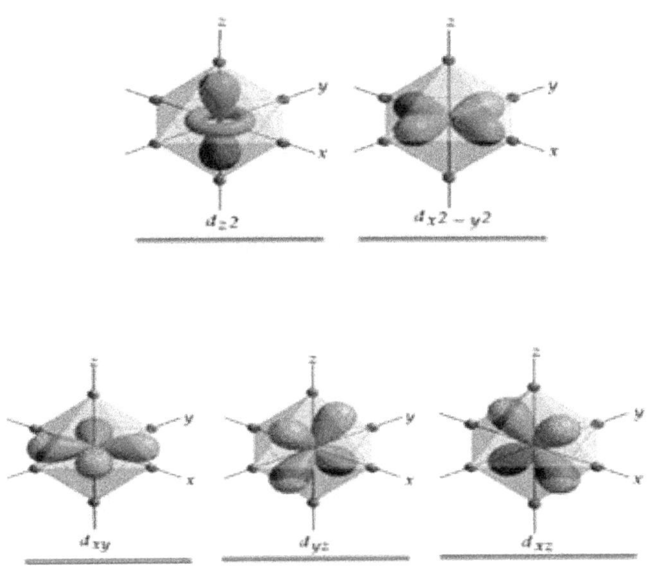

Fig. 58. Shapes of *d*- orbitals

Crystal Field Theory Assumptions

The crystal field theory is an electrostatic model which considers the metal – ligand bond to be ionic arising purely from electrostatic interactions between the metal ion and the ligand. The transition metal (central metal atom) is regarded as a positive ion and is surrounded by negative or neutral ligands which have a lone pair of electrons. If the ligand is neutral like NH_3, the negative end of the dipole in the molecule is directed towards the metal atom. The electrons on the central metal atom are under repulsive forces from the ligands; hence they occupy the d orbital furthest from the direction of the ligands.

In crystal field approach the following assumptions are made.

1. Ligands are treated as point charges.
2. There is no interaction between metal orbitals and ligand orbitals. This theory does not consider any covalent bonding (σ, π or δ) interaction or orbital overlap.
3. The d orbitals on the metal atom which were of same energy (degenerate) in an isolated atom, have their degeneracy removed by the ligands when a complex is formed. It results in splitting of d orbitals. The pattern of splitting depends on the nature of the crystal fields induced by ligands.

SPLITTING OF *d* ORBITAL ENERGIES IN OCTAHEDRAL FIELDS

If the metal ion were surrounded by a spherical electrostatic field, the energy of the *d* orbitals would be raised, but all would be raised by the same amount. As shown in Figure, an octahedral complex can be considered as a metal ion surrounded by six ligands that are located on the axes.

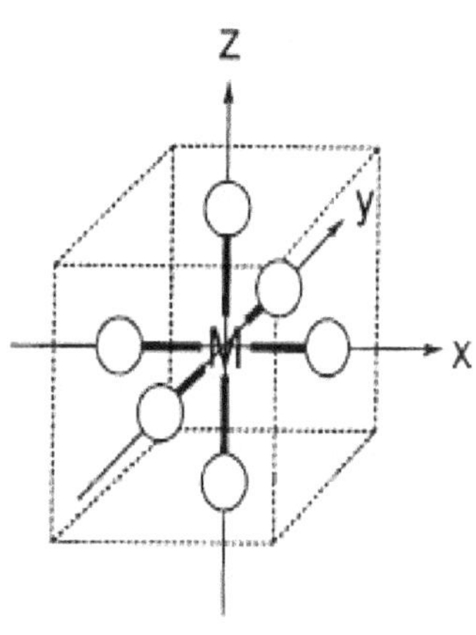

An octahedral complex with the six ligands lying on the x,y and z axes.

When six ligands surround the metal ion, the degeneracy of the *d* orbitals is removed because three of the orbitals, the d_{xy} , d_{yz} , and d_{xz} orbitals, are directed *between* the axes while the others, the d_{x2-y2} and the d_{z2}, are directed *along* the axes pointing at the ligands. Therefore, there is greater repulsion between the electrons in orbitals on the ligands and the d_{x2-y2} and the d_{z2}orbitals than there is toward the d_{xy} , d_{yz} , and d_{xz} orbitals. Because of the electrostatic field generated by the ligands, *all* of the *d* orbitals are raised in energy, but two of them are raised more than the other three.

As a result, the *d* orbitals have energies that can be represented as shown in Figure.

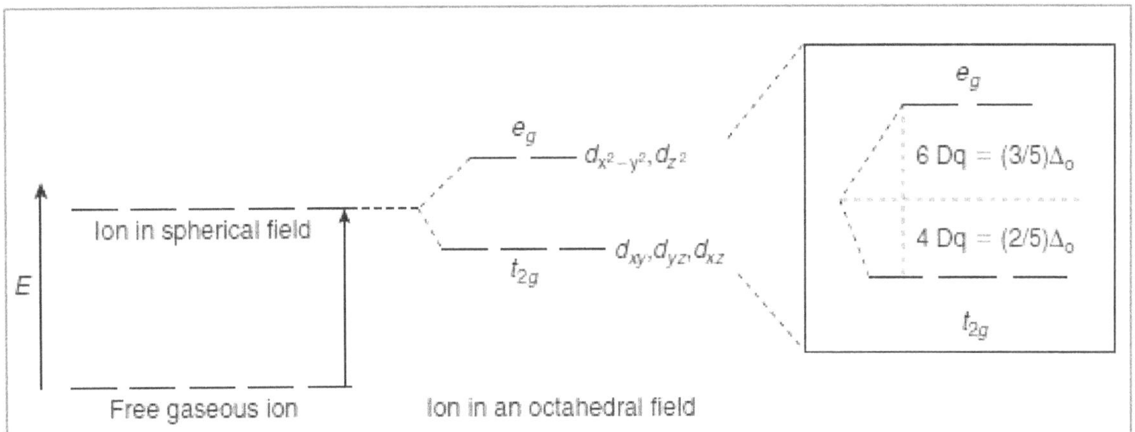

Splitting of the d orbitals in a crystal field of octahedral symmetry

The two orbitals of higher energy are designated as the e_g orbitals, and the three orbitals of lower energy make up the t_{2g} orbitals. These designations will be described in greater detail later, but the "*g*" subscript refers to being symmetrical with respect to a center of symmetry that is present in a structure that has *Oh* symmetry. The "*t* " refers to a triply degenerate set of orbitals, whereas "*e* " refers to a set that is doubly degenerate. The energy separating the two groups of orbitals is called the *crystal* or *ligand field splitting*, Δ_o. Splitting of the energies of the *d* orbitals as indicated in the figure occurs in such a way that the overall energy remains unchanged and the "center of energy" is maintained. The e_g orbitals are raised 1.5 times as much as the t_{2g} orbitals are lowered from the center of energy (Barry cetnre). Although the splitting of the *d* orbitals in an octahedral field is represented as Δ_o, it is also sometimes designated as 10 Dq, where Dq is an energy unit for a particular complex. The two orbitals making up the e_g pair are raised by 3/5 Δ_o while the t_{2g} orbitals are lowered by 2/5 Δ_o relative to the center of energy. In terms of Dq units, the two e_g orbitals are raised by 6 Dq while the three t_{2g} orbitals are 4 Dq lower than the center of energy.

The magnitude of Δ_o depends on the following factors

i) **Nature of ligand:** Studies have shown that keeping the same central metal atom, on varying the ligands attached to it, the magnitude of Δo is different in each case. Ligands which cause only a small degree of splitting are termed weak field ligands and those which cause a large splitting are called strong field ligands. The common ligands are arranged in order of their strength and the series is called the **spectrochemical series.**

Spectrochemical Series

Weak field ligands

$I^- < Br^- < S^{2-} < SCN^- < Cl^- < N_3^-, F^- <$ urea, $OH^- <$ ox, $O^{2-} < H_2O < NCS^- <$ py, $NH_3 <$ en $<$ bpy, phen $< NO_2^- < CH_3^-, C_6H_5^- < CN^- < CO$. **Strong field ligands**

ii) **Charge on the metal atom:**

The magnitude of Δo increases as the charge on central metal atom increases and in general M^{3+} complexes have double the value than that of M^{2+} complexes.

> High oxidation state favors large Δ
> - $Mn^{2+} < Ni^{2+} < Co^{2+} < Fe^{2+} < V^{2+} < Fe^{3+} < Co^{3+} < Mn^{4+} < Mo^{3+} < Rh^{3+} < Ru^{3+} < Pd^{4+} < Ir^{3+} < Pt^{4+}$

iii) Position of the metal in the transition series

The magnitude of Δ_o increases on descending down the group of transition metals. Depending upon the d electron configuration of the metal ion, and under specific environment around it, its complex acquires stability.

Illustration of CFSE

The effect of crystal field splitting is easily seen by studying the absorption spectrum of $[Ti(H_2O)_6]^{3+}$because the Ti^{3+} ion has a single electron in the 3*d* orbitals. In the octahedral field produced by the six water molecules, the 3*d* orbitals are split in energy. The only transition possible is promotion of the electron from an orbital in the t_{2g} set to one in the e_g set. This transition gives rise to a single absorption band, the maximum of which corresponds directly to the energy represented as Δ_o. As expected, the spectrum shows a single, broad band that is centered at 20,300 cm^{-1}, which corresponds directly to Δ_o.

$[Ti(H_2O)_6]^{3+}$: a d^1 complex and the electron occupies the lowest energy orbital, i.e. one of the three degenerate t_{2g} orbitals. The purple colour is a result of the absorption of light which results in the promotion of this t_{2g} electron into the e_g level. $t_{2g}^1\ e_g^0 \rightarrow t_{2g}^0\ eg^1$.The UV-Vis absorption spectrum reveals that this transition occurs with a maximum at 20300 cm-1 which corresponds to Δo 243 kJ/mol. *(1000 cm-1 = 11.96 kJ/mol or 2.86 kcal/mol or 0.124 eV.)*

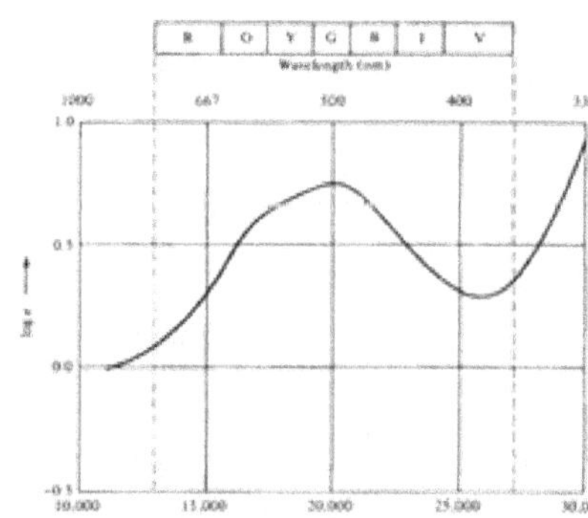

Fig. 59. UV-Visible spectrum of Titanium complex

The energy associated with this band is found to be 243kJ/mol. This energy (243 kJ/mol) is large enough to give rise to other effects when a metal ion is surrounded by six ligands. However, only for a d^1 ion is the interpretation of the spectrum this simple. When more than one electron is present in the *d* orbitals, the electrons interact by spin-orbit coupling. Any transition of an electron from the t_{2g} to the e_g orbitals is accompanied by changes in the coupling scheme when more than one electron is present.

Values of Δ_o for Octahedral Complexes of First-Row Transition Metal Ions.				
	Splittings for several ligands in cm^{-1}			
Metal ion	F^-	H_2O	NH_3	CN^-
Ti^{3+}	17,500	20,300		23,400
V^{3+}	16,100	18,500		26,600
Cr^{3+}	15,100	17,900	21,600	35,000
Fe^{3+}	14,000	14,000		34,800
Co^{3+}	13,000	20,800	22,900	33,800
Fe^{2+}		10,400	11,000	32,200
Co^{2+}		9,300	10,200	
Ni^{2+}		8,500	10,800	

Variation in Δ_o with Row in the Periodic Table.			
$[Fe(H_2O)_6]^{3+}$	$[Fe(ox)_3]^{3-}$	$[Co(H_2O)_6]^{3+}$	$[Co(NH_3)_6]^{3+}$
14,000	14,140	20,800	22,900
$[Ru(H_2O)_6]^{3+}$	$[Ru(ox)_3]^{3+}$	$[Rh(H_2O)_6]^{3+}$	$[Ru(NH_3)_6]^{3+}$
28,600	28,700	27,200	34,000
			$[Ir(NH_3)_6]^{3+}$
			41,200
Ligand field splitting (in cm^{-1}) is shown below the formula for the complexes shown.			

Table shows representative values for Δ_o in complexes containing ions of first-row metals with H_2O, NH_3, F^- and CN^- as ligands. In general, the aqua complexes of first-row transition metals have ligand field splittings of approximately 8000 to 10,000 cm^{-1} for the +2 complexes, $[M(H_2O)_6]^{2+}$, and approximately 14,000 to 21,000 cm^{-1} for the complexes of the +3 ions, $[M(H_2O)_6]^{3+}$. In most cases, there is a 50% to 100% increase in Δ_o for the +3 ion compared to the +2 ion of the same metal. For example, the value of Δ_o is 22,870 cm^{-1} for $[Co(NH_3)_6]^{3+}$ but it is only 10,200 cm^{-1} for $[Co(NH_3)_6]^{2+}$. It is interesting to note that for $[CoF_6]^{3-}$, the value of Δ_o is only about 13,000 cm^{-1}, but the energy required to force pairing of electrons in Co^{3+} is approximately 20,000 cm^{-1}. Therefore, $[CoF_6]^{3-}$, is a high-spin complex. The crystal field splitting in $[Co(NH_3)_6]^{3+}$ is 22,900 cm^{-1}, which is sufficient to force pairing, so a low-spin complex results. There is an increase of approximately 30% to 50% in Δ_o on going from a first-row transition metal to a second-row metal and another 30% to 50% increase on going from a second-row to a third-row metal when the same d^n configuration and oxidation state are involved.

Crystal field stabilization energy (CFSE)

The crystal field stabilization energy (CFSE) is the energy by which the complex is stabilized relative to the free metal atom where there is no splitting of d orbitals. It can be calculated by:

CFSE= {(n_{eg} x 3/5) – (n_{t2g} x 2/5)} Δ_0 or CFSE= {(n_{eg} x 0.6) – (n_{t2g} x 0.4)} Δ_0

where $\mathbf{n_{eg}}$ = number of electrons in e_g orbitals and $\mathbf{n_{t2g}}$ = number of electrons in t_{2g} orbitals. Consider the case of d^{10} ion where under the influence of an octahedral ligand field four electrons will move to higher energy e_g orbitals and six electrons to lower energy t_{2g} orbitals. Thus, the result for d^{10} system (eg. Zn^{2+}) is no change in energy.

CFSE = {(4 X 3/5) – (6 X 2/5)}Δ_0 = 0

A similar result would be obtained for a d^0 (e.g. Sc^{3+}) and a d^5 (e.g. Mn^{2+}, Fe^{3+}) electron configuration. In other words such ions would not be affected by the field of the ligands. Let us consider the other cases.

Taking for example $[Ti(H_2O)_6]^{3+}$ in which metal ion has d^1 configuration the CFSE is $-0.4\Delta_o$. Similarly for d^2 configuration (as in $[V(H_2O)_6]^{3+}$) the CFSE is $-0.8\Delta_o$ and in d^3 (Cr^{3+}) and d^8 (Ni^{2+}) it will be $-1.2\Delta_o$.

The ordering of the energy levels for a metal ion in an octahedral field makes it easy to visualize how high- and low-spin complexes arise when different ligands are present. If there are three or fewer electrons in the 3*d* orbitals of the metal ion, they can occupy the t_{2g} orbitals with one electron in each orbital.

If the metal ion has a d^4 configuration (e.g., Mn^{3+}), the electrons can occupy the t_{2g} orbitals only if pairing occurs, which requires that Δ_o be larger in magnitude than the energy necessary to force electron pairing, *P*. The result is a low-spin complex in which there are two unpaired electrons. If Δ_o is smaller than the pairing energy, the fourth electron will be in one of the e_g orbitals, which results in a high-spin complex having four unpaired electrons. It can be assumed that the electrons added beyond d^3 will occupy e_g orbitals in preference to pairing up in t_{2g} orbitals. This is not unreasonable, since there is a pairing energy (Table 18) associated with bringing electrons close to form a pair in an orbital. However, if the interaction between the metal ion and the ligands is sufficiently large so that $\Delta_o > PE$, it becomes more energetically favourable for the fourth electron added to occupy a t_{2g} orbital.

Ligands which produce this effect are known as **strong-field ligands** and the complexes they form are called **low-spin** complexes. Whereas ligands for which $\Delta_o < PE$ are known as **weak field ligands** and they form **high-spin** complexes.

Thus, when calculating the CESE of a strong-filed low-spin complex it is necessary to include the pairing energies associated with the additional electron pairs. In such cases CFSE for a low-spin complex ($CFSE_L$) can be written as:

$CFSE_L = \{(n_{eg} \times 3/5) - (n_{t2g} \times 2/5)\}\Delta_0 + n(PE)$ or
$CFSE_L = \{(n_{eg} \times 0.6) - (n_{t2g} \times 0.4)\}\Delta_0 + n(PE)$
Net gain in strong field ligand
where, n(PE) = number of additional electron pairs for a low-spin case. It is important to note that that for a given metal ion the pairing energy (PE) is constant.

Table 18. Crystal fiels stabilisation energy (CFSE) for complexes in octahedral field

Electronic configurations in octahedral crystal field and CFSE

d electron configuration	Arrangement in weak field ligand							Arrangement in strong field ligand							Net gain in strong field ligand
	t_{2g}			e_g		CFSE (Δ_o)	Spin only magnetic moment	t_{2g}			e_g		CFSE (Δ_o)	Spin only magnetic moment	
d^1	↑					-0.4	1.73	↑					-0.4	1.73	0
d^2	↑	↑				-0.8	2.83	↑	↑				-0.8	2.83	0
d^3	↑	↑	↑			-1.2	3.87	↑	↑	↑			-1.2	3.87	0
d^4	↑	↑	↑	↑		-0.6	4.9	↑↓	↑	↑			-1.6+PE	2.83	-1+PE
d^5	↑	↑	↑	↑	↑	0	5.92	↑↓	↑↓	↑			-2+2PE	1.73	-2+2PE
d^6	↑↓	↑	↑	↑	↑	-0.4	4.9	↑↓	↑↓	↑↓			-2.4+2PE	0	-2+2PE
d^7	↑↓	↑↓	↑	↑	↑	-0.8	3.87	↑↓	↑↓	↑↓	↑		-1.8+PE	1.73	-1+PE
d^8	↑↓	↑↓	↑↓	↑	↑	-1.2	2.83	↑↓	↑↓	↑↓	↑	↑	-1.2	2.83	0
d^9	↑↓	↑↓	↑↓	↑↓	↑	-0.6	1.73	↑↓	↑↓	↑↓	↑↓	↑	-0.6	1.73	0
d^{10}	↑↓	↑↓	↑↓	↑↓	↑↓	0	0	↑↓	↑↓	↑↓	↑↓	↑↓	0	0	0

Thus, when the configuration is d^4, two cases arise: high-spin case and low-spin case.
(i) **d^4 high spin case**

This case arises when the ligands attached to the metal are weak field ligands. The electronic configuration of d orbitals in this case is $t_{2g}^{3}\ e_{g}^{1}$ and the CFSE will be

$(3 \times -0.4\Delta o)+0.6\Delta o = -0.6\Delta o$

(ii) **d^4 low spin case**

This case arises when the ligands attached to the metal are strong field ligands. The electronic configuration of d orbitals in this case is $t_{2g}^{4}\ e_{g}^{0}$ and the CFSE will be

$(4\times -0.4\Delta o) = -1.6\Delta o+ (PE)$.

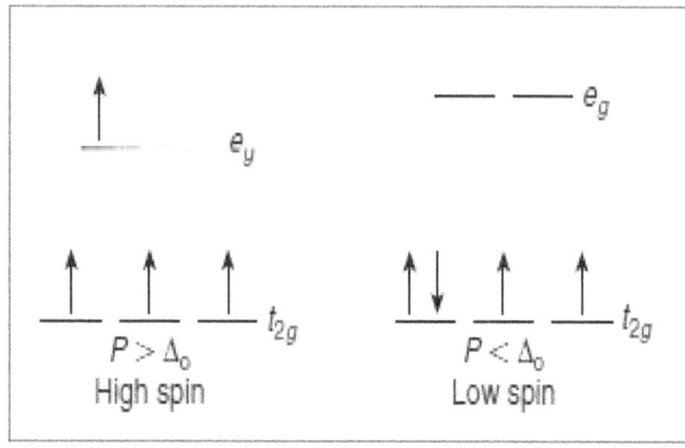

The two cases differ in the number of unpaired electrons. The CFSE in the second case is higher than in the first case. Similar situation can be observed in complexes of metal ions with d^5, d^6 and d^7 configurations. High- and low- spin configurations in an octahedral crystal field for d^6 (Fe^{2+}). Pairing energy, PE (Table 19) depends mainly on the electronic distribution of the central metal ions and its value can be obtained from the atomic spectra of the elements. The Δ_o, which depends on metal ion, ligand and the geometry of the complex, is obtained from the electronic spectra of the complex. Consideration of the data related to PE and Δo for metal ions of d^4 to d^7 configurations show that when $\Delta_o > P$, low-spin complexes are formed and when $\Delta_o < P$ then high-spin complexes are formed. The crystal field stabilization energy increases the thermodynamic stability of the complexes.

Table 19. Pairing energies of some transition metal ions

Crystal field splittings, Δ_0, and mean electron-pairing energies, PE, for some transition metal ions

Configuration	Ion	PE cm^{-1}	Ligand	Δ_0 cm^{-1}	Spin State	
					Predicated	Observed
D^4	Cr^{2+}	23,500	6 H_2O	13,900	High	High
	Mn^{3+}	28,000	6 H_2O	21,000	High	High
D^5	Mn^{2+}	25,500	6 H_2O	7,800	High	High
	Fe^{3+}	30,000	6 H_2O	13,700	High	High
D^6	Fe^{2+}	17,600	6 H_2O	10,400	High	High
			6 CN^-	33,000	Low	Low
	Co^{3+}	21,000	6 F^-	13,000	High	High
			6 NH_3	23,000	Low	Low
D^7	Co^{2+}	22,500	6 H_2O	9,300	High	High

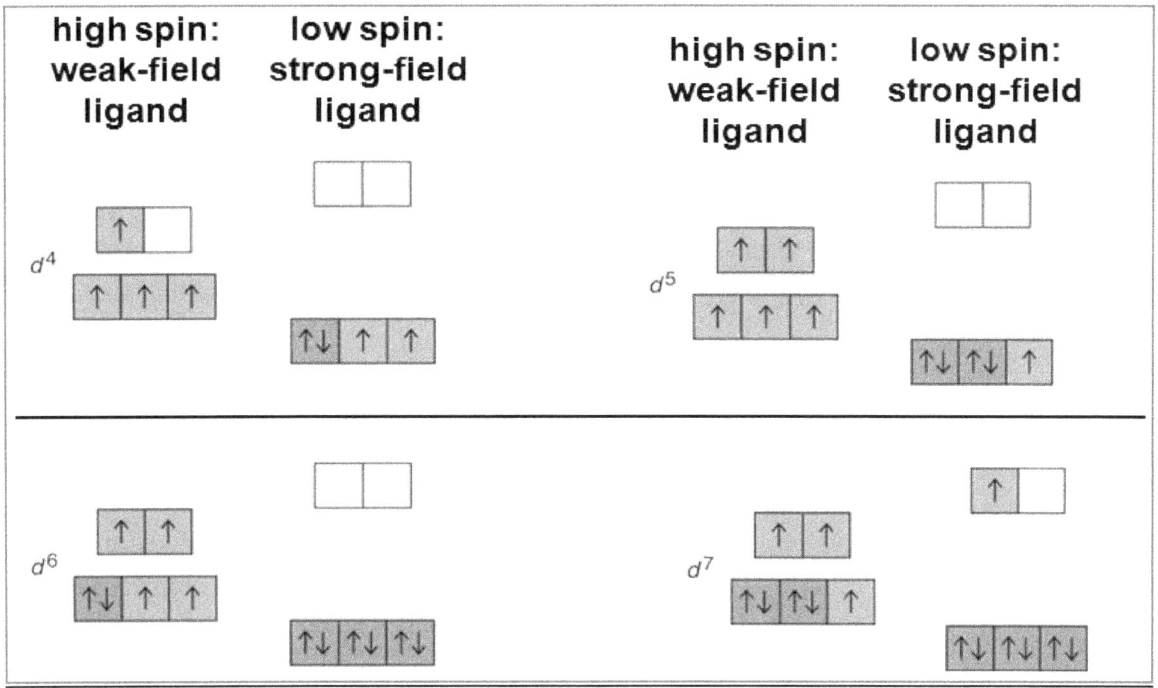

Values of Δ_o, P, CFSE and Resulting Spin State

Complex	d^n	P (cm^{-1})	Δ_o (cm^{-1})	State	CFSE
$[Cr(H_2O)_6]^{2+}$	d^4	18,800	13,900	high	$-0.6\Delta_o$
$[Fe(H_2O)_6]^{3+}$	d^5	24,000	13,700	high	0
$[Fe(H_2O)_6]^{2+}$	d^6	14,100	10,400	high	$-0.4\Delta_o + P$
$[Fe(CN)_6]^{4-}$	d^6	14,100	33,000	low	$-2.4\Delta_o + 3P$
$[CoF_6]^{3-}$	d^6	16,800	13,000	high	$-0.4\Delta_o + P$
$[Co(NH_3)_6]^{3+}$	d^6	16,800	23,000	low	$-2.4\Delta_o +3P$
$[Co(H_2O)_6]^{2+}$	d^7	18,000	9,3000	high	$-0.8\Delta_o + 2P$

Consequences of Crystal Field Splitting

In addition to the changes produced in the spectra that arise from *d* - *d* transitions, there are other effects produced as a result of the splitting of the *d* orbital energies.

1) Hydration enthalpy

The hydration process for an ion can be shown as $M^{z+}(g) + 6\ H_2O(l) \rightarrow [M(H_2O)_6]^{z+}(aq)$. The heat of hydration of an ion is related to its size and charge. However, in this case the aqua complex that is formed causes the *d* orbitals to be split in energy, and if the metal ion has electrons in the *d* orbitals, they will populate the t_{2g} orbitals, which have lower energy. This results in a release of energy over and above that produced by the hydration of an ion having a specific size and charge. High-spin (weak-field) aqua complexes normally result from the hydration of first-row transition metal ions. The actual amount of energy released will be dependent on the number of electrons in the d orbitals. For a d^1 ion, the heat of hydration will be augmented by 4 Dq.

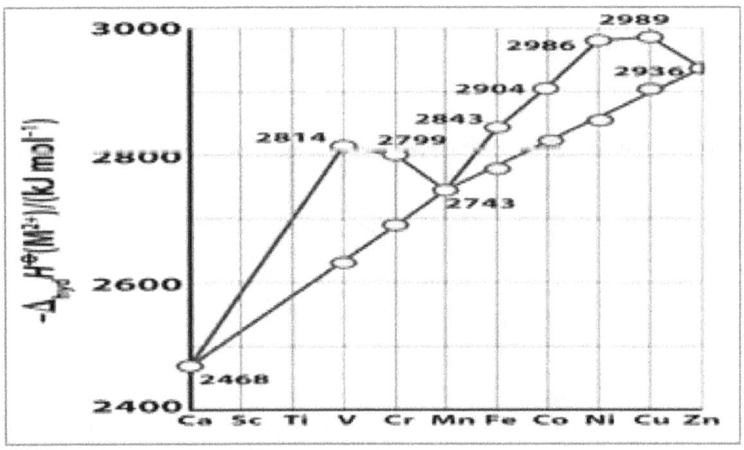

Fig. 60. Heat of hydration for 3*d*-transition metal ions

If the electron configuration is d^2, there will be 8 Dq released in addition to the heat of hydration for an ion of that size and charge.

Hydration of the metal ions produces an enthalpy change that is commensurate with the size and charge of the ion with the addition of the number of Dq units. For d^0, d^5, and d^{10} there is no additional stabilization of the aqua complex since these cases have no ligand field stabilization. The Figure 60 shows what has become known as the " double-humped " appearance that reflects the fact that the ligand field stabilization energy for the aqua complexes begins at 0, increases to 12 Dq, then drops to 0 on going from d^0 to d^5 and rcpcats thc trend on going from d^6 to d^{10}.

2) **Lattice energy**

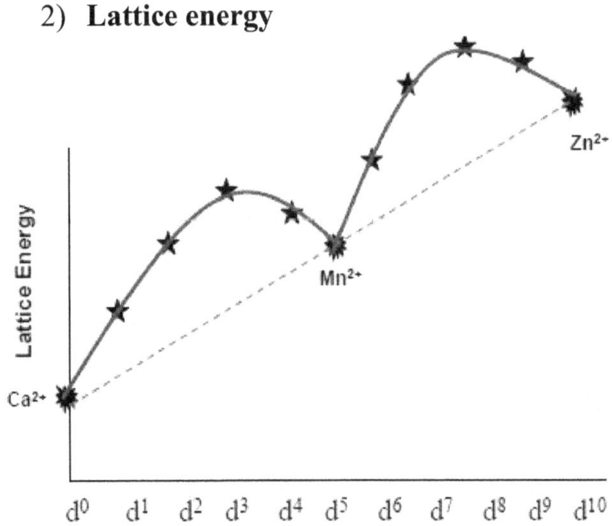

Fig. 61. Lattice energy of 3*d*-transition metal fluorides

As gaseous ions form a crystal lattice, the cations become surrounded by anions and vice versa in the solid crystal. If the cations are surrounded by six anions in an octahedral arrangement, the d orbitals will become split in energy, as illustrated earlier. When the orbitals are populated with one or more electrons, the energy released when the lattice forms will be greater by an amount that reflects the ligand field stabilization energy corresponding to the number of electrons in the d orbitals. A graph of the lattice energy for the chloride compounds of the first-row transition metals in the +2 oxidation state. The figure 61 shows the same general shape as the graph representing the hydration enthalpies of the metal ions. This is because the ligand field around the metal ion stabilizes both the hydrates and the crystal lattices in a way that depends on the number of electrons in the *d* orbitals.

SPLITTING OF *d* ORBITAL ENERGIES IN TETRAHEDRAL FIELDS

A regular tetrahedron is related to the cube with the metal ion occupying the centre and ligands occupy alternate corners of the cube. The e_g orbitals point along x, y and z axes and t_{2g} orbitals point in between the axes. The three axes point to the centres of the faces of the cube. Therefore the direction of approach of the ligands does not coincide with either of the two set of orbitals. The angle between an e_g orbital, the central metal and the ligand is half of the tetrahedral angle i.e., $54^o\ 44'$. The angle between a t_{2g} orbital, the central metal and the ligand is $35^o\ 16'$. Thus, the t_{2g} orbitals are nearer to the ligands as compared to the e_g orbitals. Although the approach of the ligands raises the energy of all the d orbitals but since the t_{2g} orbitals are nearer to the ligands they are raised more in energy. Therefore the crystal field splitting is opposite to that in case of octahedral complexes. The t_{2g} orbitals in tetrahedral case are raised by $0.4\Delta_t$ above the zero point energy while the other set of orbitals are $0.6\Delta_t$ below the average.

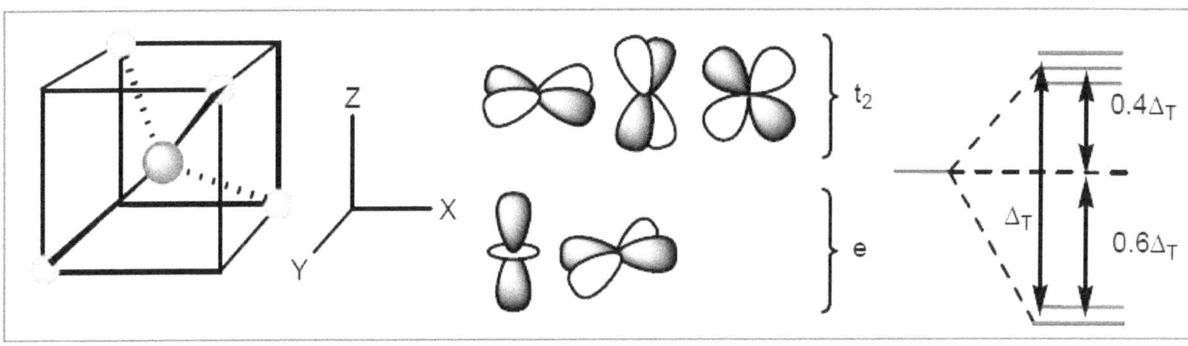

The magnitude of crystal field splitting in tetrahedral complexes (Δt) is considerably less than in octahedral complexes. Since there are only four ligands, the ligand field is only 2/3 times of the octahedral field and since the orbitals do not coincide with the direction of the ligands, the crystal field splitting is further reduced by a factor of 2/3.Therefore the crystal field splitting in tetrahedral complexes (Δt) is approximately 4/9 of the octahedral field (Δo). Unlike the octahedral complexes, where a strong field ligand causes a larger splitting compared to weak field ligand, the tetrahedral Δt is always very small and energetically not favourable to pair electrons. Therefore all tetrahedral complexes are high spin complexes.

There are several differences between the splitting in octahedral and tetrahedral fields. Not only are the two sets of orbitals inverted in energy, but also the splitting in the tetrahedral field is much smaller than that produced by an octahedral field. First, there are only four ligands producing the field rather than the six ligands present in the octahedral complex. Second, none of the *d* orbitals point directly at the ligands in the tetrahedral field. In an octahedral complex, two of the orbitals point directly toward the ligands and three point between them. As a result, there is a maximum energy splitting effect on the *d* orbitals in an octahedral field. In fact, it can be shown that if identical ligands are present in the complexes and the metal-to-ligand distances are identical, $\Delta_t = (4/9)\ \Delta_o$.

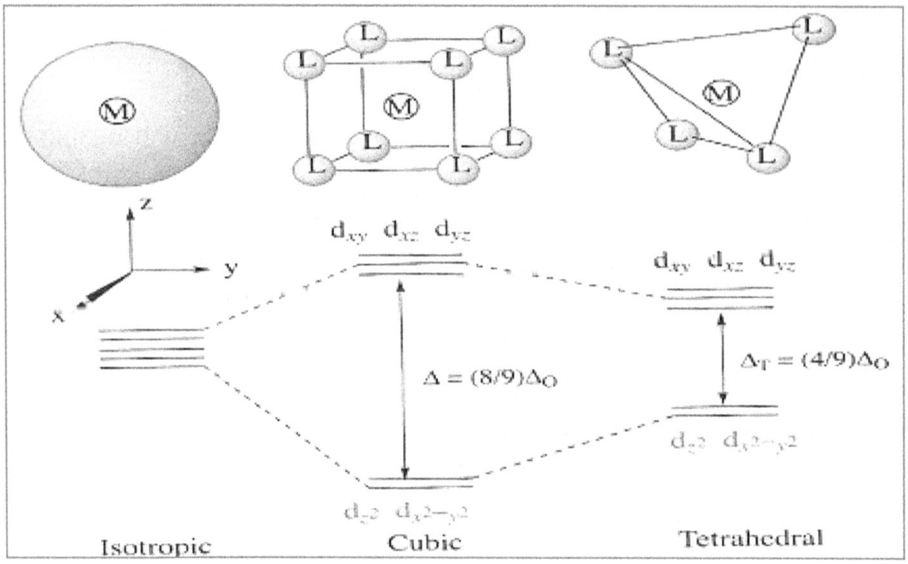

The result is that *there are no low-spin tetrahedral complexes* because the splitting of the *d* orbitals is not large enough to force electron pairing. Third, because there are only four ligands surrounding the metal ion in a tetrahedral field, the energy of all of the *d* orbitals is raised less than they are in an octahedral complex. The subscripts " *g* " do not appear on the subsets of orbitals because there is no center of symmetry in a tetrahedral structure.

The CFSE for tetrahedral complexes and its comparison with the octahedral CFSE is given in Table. The CFSE for d^0, d^5 and d^{10} configurations is zero in both octahedral and tetrahedral cases. In all other cases the octahedral CFSE is greater than that of tetrahedral CFSE. Overall octahedral complexes are more stable and more common than tetrahedral complexes, partly because of the formation of six bonds instead of four and also because of larger CFSE.

Δ_T for representative tetrahedral complexes

Complex	Δ_T/cm^{-1}
VCl_4	9010
$[CoCl_4]^{2-}$	3300
$[CoBr_4]^{2-}$	2900
$[CoI_4]^{2-}$	2700
$[Co(NCS)_4]^{2-}$	4700

Tetrahedral complexes are favoured when,

(i) the ligands are large and bulky and could cause crowding in an octahedral complex,

(ii) the ligands are weak field, so that the loss in CFSE is not that important,

(iii) the central metal has a low oxidation state and as a result the magnitude of Δ_t is small, and

(iv) the electronic configuration of the central metal is d^0, d^5 or d^{10} as there is no CFSE in these cases or d^1 or d^6 where the loss in CFSE is small.

Table 20. Crystal fiels stabilisation energy (CFSE) for complexes in tetrahedral field

Electronic configurations in tetrahedral field and CFSE										
d electron configuration	Arrangement of electrons					Spin only magnetic moment (in B.M.)	Tetrahedral CFSE(Δ_t)	Tetrahedral CFSE on octahedral CFSE scale (Δ_t =4/9 Δ_o)	Octahedral CFSE	
	e_g		t_{2g}						Weak field	Strong field
D^1	↑					1.73	-0.6	-0.27	-0.4	-0.4
D^2	↑	↑				2.83	-1.2	-0.53	-0.8	-0.8
D^3	↑	↑	↑			3.87	-0.8	-0.36	-1.2	-1.2
D^4	↑	↑	↑	↑		4.90	-0.4	-0.18	-0.6	-1.6
D^5	↑	↑	↑	↑	↑	5.92	0	0	0	-2.0
D^6	↑↓	↑	↑	↑	↑	4.90	-0.6	-0.27	-0.4	-2.4
D^7	↑↓	↑↓	↑	↑	↑	3.87	-1.2	-0.53	-0.8	-1.8
D^8	↑↓	↑↓	↑↓	↑	↑	2.83	-0.8	-0.36	-1.2	-1.2
D^9	↑↓	↑↓	↑↓	↑↓	↑	1.73	-0.4	-0.18	-0.6	-0.6
D^{10}	↑↓	↑↓	↑↓	↑↓	↑↓	0	0	0	0	0

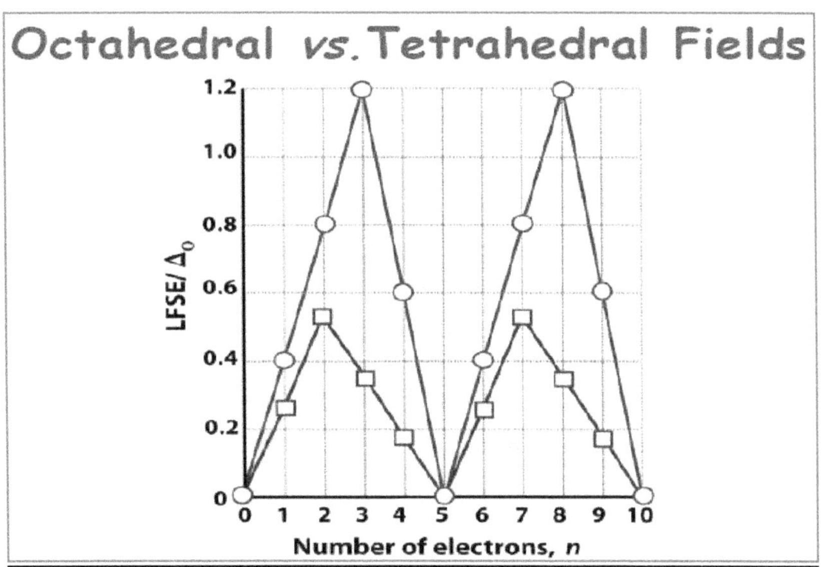

1.10 JAHN-TELLER DISTORTION OR J-T EFFECT

Tetragonal distortion of octahedral complexes (Jahn-Teller distortion or J-T Effect)

The stereochemistry or shape of transition metal complexes is determined by the tendency of electron pairs to occupy positions as far away from each other as possible. If the d electrons are symmetrically arranged with respect to an octahedral field they will repel all the six ligands equally and a regular octahedral geometry is achieved. The various symmetrical electronic arrangements are observed in d^0 (strong and weak field), d^3 (strong and weak field), d^5 (weak field), d^6 (strong field), d^8 (weak field) and d^{10} (strong and weak field).In all other cases the electronic arrangements are unsymmetrical.

As the t_{2g} orbitals point in between the ligands, the asymmetric filling of these orbitals has little effect on the shape of the complex. However, the e_g orbitals point directly towards the ligands and as a result the asymmetric filling of these orbitals leads to more repulsion of some ligands compared to others. This leads to a significant deviation from octahedral geometry. The reason for this is that in an octahedral environment if e_g orbitals, d_{x2-y2} and d_{z2} are unsymmertically filled (for example $t_{2g}^3\ e_g^1$, $t_{2g}^6\ e_g^1$ and $t_{2g}^6\ e_g^3$) their degeneracy is removed and the two orbitals become unequal in energy. Out of the two orbitals, the d_{z2} orbital has its two lobes pointing towards z and –z axes while the d_{x2-y2} has four lobes pointing towards x, y, -x and –y axes. In order to minimize the repulsion with the ligands the single e_g electron will occupy the d_{z2} orbital.

If bot the e_g orbitals are symmetrically filled - all ligands are repelled equally. **Result:** regular octahedron

If asymmetrically filled - some ligands are repelled more than the other . **Result:** Distorted octahedron

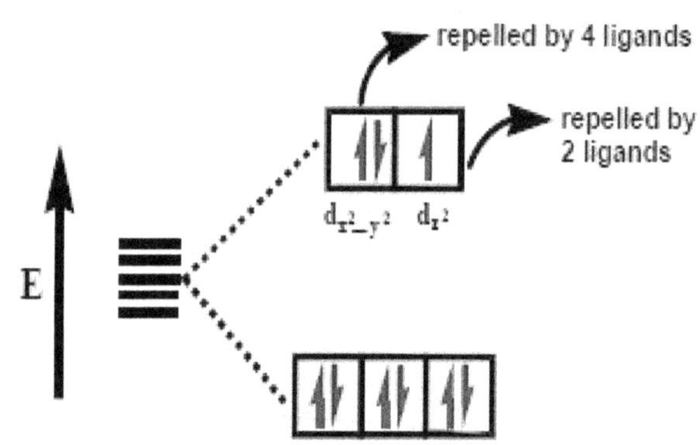

Therefore two ligands approaching along the z and -z axes will experience greater repulsion than the four ligands along the other four axes. This situation results in tetragonal distortion or strictly speaking **tetragonal elongation** of octahedral complexes with four short and two long bonds.

In case where the d_{x2-y2} orbital contains the extra electron, then the elongation will occur along the x and y axes and, therefore, ligands approach more closely along the z axis. Thus there will be four long bonds and two short bonds. This means compression of octahedron along the z-axis and is called **tetragonal compression**. The tetragonal elongation is much more common.

Thus one can come to a conclusion that, when there is unequal filling of d_{z2} and d_{x2-y2} orbitals, distortion occurs. This is known as **Jahn-Teller distortion (Jahn-Teller Effect)**.

The **Jahn-Teller Theorem** was published in 1937 and states:

"any non-linear molecular system in a degenerate electronic state will be unstable and will undergo distortion to form a system of lower symmetry and lower energy thereby removing the degeneracy"

Tetragonal elongation: Consider e_g configuration: $(d_{z2})^2 (d_{x2-y2})^1$

Ligands along z, -z will be repelled more and bonds elongated. i.e. the octahedron will be elongated along the z axis.

Tetragonal compression: **Consider e_g configuration: $(d_{z2})^1 (d_{x2-y2})^2$**

Ligands along x, -x, y, -y will be repelled more and bonds elongated. i.e., the octahedron will be compressed along the z axis.

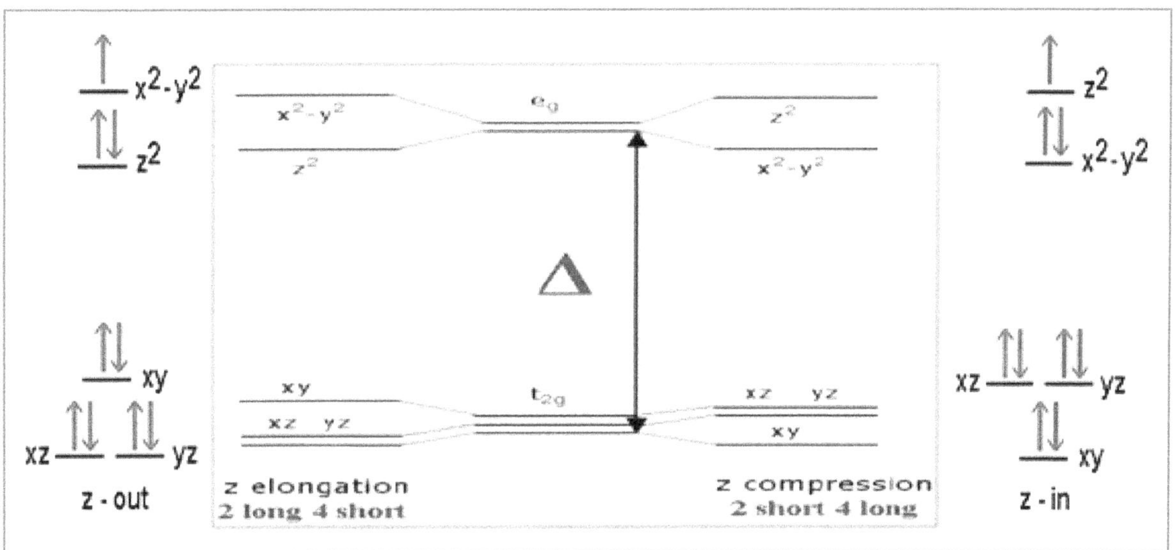

Compounds containing approximately octahedral Cu^{2+} (d^9- t_{2g}^6 e_g^3), Mn^{3+} (d^4 - $t_{2g}^3 e_g^1$) and L.S. Ni^{3+} (d^7 - $t_{2g}^6 e_g^1$) often display distorted coordination environments as the distortion breaks the degeneracy of the octahedral ground state.

Many Cu(II) complexes (e.g. $[Cu(NO_2)_6]^{4-}$), where Cu^{2+} has t_{2g}^6 e_g^3 configuration, show tetragonally distorted octahedral structures. Here also to minimize repulsion with the ligands, two electrons occupy the d_{z2} orbital and one occupies the d_{x2-y2} orbital. One experimental verification of this phenomenon comes in the form of bond lengths. In $CuCl_2$, the Cu^{2+} is surrounded by six Cl^- ions, and the equatorial Cu – Cl bonds are 230 pm in length, whereas the axial bonds measure 295 pm. Bond distances in many other copper(II) compounds could also be cited to illustrate Jahn-Teller distortion. Other examples of this distortion include CrF_2 and MnF_3 where metal ions of d^4 configuration are octahedrally surrounded by six weak field ligands, F-ions.

Tetragonal elongation - Jahn-Teller Effect: Schematic representation

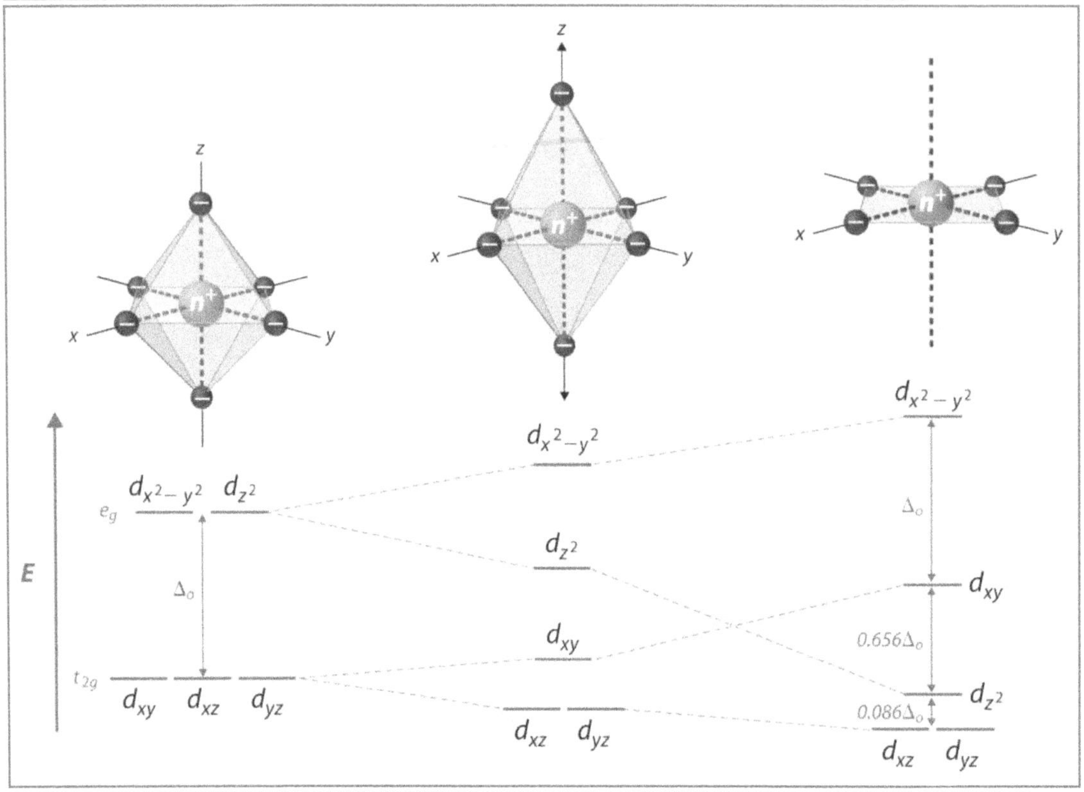

Jahn-Teller distorted complexes

$CuBr_2$	4 Br at 240pm 2 Br at 318pm
$CuCl_2.2H_2O$	2 O at 193pm 2 Cl at 228pm 2 Cl at 295pm
$CsCuCl_3$	4 Cl at 230pm 2 Cl at 265pm
CuF_2	4 F at 193pm 2 F at 227pm
$CuSO_4.4NH_3.H_2O$	4 N at 205pm 1 O at 259pm 1 O at 337pm
K_2CuF_4	4 F at 191pm 2 F at 237pm
CrF_2	4 F at 200pm 2 F at 243pm
$KCrF_3$	4 F at 214pm 2 F at 200pm
MnF_3	2 F at 209pm 2 F at 191pm 2 F at 179pm

Structures of complexes with J-T distortion

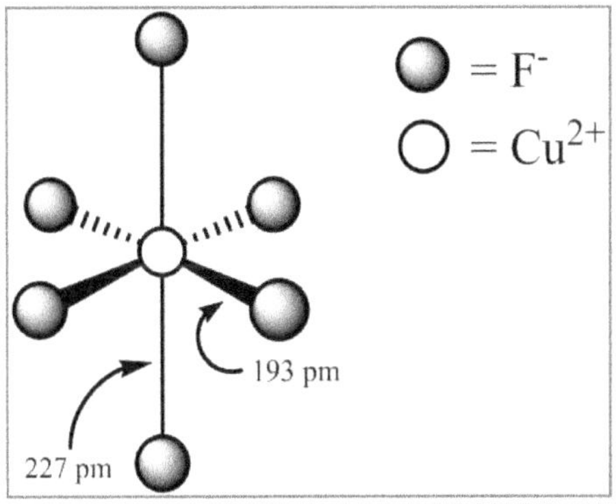

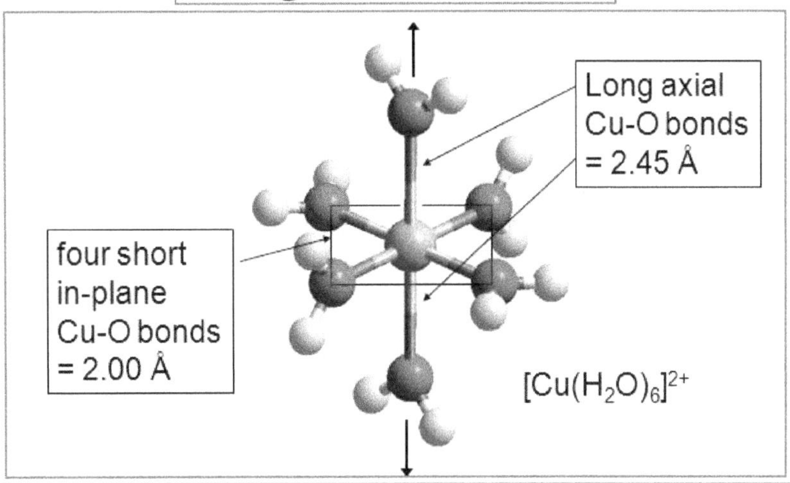

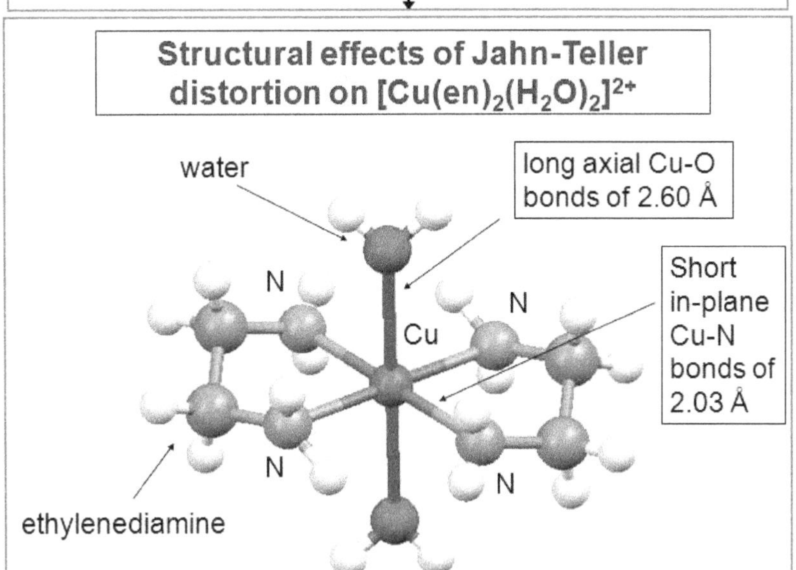

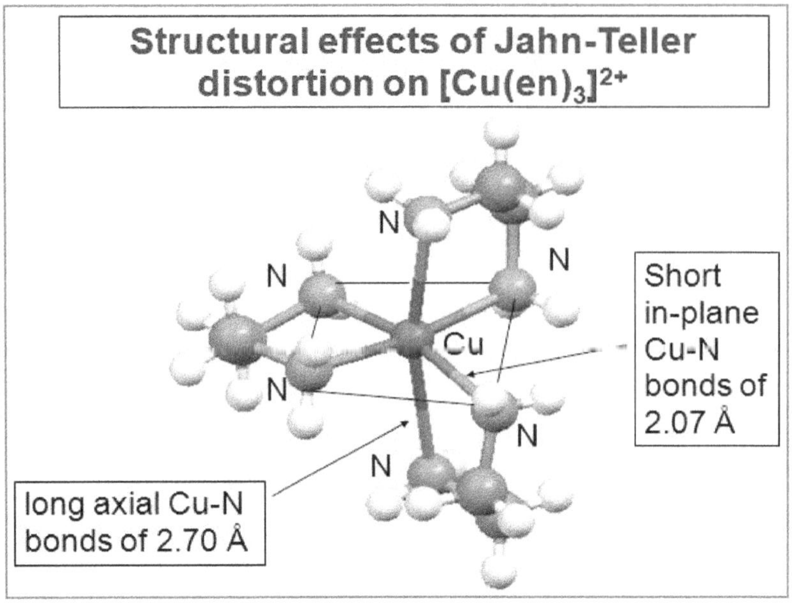

Square planar complexes – the low-spin d^8 metal ions
All high-spin d^8 metal ions are octahedral (or tetrahedral), but the low-spin d^8 metal ions are all square planar.

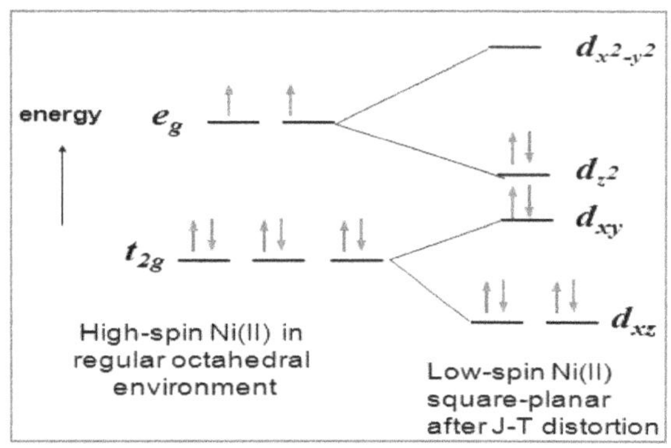

Important examples of square-planar low-spin d^8 metal Ions are Ni(II), Pd(II), Pt(II), Au(III), Co(I), Rh(I), and Ir(I). Asplitting of the d sub-shell in Ni(II) low-spin square-planar complexes is as shown above.

High-spin Ni(II) – only one way of filling the e_g level – not degenerate, no J-T distortion

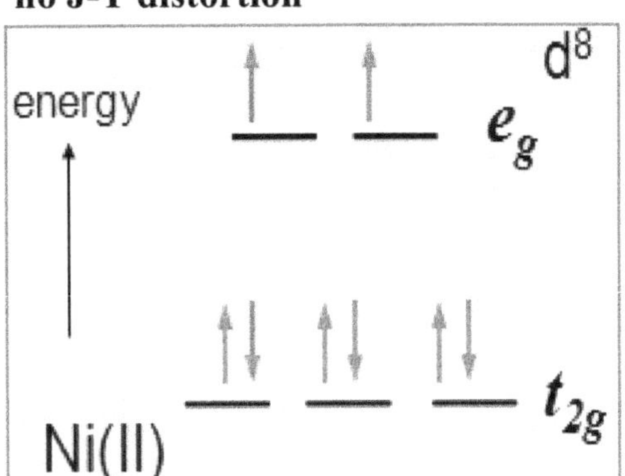

Cu(II) with its d^9 configuration is degenerate and has J-T distortion

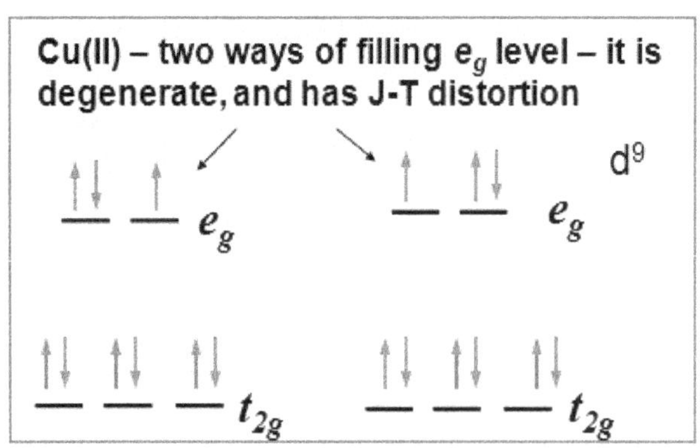

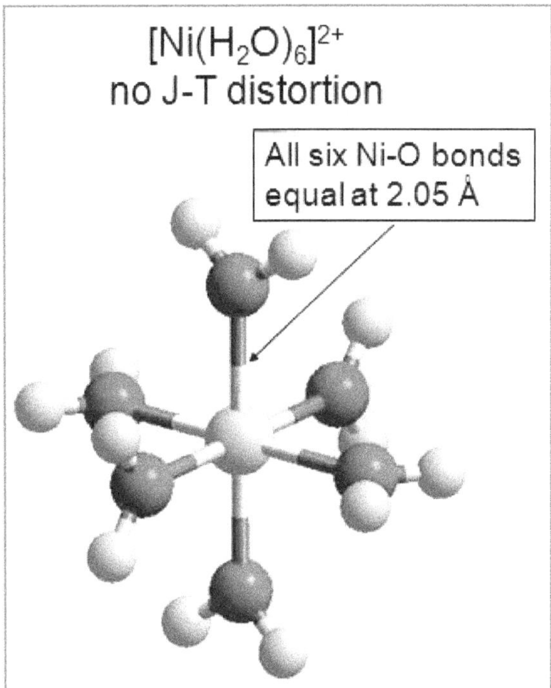

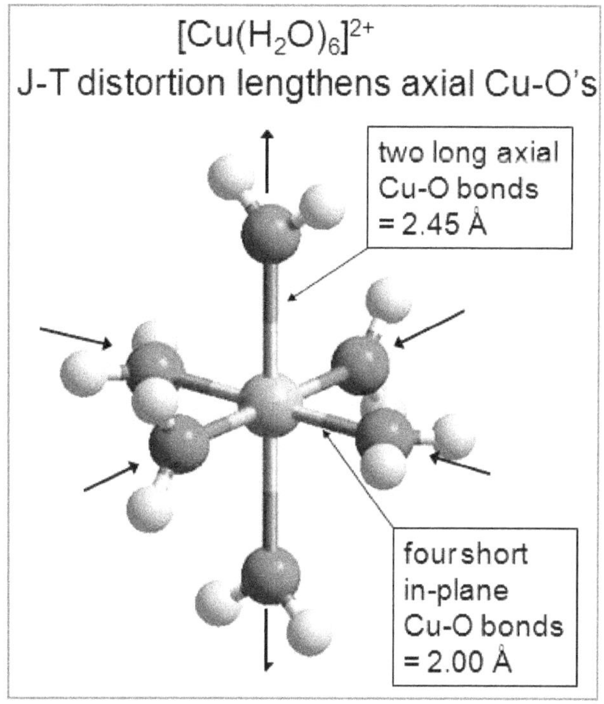

Splitting of d-orbitals in square planar complexes

If the central metal atom in a complex has a d^8 configuration, six electrons will occupy t_{2g} orbitals and the rest two will go to the e_g orbitals irrespective of whether ligand is weak or strong field.

Now taking the case of very strong field ligand, the single electron in d_{x2-y2} orbital is repelled by four ligands while electron in d_{z2} orbital is repelled by only two ligands. As a result the energy of d_{x2-y2} orbital increases relative to the other orbital. If ligand field is strong enough, the difference in energy between these two orbitals becomes larger than the energy needed to pair up electrons. Therefore pairing up of electrons in d_{z2} orbital takes place leaving the other orbital empty. Now four ligands along x, y, -x and –y axes can approach metal atom without any difficulty but ligands approaching along z and –z axes meet very strong repulsive forces. As a result only four ligands succeed in bonding to the metal and a square planar complex is formed. Square planar complexes are formed by d^8 ions with strong field ligands, e.g.$[Ni(CN)_4]^{2-}$. The CFSE is large for heavier elements and in highly charged species, and as a consequence all complexes of Pt^{2+}, Pd^{2+} and Au^{3+} are square planar, whether with strong or weak field. Square planar is also common for complexes of the d^9 ion, Cu^{2+} e.g.$[Cu(py)_4]^{2+}$. For this only one electron has to be placed in the high energy d_{x2-y2} orbital. This geometry also arise in d^4 ions in weak field (e.g. mononuclear S-coordinated complex: $[Mn(S_2C_6H_3Me)_2]^-$).

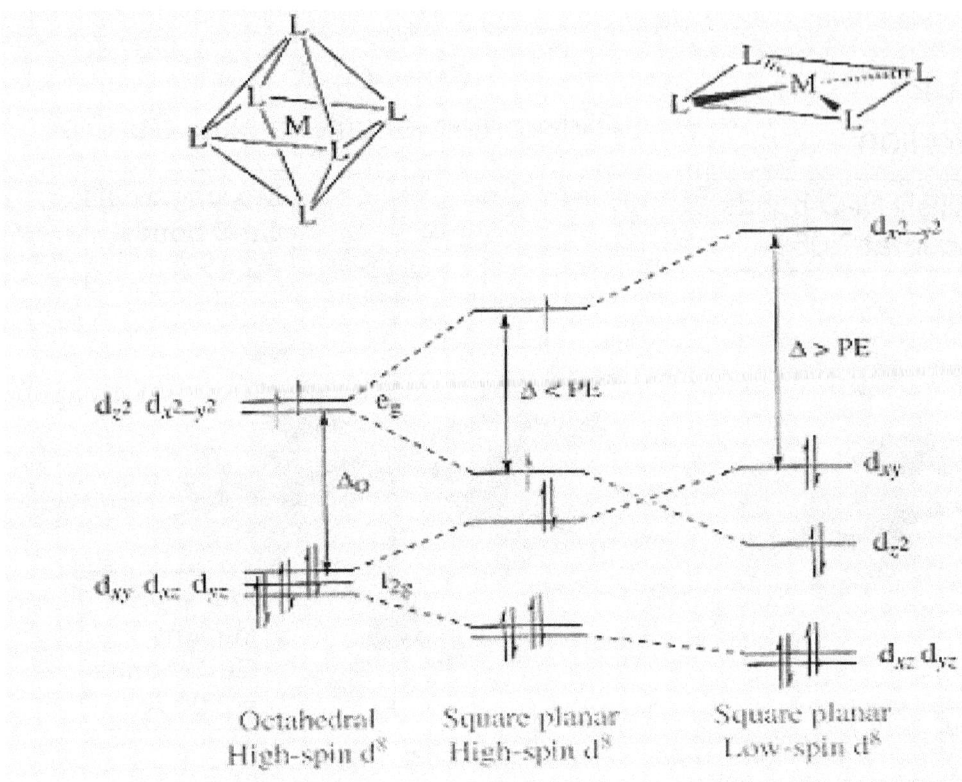

SPLITTING OF *d* ORBITAL ENERGIES IN SQUARE PLANAR FIELDS

As ligands move away along the z-axis, d-orbitals with a z-component will fall in energy.

□ The d_{z2} orbital falls the most, as its electrons are concentrated in lobes along the z-axis.

□ The d_{xz} and d_{yz} orbitals also drop in energy, but not as much.

□ Conversely, the d_{x2-y2} and the d_{xy} orbitals increase in energy. The splitting diagram for square planar complexes is more complex than for octahedral and tetrahedral complexes, and is shown below with the relative energies of each orbital.

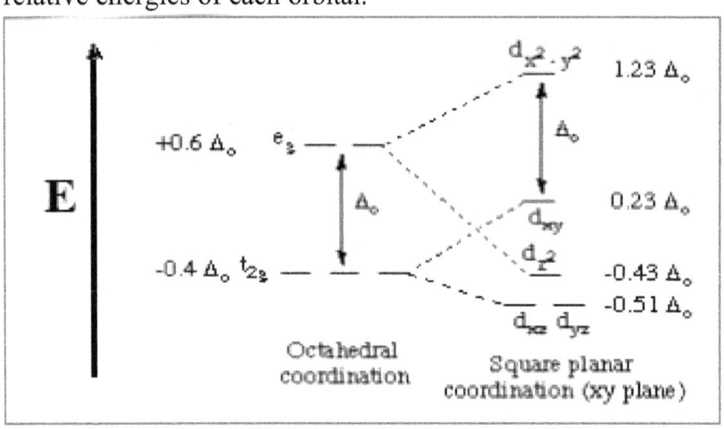

Crystal Field Stabilization Energy in Square Planar Complexes

Square planar coordination is **rare** except for **d^8 metal ions**. Among the d^8 metal ions exhibiting square planar coordination are nickel(II), palladium(II), platinum(II), rhodium(I), iridium(I), copper(III), silver(III), and gold(III). Copper(II) and silver(II), both d^9 ions, are occasionally found in square planar coordination.

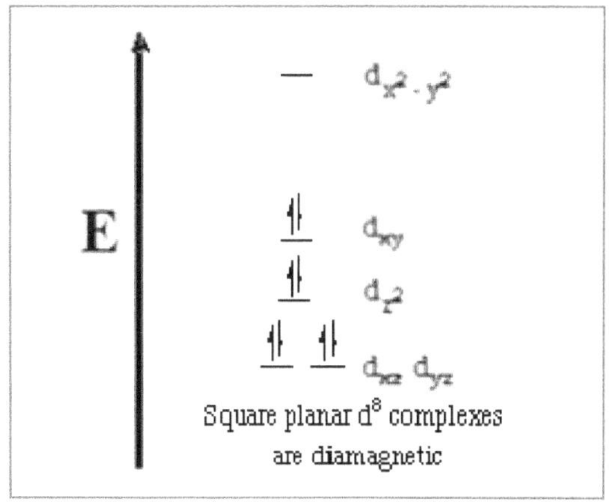

All known square **planar complexes** of **d^8 ions are diamagnetic**, because the highest-energy orbital (d_{x2-y2}) is greatly destabilized, and pairing in the d_{xy} orbital is more favorable than placing an unpaired electron in the d_{x2-y2} orbital.

The crystal field stabilization energy for a diamagnetic square planar d^8 metal complex is readily calculated by the usual method:

$$\begin{aligned} CFSE_{square\ planar} &= 4\text{ electrons }(-0.51\,\Delta_o) + 2\text{ electrons }(-0.43\,\Delta_o) \\ &\quad + 2\text{ electrons }(+0.23\,\Delta_o) + P \\ &= -2.44\,\Delta_o + P \end{aligned}$$

The pairing energy correction is included because a free d^8 ion has 2 unpaired electrons, but a square planar d^8 complex has no unpaired electrons.

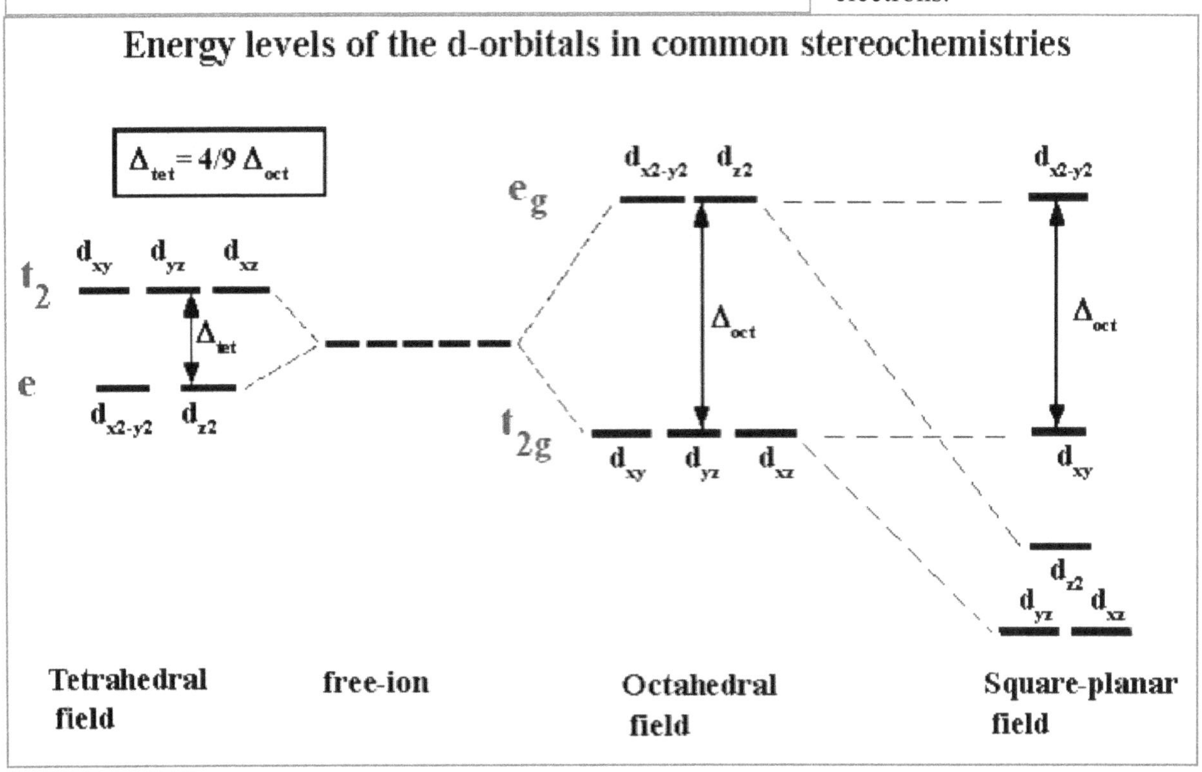

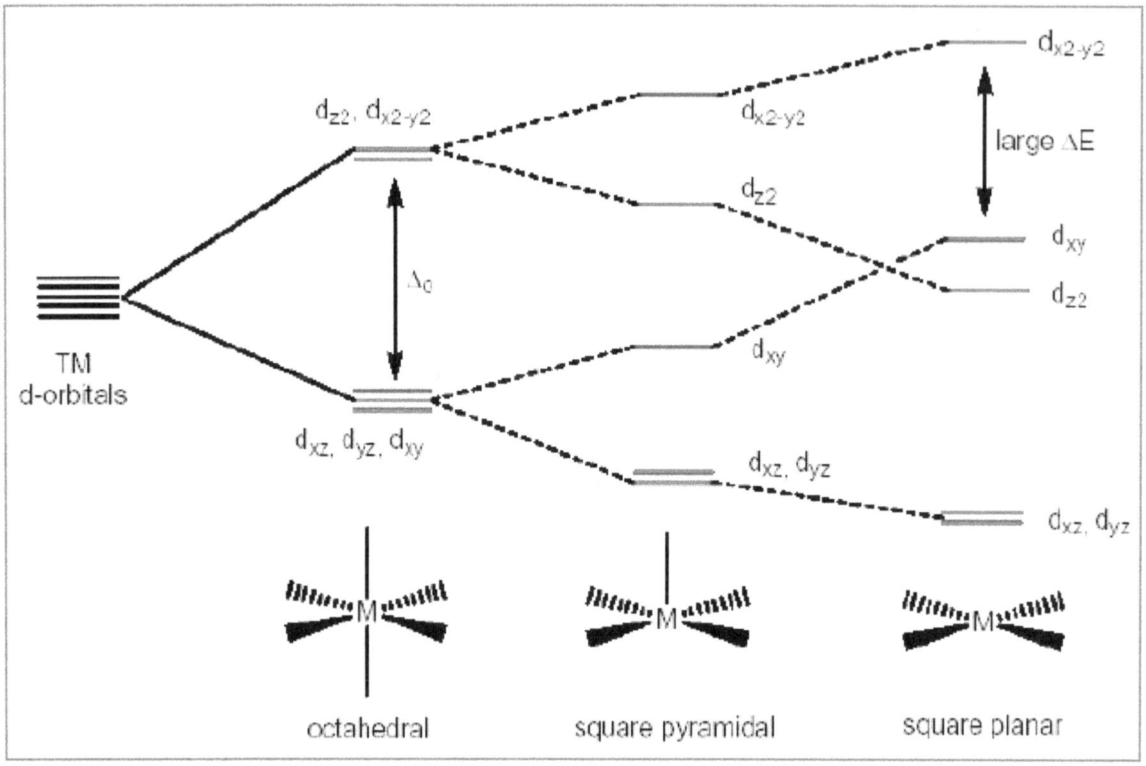

Limitations of crystal filed theory

Although the CFA successfully explains many of the properties of transitional metal complexes such as their electronic spectra and colour, it fails to account for some important observations. In particular, spectroscopic measurements have been used to define a relative order of ligands in terms of the magnitude of the crystal field splitting they induce. This order, known as spectrochemical series, is based on measurement with a variety of metal ions and so represents an averaged ordering of ligands which may not be exactly correct for any given metal ion but represents the general picture. Since the CF model is an ionic model based on the electrostatic effect of the ligands on the energies of the metal ion d orbitals, it is surprising to find that anionic ligands such as Br^-, F^- and OH^- should be weaker-field than neutral ligand such as CO, NH_3 and PR_3. This anomaly arises because the CFA model does not take account of covalence in metal-ligand bonding. Modifications to the CFA model in the ligand field approach (LFT) model take accounts of these facts. Molecular orbital Theory (MOT) takes into account both ionic as well as covalent contributions in the formation of complexes. This extensively involves symmetry and group theory applications.

1.11 MOLECULAR ORBITAL THEORY

Crystal field theory adequately accounts for large quantity of data on transition metal complexes. But CFT has serious limitations. There are several experimental and semitheoretical arguments that can be presented against the assumption that the splitting between metal and ligand is ionic with no covalent character. Indirect evidence that the electrons are shared between the ligands and the central metal ions comes from the nephelauxetic effect. It is found that the electron-electron repulsion in complexes is somewhat less than that in the free ion.

Additional evidence for covalency in metal-ligand bonds is provided by electron paramagnetic resonance. The EPR spectra of many complexes, however, show hyperfine splitting patterns that arise from the interaction of the unpaired metal electron with magnetic nuclei on the ligands. This clearly indicates that the electron is at least partially delocalized over the ligands.

Bonding in Transition metal complexes can broadly discussed under the following two considerations,

A. Geometry

ML_6 -O_h (octahedral)
ML_4 -T_d vs D_{4h} (tetrahedral vs. square planar)

B. Ligand Type

π – acceptors- CO, NO^+, CNR, CN^-
π – donors- halides (X- = Cl, Br, I), NH_2^- (amide), NR_2^-, O_2^-, OR^-, SR^-
σ – donors- H^-, NH_3

many ligands are a combination of donor types, but the "pure" donor diagrams can be considered.
M O theory uses **symmetry-adapted linear combinations** (SALCs) of metal-ligand interactions and that relies on some application of symmetry elements and group theory to produce a viable bonding model.

Metal orbitals

In a strictly octahedral environment, the metal orbitals divide by symmetry into four sets, which are given labels that describe their symmetry (subscripts g or u, 1 or 2), and their multiplicity (a or b, e, t). The resulting Symmetry Labels (also called Mulliken Symbols) are used to now describe ***sets of orbitals in a complex***, rather than just the atomic orbital. Therefore, orbitals or groups of **metal orbitals** that are labelled by a particular Mulliken Symbol can only interact with a **ligand set of orbitals** that is described by the same label.

Metal Orbital	Symmetry Label	Degeneracy
s	a_{1g}	1
p_x, p_y, p_z	t_{1u}	3
d_{xy}, d_{yx}, d_{xz}	t_{2g}	3
d_{x2-y2}, d_{z2}	e_g	2

Among the metal orbitals, only the t_{2g} set (dxy, dxz, dyz) does not possess suitable symmetry to undergo σ-bonding interactions with the ligand SALCs. The metal t_{2g} set remains non-bonding in this scenario.

Metal -ligand σ – bonding

Six symmetry-adapted linear combinations of the six ligand σ symmetry orbitals can also be formed. These combinations are shown below, where we are only considering ligand s orbitals to keep things simple. Bear in mind that these are not specifically s or p atomic orbitals any more, but only the orbitals, atomic or molecular, which exhibit σ symmetry with respect to the metal-ligand bond axis. We are now treating all six ligands as a group where six metal orbitals can be broken down into three groups according to their symmetry (**a_{1g}, t_{1u} and e_g**).

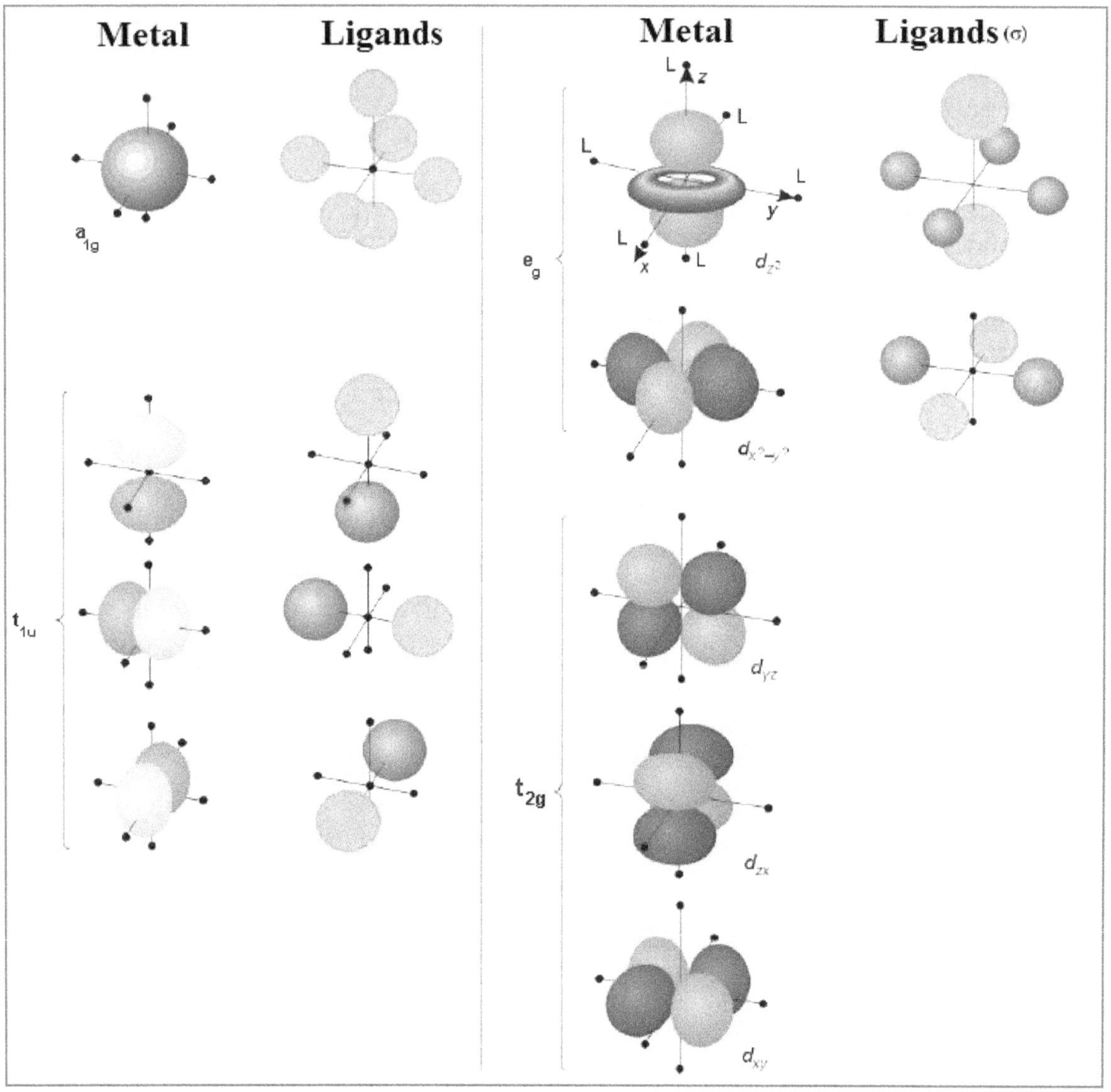

Fig. 62. Shapes of Molecular orbitals of ligands

Molecular orbitals for the resulting complex are formed by combining SALCs and metal orbitals of the same symmetry type. Since we have six metal orbitals that can be broken down into three groups according to their symmetry (**a_{1g}, t_{1u} and e_g**) and six ligand SALCs, these combine to form 6 bonding and 6 antibonding MOs. The resulting MO diagram can now be populated with electrons according to the Aufbau process, Pauli principle and Hund's rule. For the resulting energy levels, the greatest contribution to the molecular orbital of lowest energy is from atomic orbitals of lowest energy. For most ligands, particularly those of period 2 and 3 (n quantum number ≤ 3) in the main group, the ligand σ orbitals are derived from AOs with energies much lower than those of the metal d-orbitals. Thus, the six bonding MOs of the complex are mostly ligand-orbital in character. These six bonding orbitals can accommodate the 12 electrons provided by the six ligand lone pairs.

The d electrons of the metal centre, if any, occupy the lower energy, non-bonding metal t_{2g} set and the higher energy, antibonding e_g set. Therefore, the metal based d electrons tend to reside largely on the metal atom. In summary, the frontier orbitals of the complex are the non-bonding, entirely metal based t_{2g} orbitals and the antibonding e_g orbitals, which are mainly metal in character anyway. Therefore, because most bonding (and therefore "chemistry") occurs in the valence (or "frontier") orbitals of a compound, most reactions of metal complexes occur at the metal centre. The octahedral ligand field splitting parameter, Δ_o, in this approach is called the HOMO-LUMO separation, or HOMO-LUMO gap. It is approximately the splitting of the metal d orbitals caused by the ligands, whereas in CFT Δ_o is purely a metal d orbital separation.

MO diagram for ML_6 (σ – donor) Octahedral complex

The diagram below shows how the SALCs of the ligand orbital sets combine with the metal orbitals to produce molecular orbital energy levels (the MO diagram) of an octahedral ML_6 complex. The t_{2g} and e_g* (which is now of antibonding symmetry with respect to the M-L bond) are empty in this diagram, suggesting the metal ion, M, has a d^0 electron configuration. With several d orbitals (as well as s and p orbitals) available in the valence shell, enough delocalized MOs can be constructed to accommodate all the electrons needed to bind the ligands. Also, since lanthanides have f orbitals as well, they can accommodate even more ligand electrons and, therefore, can easily have coordination numbers higher than 6.

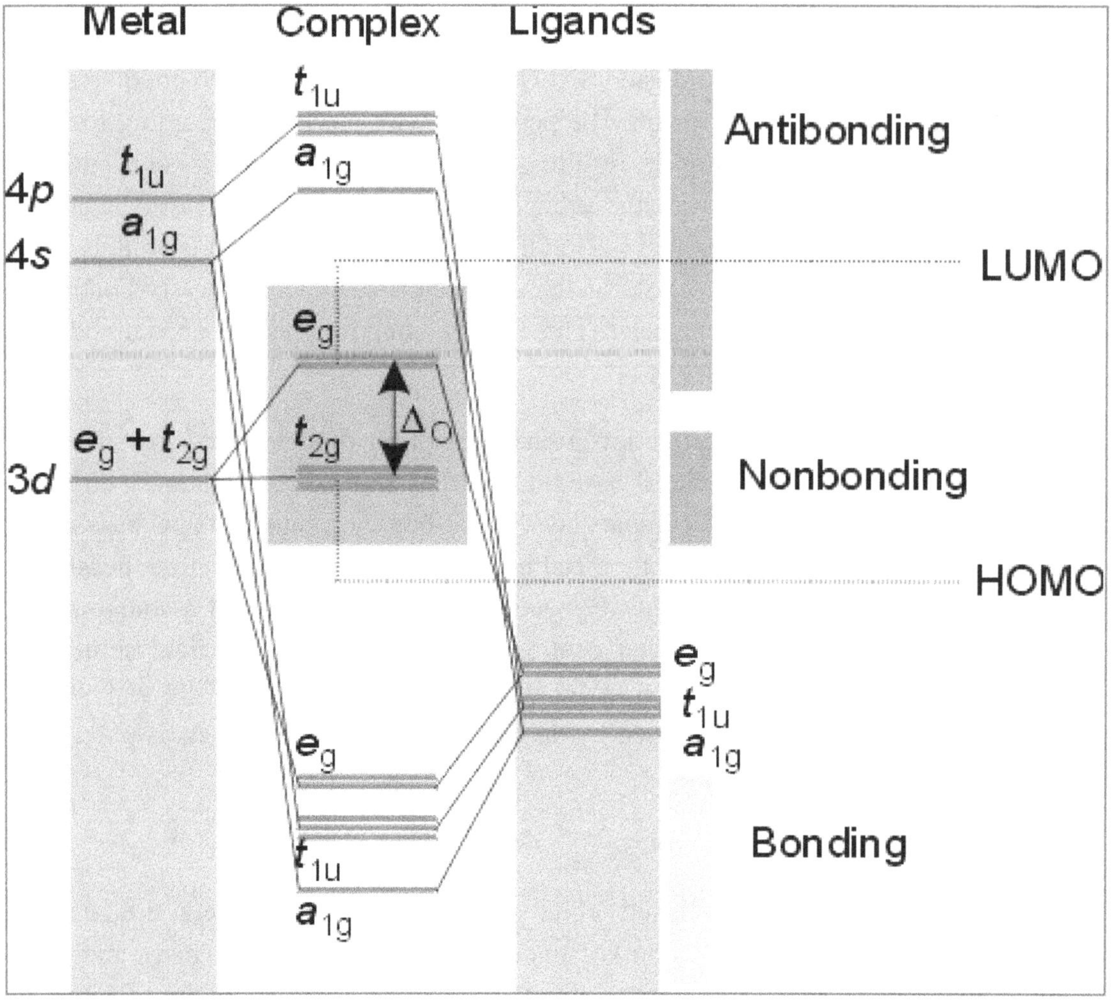

Therefore, six group orbitals have been obtained from six ligand σ orbitals. The combination with the six central ion orbitals of the same symmetry will give rise to 12 MO: six bonding and six antibonding. The t_{2g} orbitals of the central ion (d_{xy}, d_{xz}, d_{yz}) do not have adequate symmetry and do not play a role in this type of bond. They remain, therefore, as nonbonding orbitals.

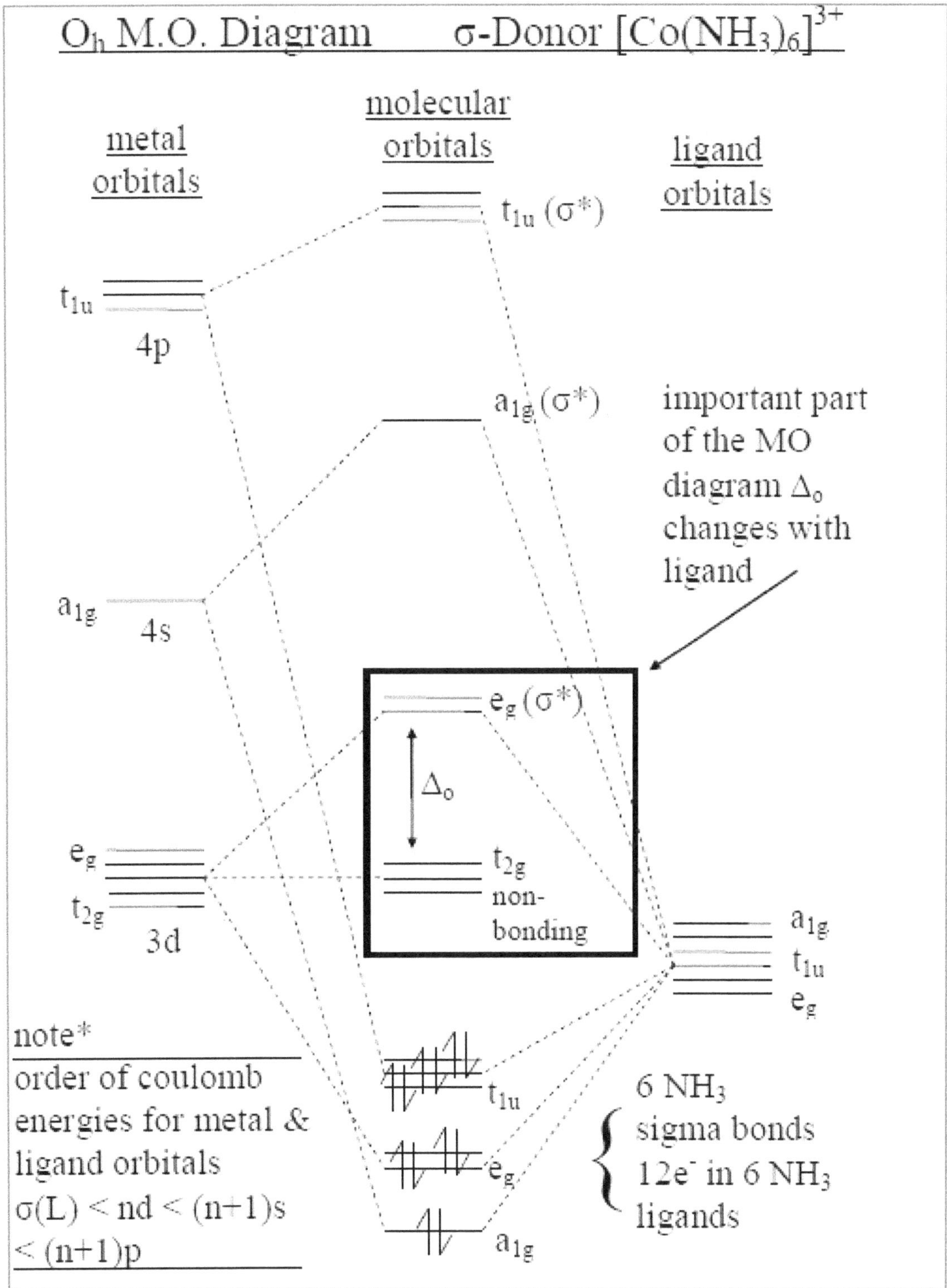

Metal –ligand π – bonding

If the ligands in a complex have orbitals with local π symmetry with respect to the M-L axis (a t_{2g} ligand SALC), then they may form π MOs with the t_{2g} set of metal orbitals. These ligand π orbitals may be populated or vacant and, therefore, may act as electron donors or acceptors as shown in the diagram below. The effect of this bonding on Δo depends on where the electron density is going. For π-acceptor ligands, the bonding is SYNERGIC: σ-donation to the metal strengthens π-back bonding to the ligand, and π-donation from the metal to the ligand strengthens the σ-donor component of bonding. This is because σ-donation leads to increased electron density on the metal, which allows increased π- back donation. Conversely, π-back donation reduces the amount of electron density on the metal, which allows more σ-donation from the ligand to the metal.

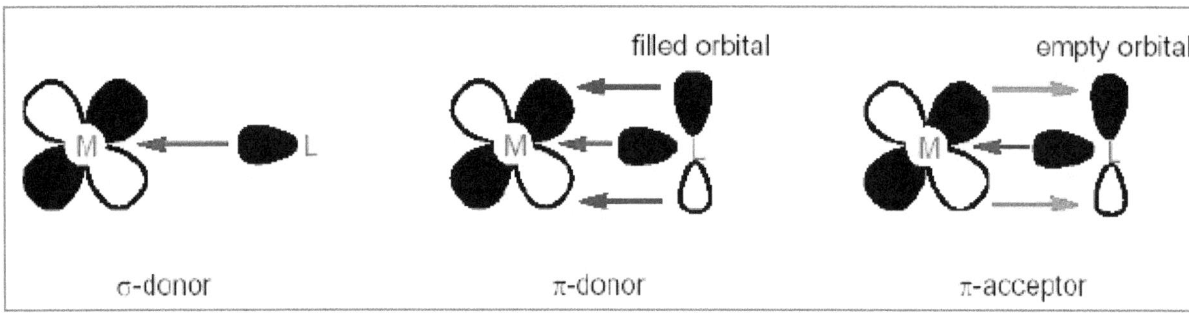

A π-acceptor ligand is a ligand that has filled π orbitals, which are much lower in energy than the metal t_{2g} orbitals. It also has empty π* orbitals which are close enough in energy to the metal d orbitals to allow population. Examples of ligands that possess orbitals of this appropriate energy are CO and N_2.

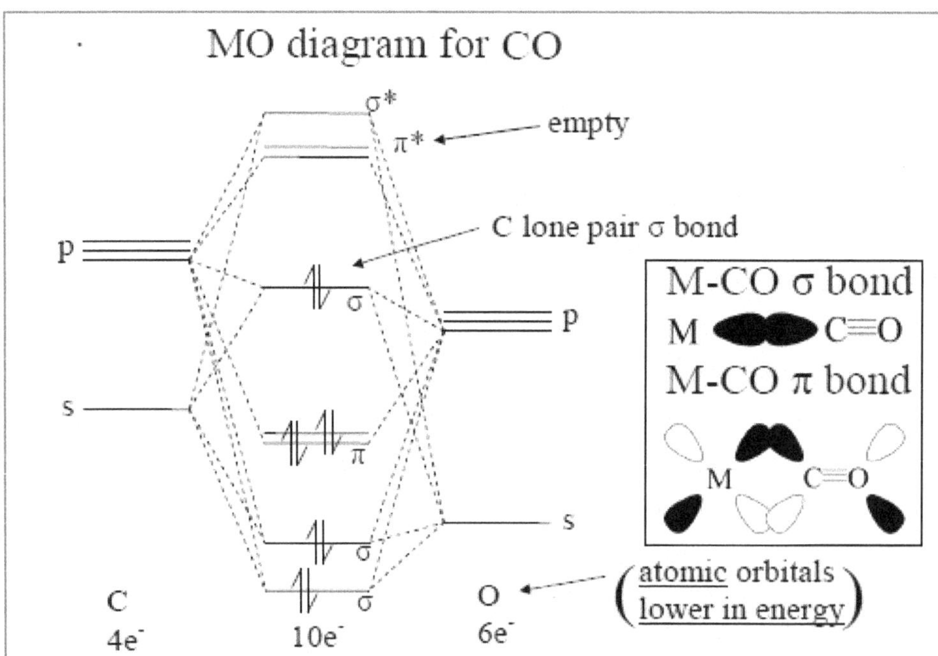

A π donor ligand has filled orbitals that have π symmetry around the M-L axis. The energies of these orbitals are similar to those of the metal d orbitals and the ligand has no vacant low energy π orbitals (strictly speaking these are of π* symmetry). Examples of π donor ligands include Cl^-, OH^-, NR_2^- and H_2O. These all turn out to be low in the spectrochemical series (weak field ligands) that lower Δ_o. The full π orbitals of the π donor ligands lie lower in energy than the partially filled d orbitals of the metal. This means that when they form molecular orbitals with the metal t_{2g} orbitals, the bonding combination lies lower than the ligand orbitals and the antibonding combination lies above the energy of the d orbitals of the free metal (figure below on the left). The result is that ligand electrons occupy the bonding t_{2g} MO, while the metal electrons now occupy the antibonding t_{2g} MO set, which has been raised in energy.

The bonding π orbital of MO is populated and is largely localized on the O atom because O is more electronegative than C, hence the lower energy π bonding orbital is mostly O in character. The antibonding π* orbital of CO has its largest amplitude on C and has appropriate symmetry and energy to overlap the t_{2g} orbitals of the metal. Because of the resulting energy correlation, the π donor ability of CO is very small, and its bonding to metals is dominated by its π acceptor character along with its σ donor ability. Because these π acceptor ligands are typically of antibonding character with respect to the ligand atoms, they lie higher in energy than the metal d orbitals, they form molecular orbitals of t_{2g} symmetry which are largely of metal d orbital in nature. These bonding combinations lie slightly lower in energy than the metal d orbitals themselves. As a result, Δo is increased due to the "pushing down" in energy of the bonding t_{2g} symmetry molecular orbitals.

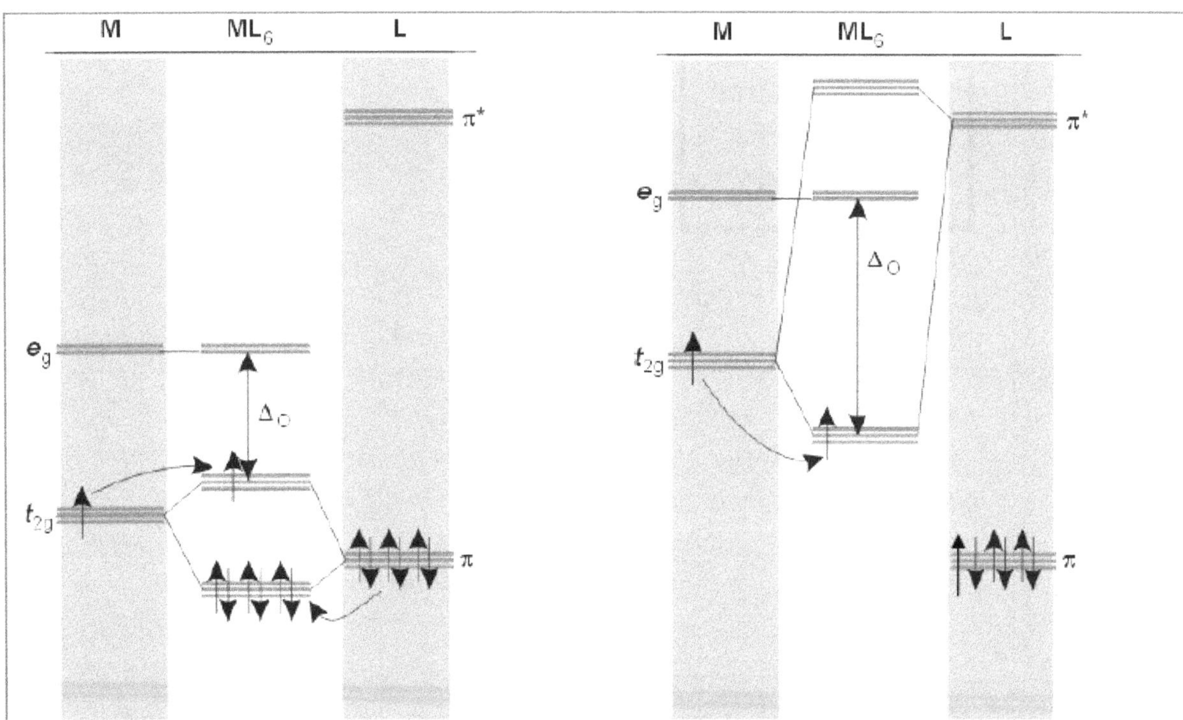

Ligands that are π donors decrease Δ_o. Ligands that are π acceptors increase Δ_o.

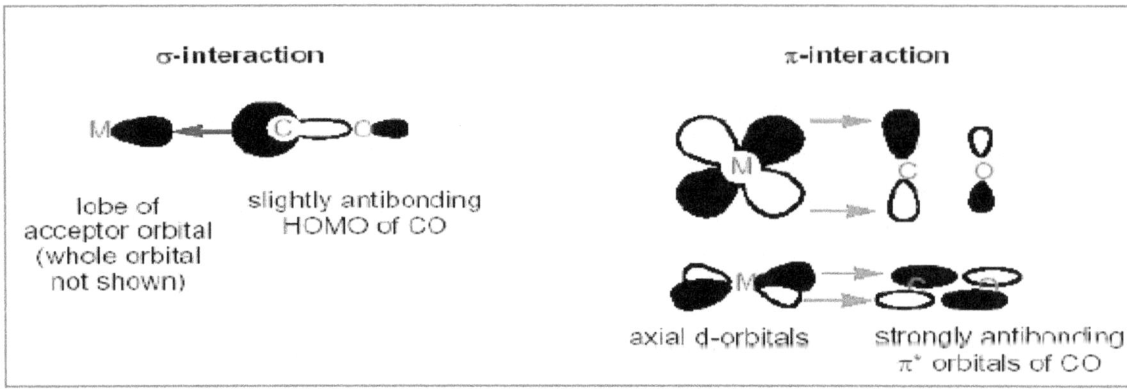
σ-interaction
π-interaction
M
C
O
lobe of acceptor orbital (whole orbital not shown)
slightly antibonding HOMO of CO
axial d-orbitals
strongly antibonding π* orbitals of CO

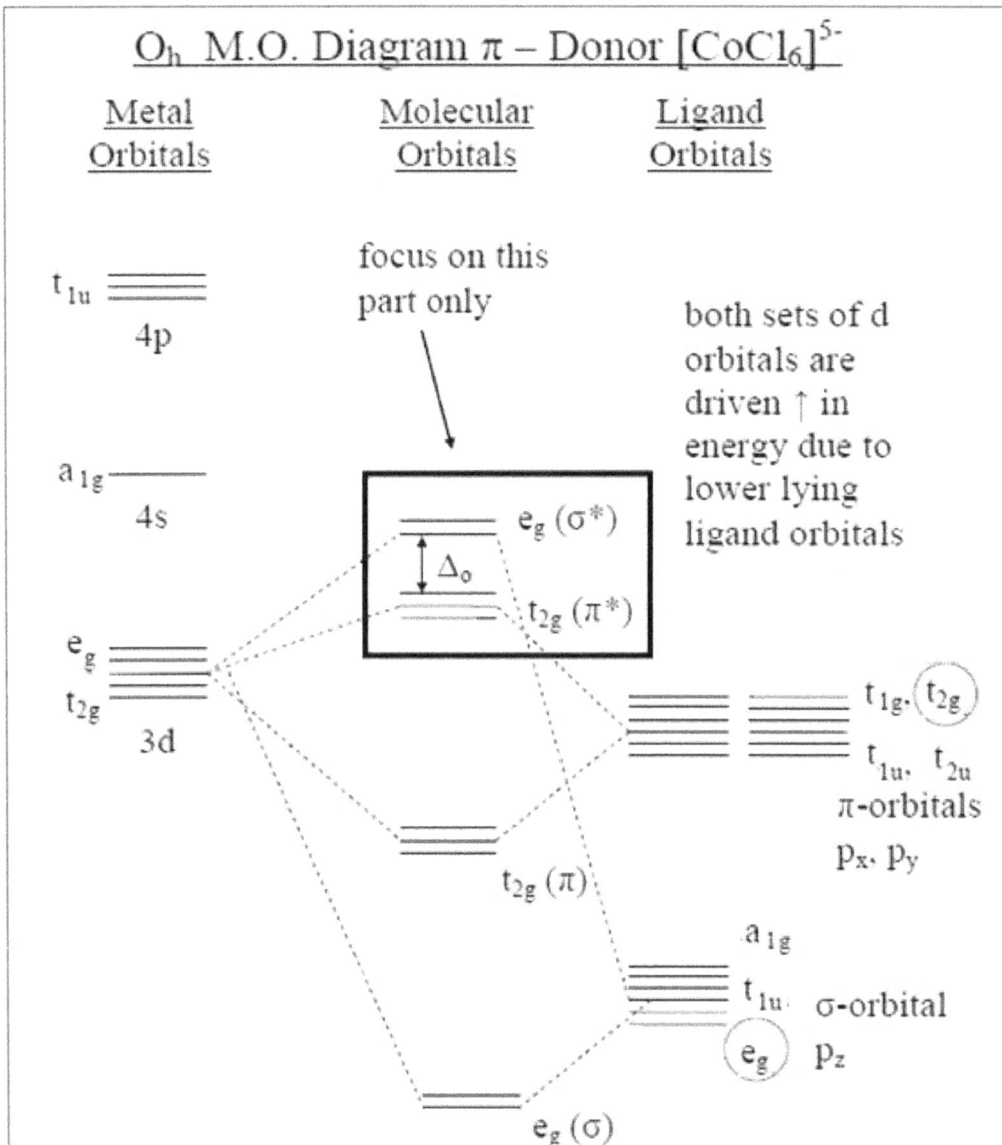
Oh M.O. Diagram π – Donor [CoCl6]5-
Metal Orbitals
Molecular Orbitals
Ligand Orbitals
focus on this part only
both sets of d orbitals are driven ↑ in energy due to lower lying ligand orbitals
t1u
4p
a1g
4s
eg
t2g
3d
eg (σ*)
Δo
t2g (π*)
t2g (π)
eg (σ)
t1g, t2g
t1u, t2u
π-orbitals
px, py
a1g
t1u
σ-orbital
eg
pz

Below is the resulting molecular orbital (MO) diagram for an octahedral complex with strong π-acceptor ligands (e.g. $Cr(CO)_6$). The Cr^0 centre is a $3d^6$ configuration. The 6 CO ligands each donate two electrons to the metal centre via σ bonding. The π bonding symmetry orbitals of the ligands are also populated, but cannot effectively act as π donors to the metal centre, partially because of their energies being much lower than the appropriate symmetry metal t_{2g} set, but also because the metal d^6 centre has its t_{2g} set populated by metal based electrons; electron-electron repulsion further hinders any π donor ability from the CO ligand.

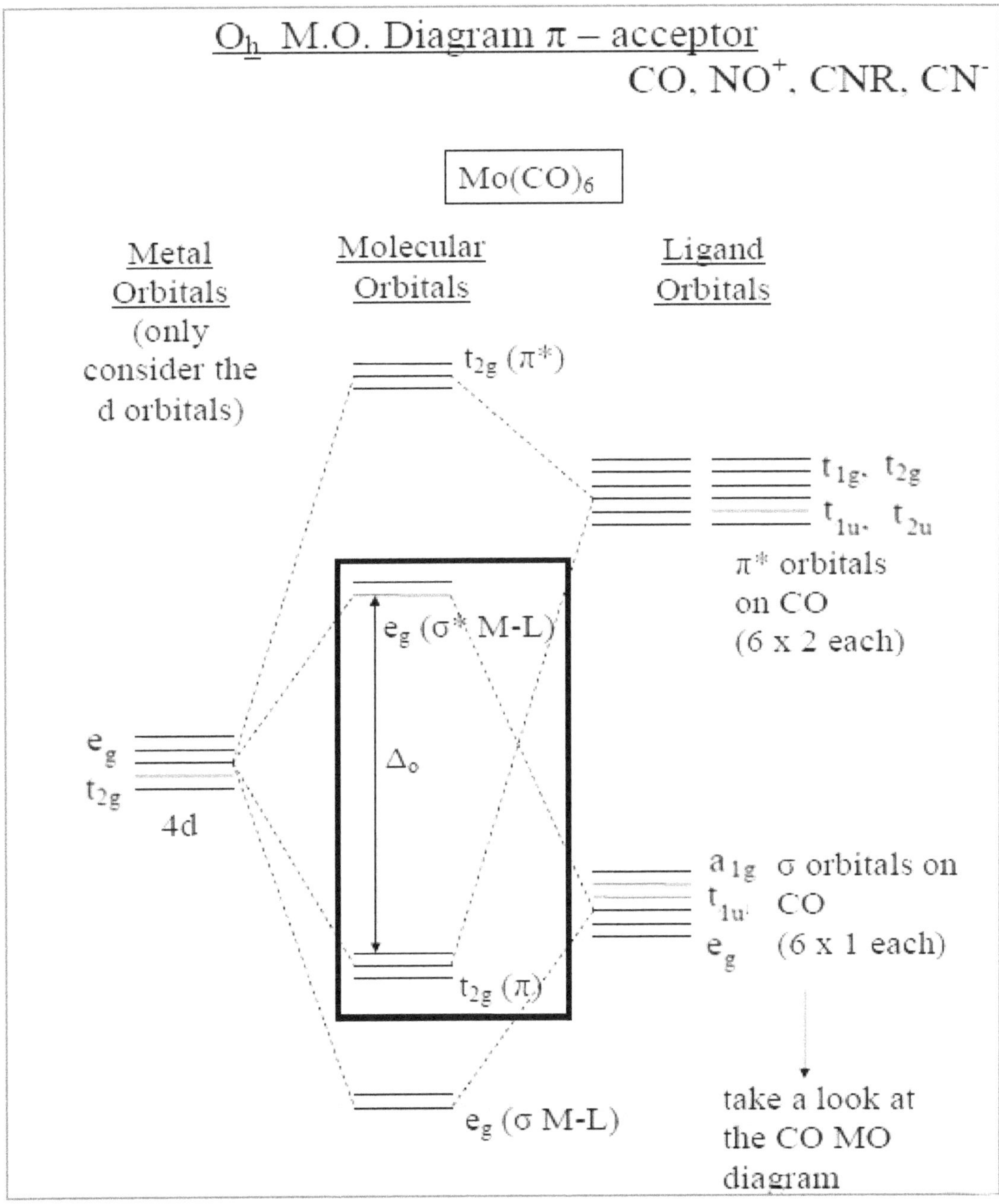

The antibonding π^* orbitals of the CO ligand are vacant and are low enough in energy to undergo orbital overlap with the populated metal t_{2g} set. Therefore, the metal "back-donates" this electron density into the CO π^* orbitals. In total, this complex has 18 electrons that participate in bonding. In transition metal complexes, this "18 electron rule" is similar to the "octet rule" used in main group chemistry. Complexes possessing 18 bonding electrons are particularly stable.

Molecular Orbitals of a Tetrahedral Complex

The steps to be followed here are the same as in the case of octahedral complexes. They must first be carried out for σ interactions and then for π interactions. In the T d point group the four σ orbitals of the ligands can be grouped according to the group orbitals $a_1 + t_2$.

For π - group orbitals, although this is not immediately obvious due to their orientation, these orbitals belong to the symmetry species $t_1 + t_2 + e$. The t_1 group orbitals do not correspond to any central ion orbital and the t_2 group form σ bonds. The orientation and the MOs with e symmetry, as well the energy diagrams including the σ and π bonds are difficult to visualize due to the presence of mixed orbitals.

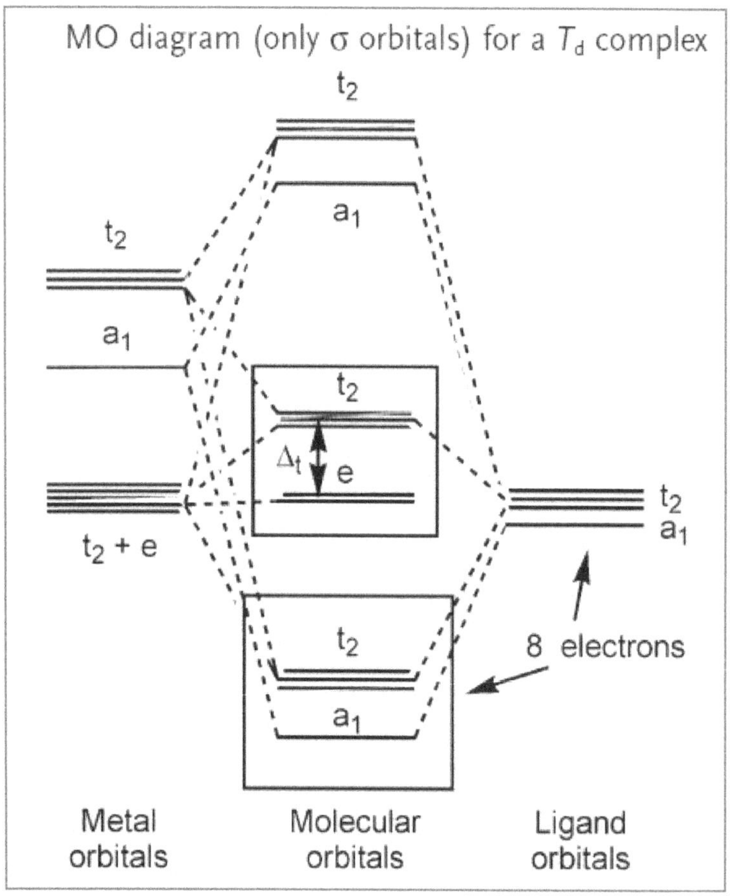

Molecular Orbitals of a Square - planar Complex

In the D_{4h} point group the four σ orbitals of the ligands can be grouped according to the group orbitals a_{1g} + e_u + b_{1g}. There are two kinds of π group orbitals: those in the molecular plane and those which are perpendicular to it. The first four belong to the symmetry species a_{2g} + e_u + b_{2g}, while the four perpendicular ones belong to the species a_{2u} + e_g + b_{2u}. The corresponding π_z MO orbitals can be easily deduced by the reader following an analogous procedure. The corresponding – and simplified – energy diagram for the MO obtained is shown schematically in the figure given below,

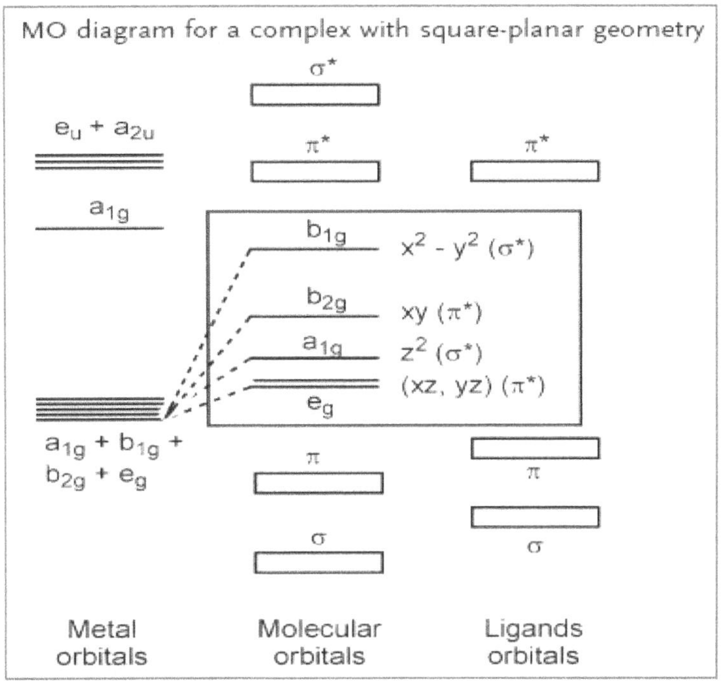

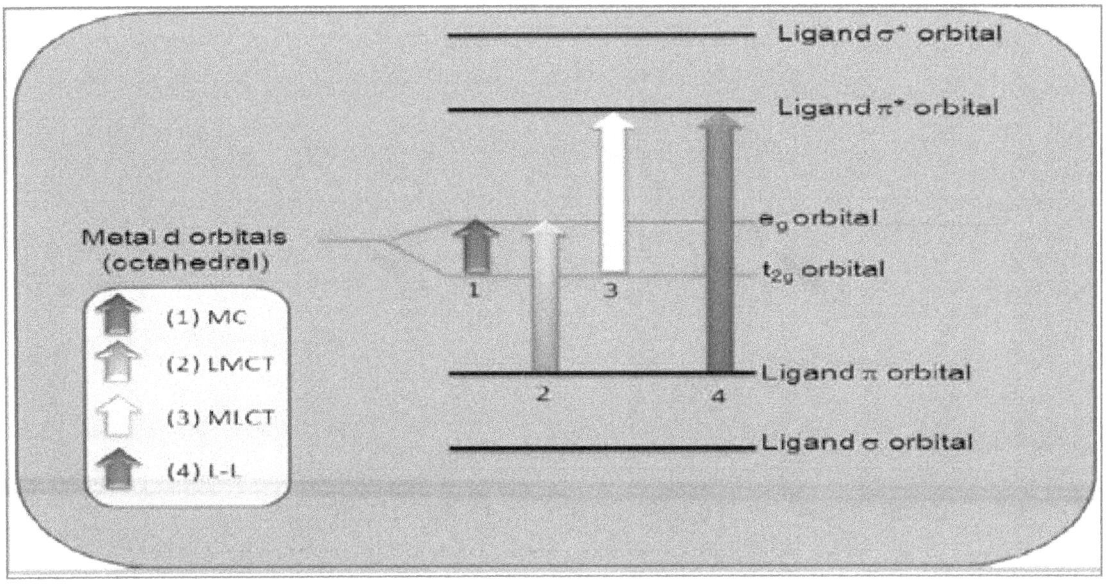

1.12 SUBSTITUTIONS IN SQUARE PLANAR COMPLEXS

The majority of square planar complexes are those that contain d^8 metal ions, of which the most common examples are Ni^{2+} , Pd^{2+}, and Pt^{2+}, although some complexes containing Au^{3+} have also been studied. As a general trend, the rate of substitution in these complexes is
$Ni^{2+} > Pd^{2+} > Pt^{2+} < Au^{3+}$
For the first three metal ions in the series, the rate is inversely related to the ligand field stabilization energy, as would be expected. The fact that Au^{3+} complexes undergo substitution much faster than those of Pt^{2+} comes from the fact that Au^{3+} has a higher charge, which gives rise to greater attraction for a potential ligand. Complexes of Pt^{2+} have been extensively studied, and much of the discussion will be concerned with those complexes.

For Square Planar complex; $ML_3X + Y \rightarrow ML_3Y + X$
X and Y can be any pair of ligands.

- Rate = k[ML_3X][Y] i.e. second order kinetics so it depends on the nature of both the leaving group X and the entering group Y.
- Steric crowding slows down the reaction.
- Evidence for an **associative mechanism**.

Mechanism for Square Planar Ligand Substitution

- For square planar **BOTH bond-breaking and bond making are important** in the reaction mechanism (i.e. an associative mechanism).

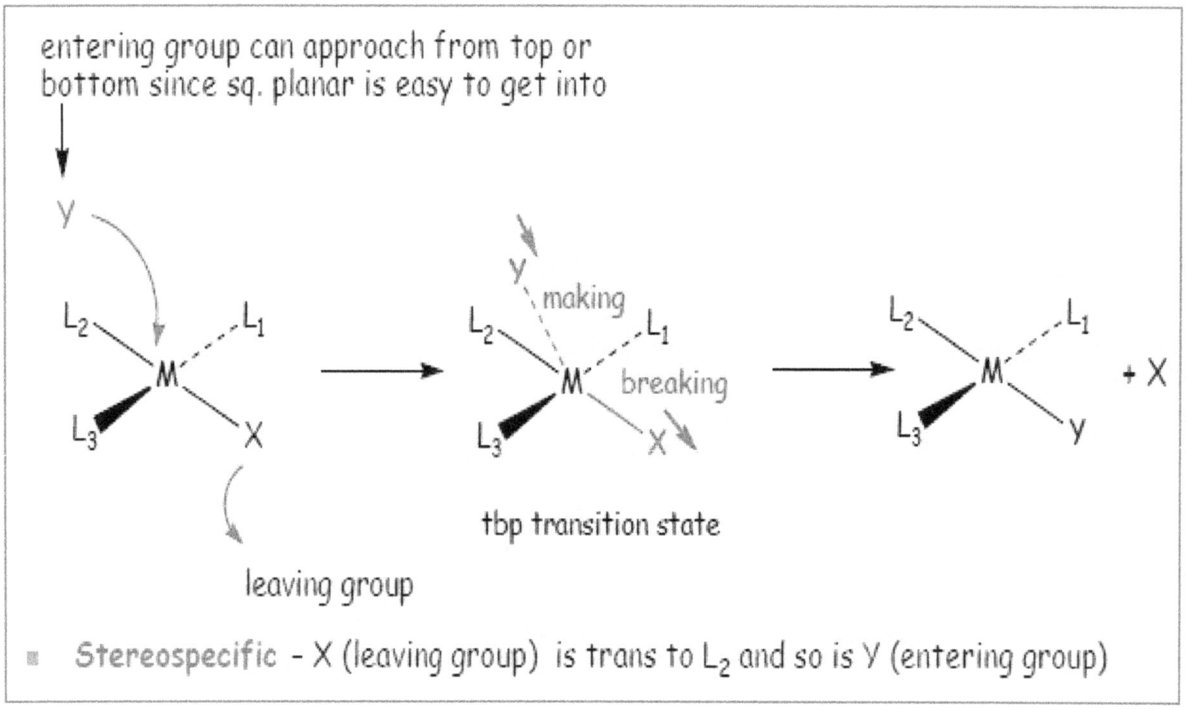

Initial attack by the entering group at a square planar Pt(II) centre is **from above or below the plane**. Nucleophile Y then coordinates to give a **trigonal bipyramidal intermediate species** which loses X with **retention of stereochemistry**.

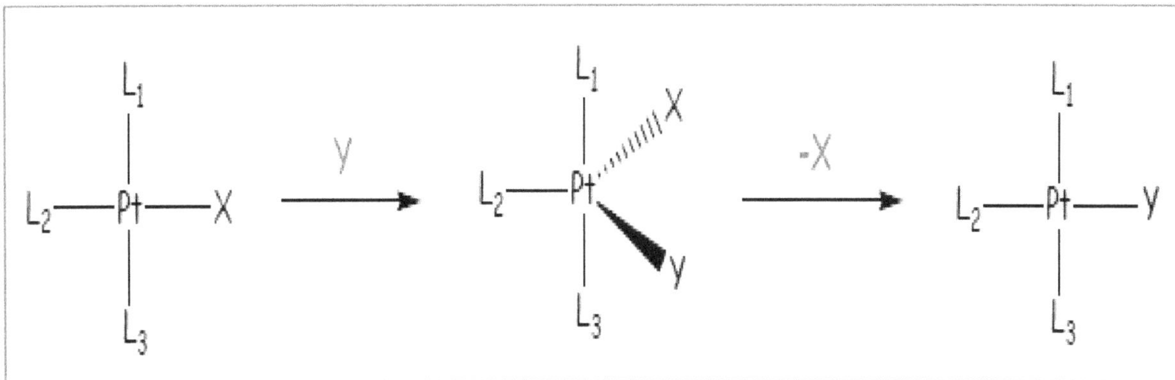

Examples:

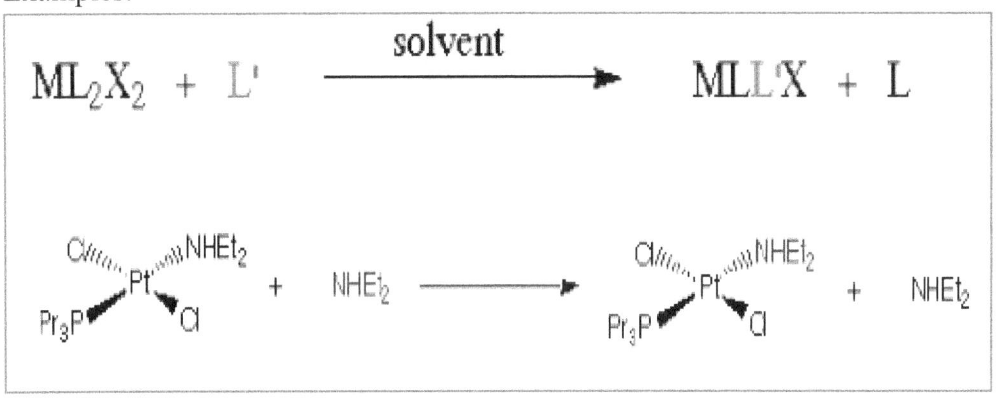

Factors which affect the rate of substitution

i). Role of the Entering Group:

The rate of substitution is proportional to the nucleophilicity of entering group

- i.e. for most reactions of Pt(II), the rate constant increases in the order:

$H_2O < NH_3 = py < Br^- < I^- < CN^-$

- The ordering is consistent with Pt(II) being a soft metal centre.

ii). The Role of The Leaving Group:

For the reaction

$[Pt(dien)X]^+ + py \rightarrow [Pt(dien)(py)]^+ + X^-$

in H_2O at 25^0C the sequence of lability is;

$H_2O > Cl^- > Br^- > I^- > N_3^- > SCN^- > NO_2^- > CN^-$ with a spread of over 10^6 in rate across series.

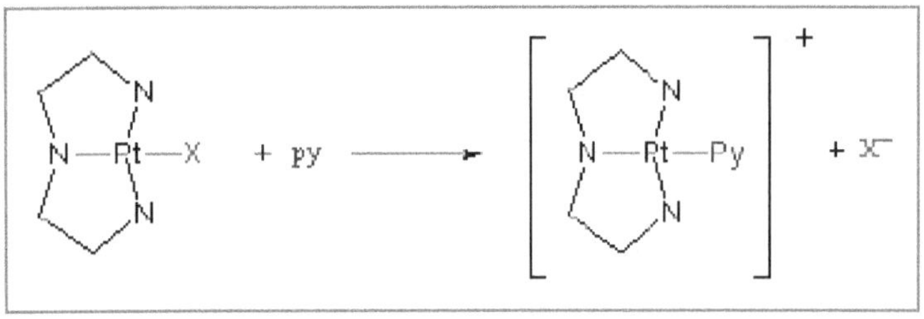

The leaving group does not affect the nucleophilic discrimination factors only the intrinsic reactivity. The series tend to parallel the strength of the Metal-L bond.

iii). The Nature of the Other Ligands in the Complex:

iv). Effect of the Metal Centre:

TRANS EFFECT

It is observed that during the substitution reactions of square planar metal complexes, some ligands preferentially direct the substitution **trans** to themselves. i.e., the choice of leaving group is determined by the nature of ligand trans to it.

*The **Trans effect** can be defined as the effect of a ligand over rate of substitution of another ligand positioned trans to it in the square planar complexes.*

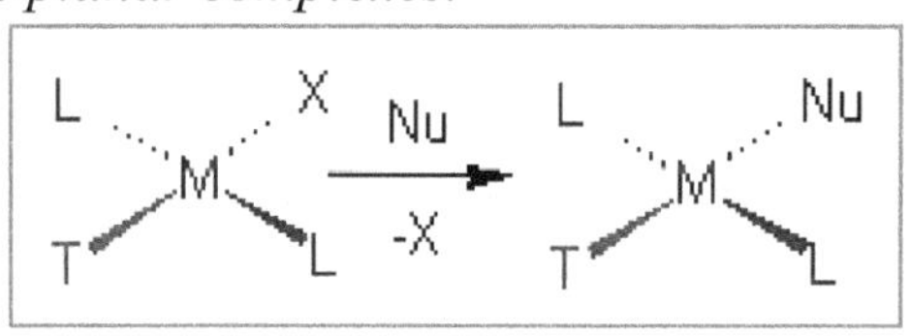

'T' is the trans directing group and 'Nu' is the nucleophilic ligand which preferentially substitutes the ligand 'X' which is trans to ligand 'T'.

* The Trans effect was first recognized by Ilya Ilich Chernyaev, a russian chemist in square planar complexes of Platinum(II).

* However the Trans effect is also observed in octahedral complexes.

* The true Trans effect is a kinetic effect.

ILLUSTRATIONS OF TRANS EFFECT

The Trans effect can dictate the product formed in the substitution reactions. The classic example of Trans effect is the synthesis of **cisplatin**, cis-diamminedichloridoplatinum(II). It is prepared by substituting the two chloro groups of $PtCl_4^{2-}$ by ammonia molecules.

In the first step, any of the chloro group is substituted by ammonia randomly. But in the second step, the ammonia group preferentially substitutes the chloro group cis to the first ammonia. This can be attributed to the fact that the Cl^- has a larger trans effect than NH_3.

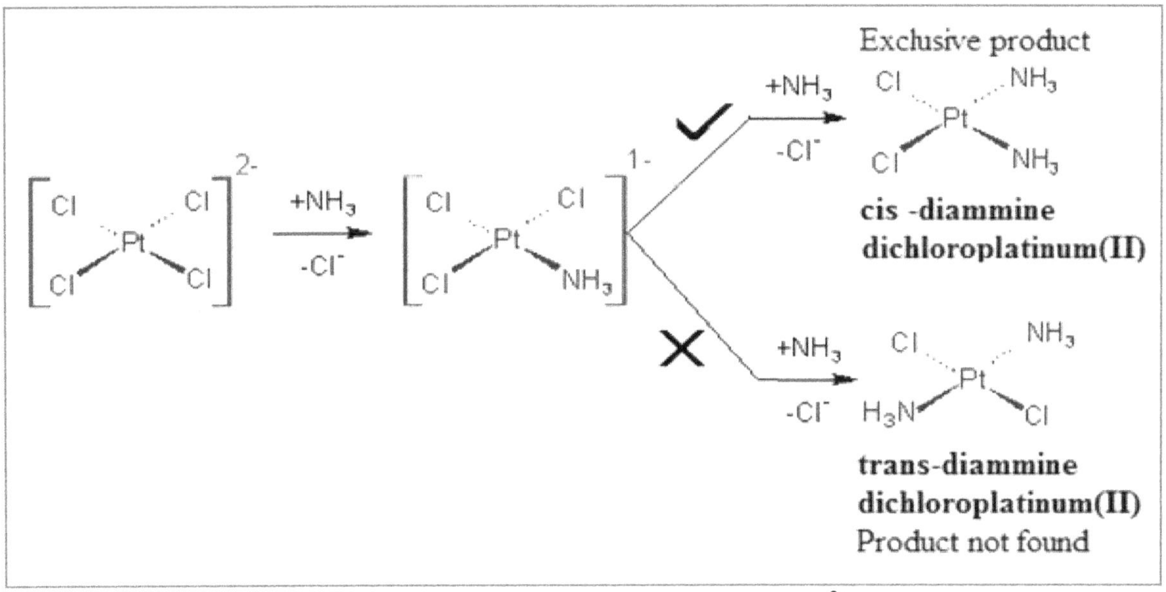

Whereas, the trans product is obtained by starting from $Pt(NH_3)_4^{2+}$. In this case the second Cl group is substituted preferentially at trans position to the first one.

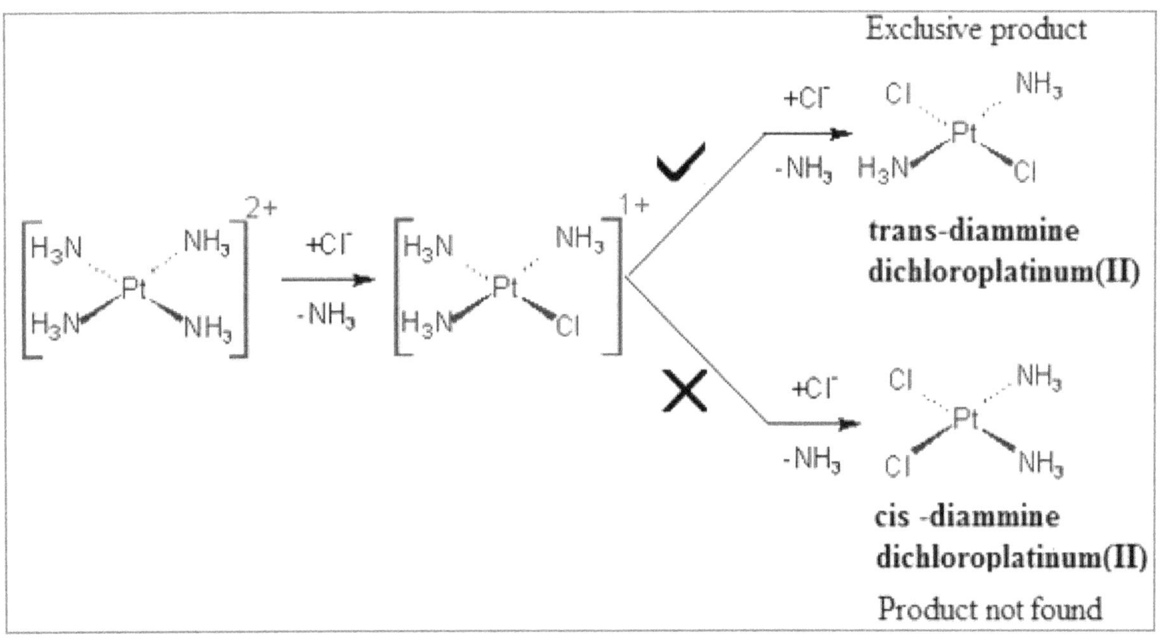

Mechanism:

The above kinetic data along with preferential substitution at trans position of certain ligands suggest the possible formation of *5-coordinate trigonal planar transition state (or intermediate)*.

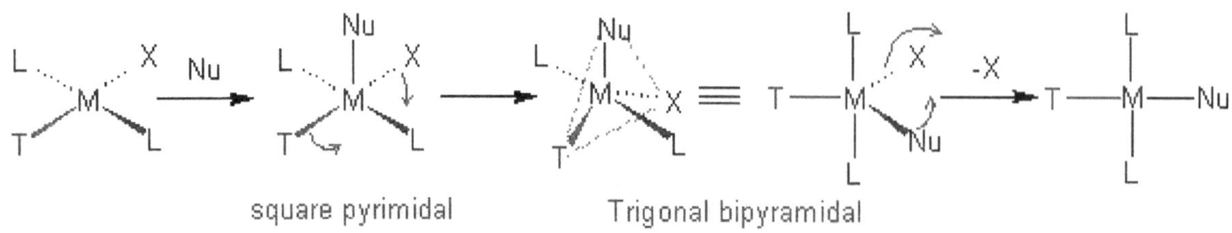

In the final compound, the incoming Nu occupies the trans position to ligand T. Also observe the retention of configuration.

In the 5-coordinate Trigonal bipyramidal transition state, the electrostatic repulsion is decreased due to removal of electron density in the equatorial plane. The removal of electron density if facilitated by the π-interaction of the trans directing ligand, T as shown below.

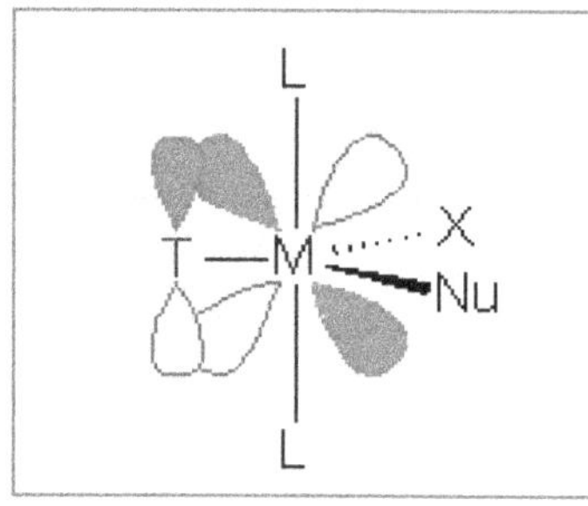

* The square planar substitution reactions occur slowly due to loss of CFSE during the formation of trigonal bipyramidal complex from square planar one. The loss of CFSE is increased down the group. Hence the square planar substitutions of 4d and 5d series are slower. This is why most of the square planar substitution kinetic studies are done on Pt(II) complexes.

* The rough order of trans-directing effect of various ligands is:
NO^+, CO, CN^-, C_2H_4 > PR_3, H^- > Me^- > Ph^- > NO_2^-, I^-, SCN^- > Br^- > Cl^- > Py, NH_3, OH^-, H_2O

Other contributing factors to the trans-effect

□ In the trigonal plane of the 5-coordinate transition state or intermediate, a π- bonding interaction can occur between a metal d-orbital (e.g. dxy) and suitable orbitals (p atomic orbitals, or molecular orbitals of p-symmetry) of ligand L2 (the ligand trans to the leaving group) and Y (the entering group).

□ These 3 ligands and the metal centre can communicate electronically through π- bonding only if they all lie in the same plane in the transition state or intermediate.

□ This implies the 5-coordinate species must be trigonal pyramidal.

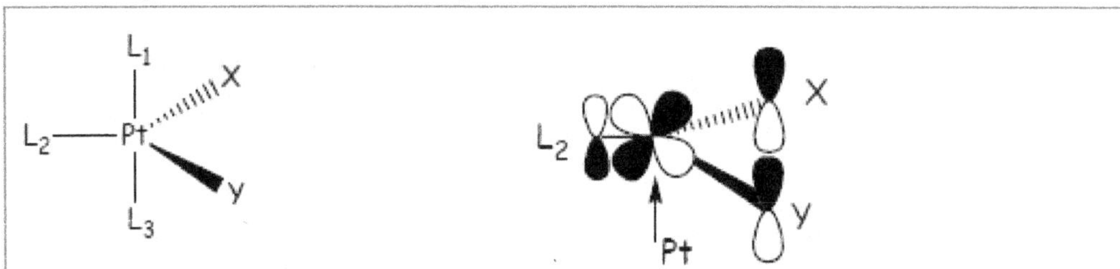

trigonal bipyramidal transition state or intermediate

π-bonding in the trigonal plane

1.13 RACEMISATION IN COMPLEXES

An optically active compound of this type can sometimes be converted into the racemic mixture even in the solid state. In some cases the reaction is induced thermally, but racemization has also been brought about by high pressure. Although other types of complexes can exist as optical isomers, it has been the study of racemization in octahedral complexes that has been most extensive. Two types of processes have been identified, those in which ligands become detached from the metal and those in which there is no metal-ligand bond rupture (intramolecular).

If one end of a chelate ring on an octahedral complex is detached from the metal, the five-coordinate transition state can be considered as a fluxional molecule in which there is some interchange of positions. When the chelate ring reforms, it may be with a different orientation that could lead to racemization.

Possible mechanisms for racemization of [M(bidentate ligand)3]$^{x+/-}$
1. Trigonal Twist mechanisms (Bailar twist or Ray-Dutt twist):
Octahedral Complexes

- Octahedral complexes are generally far more stereochemically rigid than 5-coordinate complexes.
- However, chiral molecules such as $[Ni(en)_3]^{2+}$, $[M(bipy)_3]^{2+}$ (M= Ni, Fe) and $[M(ox)_3]^{3-}$ (M = Ni, Cr, Co, Rh) do undergo racemization at temperatures between 0°C and 200°C.

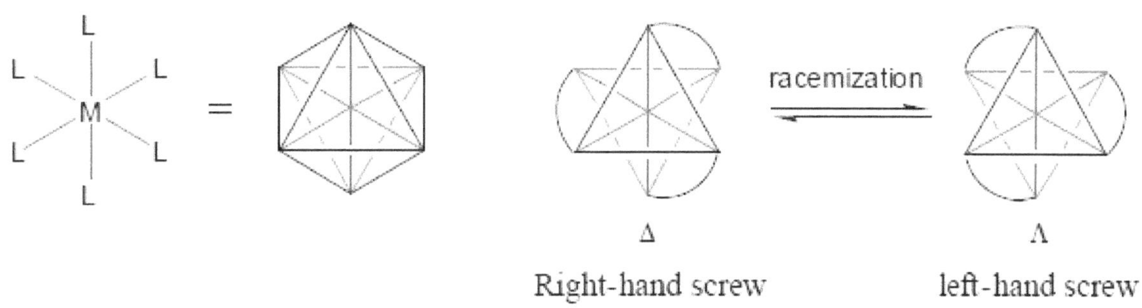

Fig. 63. Racemic structures of octahedral complexes

Numerous racemization reactions occur in which there apparently no bond is breaking as the transition state forms. Although there have been various schemes proposed to account for these observations, the two most likely and important models involve twist mechanisms.

In a regular octahedral structure, C_3 axes pass through the structure and exit from the center of the triangular faces in the"upper" and "lower" halves of the structure. If one triangular face is rotated relative to the other around a C_3 axis, a trigonal prism results. However, for a trischelate complex there are two ways (actually there are four, but most indications are that two are most important) in which the rings (that remain attached to the metal) can be oriented. In one way, the rings span the rectangular faces along the three longitudinal edges. This "trigonal twist" is known as the *Bailar twist as shown in Fig. 63* (Bailar, 1958).

In a second type of trigonal twist (Fig. 64), the transition state is a trigonal prism in which one of the chelating groups lies along the longitudinal edge of a rectangular face but the other two are attached along edges where the triangular faces join the rectangular faces. This mechanism is known as the *Ray and Dutt twist*. A twist mechanism is thought to be operative for the racemization of $[Ni(ox)_3]^{3-}$.

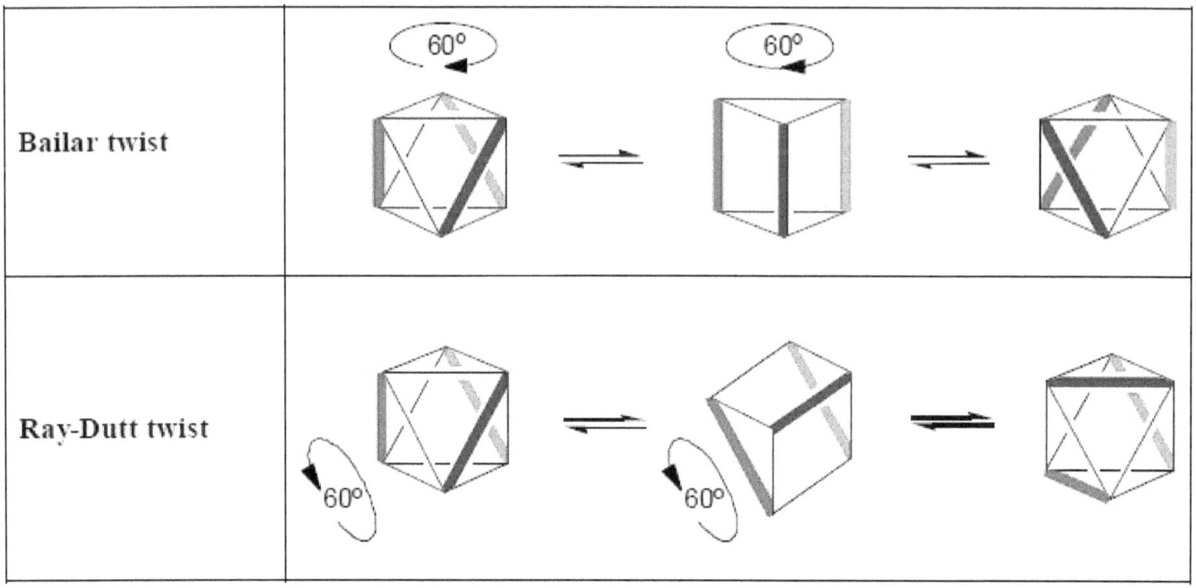

Fig. 64. Trigonal twist structures for octahedral complexes

1. **Dissociation of one end of a bidentate ligand** to give a 5-coordinate species that can undergo rearrangement within the time that the donor atom remains uncoordinated. This mechanism is thought to be operative for the isomerization of $[Rh(ox)_3]^{3-}$.

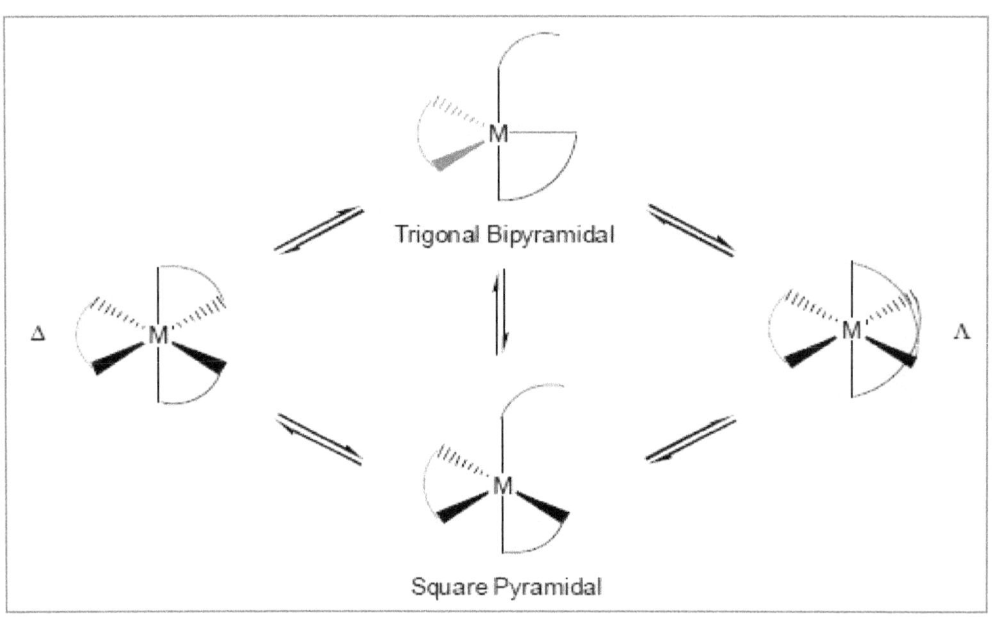

Tetrahedral and Square Planar Complexes

- Tetrahedral and square planar complexes are generally rigid – *i.e.* isomers (optical isomers for tetrahedral complexes, *cis*- and *trans*-isomers for square planar complexes) do not readily interconvert.

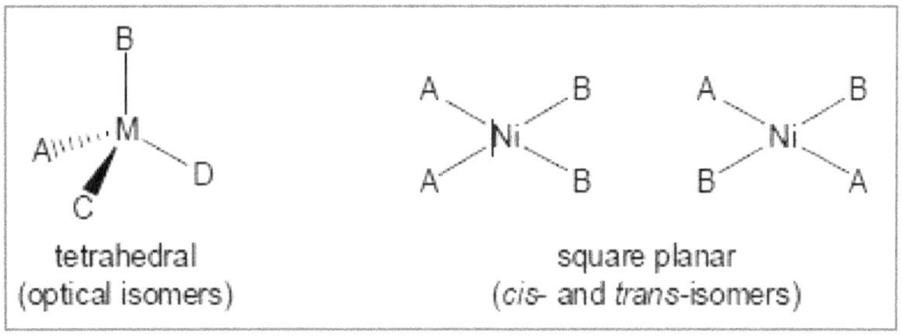

Exceptions:

- In the presence of strongly donating solvents or suitably coordinating counter-anions, 4-coordinate complexes can sometimes undergo ligand redistribution *via* 5- or 6-coordinate intermediates.

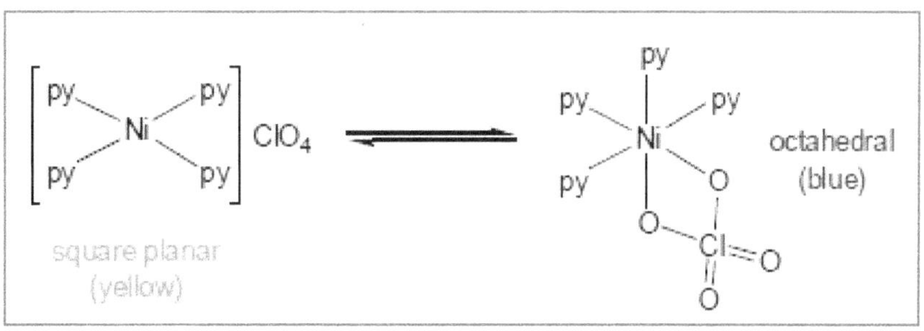

- Most complexes are either tetrahedral (non-d^8) or square planar (RhI, IrI, PdII, PtII, AuIII). However, nickel(II) complexes can be either square planar or tetrahedral, and often undergo facile square planar / tetrahedral interconversion.

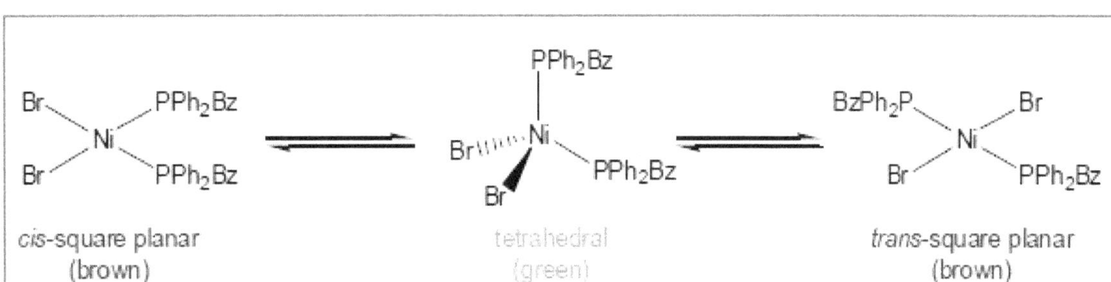

1.14 ELECTRON TRANSFER REACTIONS

Taube won the Nobel Prize (1983) for the study of electron transfer reactions in transition metal complexes, classifying such reactions into two mechanisms. The mechanism of electron transfer in which a bridging ligand is shared between two metals is called the **inner-sphere mechanism,** and the one involving a direct transfer of electrons between two metals without a bridging ligand is called the **outer-sphere mechanism.**

An aqueous solution containing complexes of two different metal ions may make it possible for a redox reaction to occur. In such cases, electrons are transferred from the metal ion being oxidized to the metal ion being reduced. For example,

$$Cr^{2+}(aq) + Fe^{3+}(aq) \rightarrow Cr^{3+}(aq) + Fe^{2+}(aq)$$

In this reaction, an electron is transferred from Cr^{2+} to Fe^{3+} and such reactions are usually called electron transfer or electron exchange reactions. Electron transfer reactions may also occur in cases where only one type of metal ion is involved. For example, the reaction

$$[{}^{*}Fe(CN)_6]^{4-} + [Fe(CN)_6]^{3-} \rightarrow [{}^{*}Fe(CN)_6]^{3-} + [Fe(CN)_6]^{4-}$$

represents an electron transfer from $[{}^{*}Fe(CN)_6]^{4-}$ (where *Fe is a different isotope of iron) to Fe^{3+} in $[Fe(CN)_6]^{3+}$. This is an electron transfer in which the product differs from the reactants only in that a different isotope of Fe is contained in the +2 and +3 oxidation states.

Electron transfer between metal ions contained in complexes can occur in two different ways, depending on the nature of the metal complexes that are present. If the complexes are inert, electron transfer occurring faster than the substitution processes must occur without breaking the bond between the metal and ligand. Such electron transfers are said to take place by an *outer sphere* mechanism. Thus, each metal ion remains attached to its original ligands and the electron is transferred through the coordination spheres of the metal ions.

Outer-sphere mechanism When $[Fe(phen)_3]^{3+}$ (phen is orthophenanthroline) is reduced by $[Fe(CN)_6]^{4-}$, no ligand bridge forms between the metals and an electron moves from the HOMO of Fe(II) to the LUMO of Fe(III) in a very short and direct contact between the complexes. As the result of the electron transfer, $[Fe(phen)_3]^{2+}$ and $[Fe(CN)_6]^{3-}$ form. This kind of reaction is a redox one via the outer-sphere mechanism, and is characteristic of a complex system that has a very slow ligand substitution rate compared with the speed of electron transfer, especially in systems that have the same ligands but different oxidation-numbers, for example, $[Fe(CN)_6]^{3-}$ to $[Fe(CN)_6]^{4-}$ has a high rate of electron transfer. R. A. Marcus won the Nobel Prize (1992) for his study of this outer-sphere electron transfer mechanism.

In the second case, the ligand replacement processes are more rapid than the electron transfer process. If this is the case (as it is with labile complexes), a ligand may leave the coordination sphere of one of the metal ions and be replaced by forming a bridge utilizing a ligand already attached to a second metal ion. Electron transfer then occurs through a bridging ligand, and this is called an *inner sphere* mechanism.

Inner-sphere mechanism When $[CoCl(NH_3)_5]^{2+}$ is reduced by $[Cr(OH_2)_6]^{2+}$, an intermediate complex, $[(NH_3)_5Co\text{-}Cl\text{-}Cr(OH_2)_5]^{4+}$, is formed in which the chlorine atom forms a bridge between cobalt and chromium. As a result of an electron transfer from chromium to cobalt through chlorine, $[Co(NH_3)_5Cl]^{+}$, in which cobalt is reduced from a trivalent to a divalent oxidation state and $[Cr(OH_2)_6]^{3+}$, in which chromium is oxidized from a divalent to a trivalent oxidation state, are formed. This kind of reaction is a redox reaction via the inner-sphere mechanism. The anions other than halogens suitable for such bridge formation are SCN^-, N_3^-, CN^-, *etc.*

In an inner sphere reaction, before the electron transfer can take place a ligand bridge is formed between the two reacting metal centers. Consider,

$$Fe(CN)_6^{3-} + Co(CN)_5H_2O^{3-} \rightarrow Fe(CN)_6^{4-} + Co(CN)_5H_2O^{2-}$$

Recalling that the cycanide ligand has a 1- charge, the reaction involves changes in oxidation states from

$$(Fe^{3+}, Co^{2+}) \rightarrow (Fe^{2+}, Co^{3+})$$

The first step of the reaction involves loss of the weakly bonded H2O ligand and formation of a CN bridege. This creates a relatively long-lived intermediate,

$$Fe^{III}(CN)_6^{3-} + Co^{II}(CN)_5H_2O^{3-} \rightarrow (CN)_5Fe^{III}\text{-}C{=}N\text{-}Co^{II}(CN)_5^{6-} + H_2O$$ *NOTE: should be a triple bond*

The second step is the electron transfer step,

$$(CN)_5Fe^{III}\text{-}C{=}N\text{-}Co^{II}(CN)_5^{6-} \rightarrow (CN)_5Fe^{II}\text{-}C{=}N\text{-}Co^{III}(CN)_5^{6-}$$ *NOTE: should be a triple bond*

Finally, the bridged intermediate dissociates and a H2O ligand fills the vacant coordination slot on the Co atom,

$$(CN)_5Fe^{II}\text{-}C{=}N\text{-}Co^{III}(CN)_5^{6-} + H_2O \rightarrow Fe(CN)_6^{4-} + Co(CN)_5H_2O^{2-}$$

1.15 EXERCISE QUESTIONS

Exercise Questions: Type 1

1. What is meant by anionic complex? Give an example.
2. Write the structure of any one hexadentate ligand?
3. Explain very briefly the chelate effect in coordination compounds.
4. Calculate the EAN for the following complexes;
   i. $K_3[Fe(C_2O_4)_3]$
   ii. $[Co(NH_3)_6]^{3+}$
5. Write the IUPAC name for the following complexes;
   i. $K_2[Cr(CN)_2O_2(O_2)Cl]$
   ii. $Ni(CO)_2(PPh_3)_2$
6. Explain ionisation isomerism with suitable examples.
7. Write the structures of cis and trans- tetramminedichlorocobalt isomers.
8. As per VBT, comment on the hybridisation and structure of the following complexes;
   i. $[Mn(NO_2)_6]^{3-}$
   ii. $[Ni(CN)_5]^{3-}$
9. Write the postulates of crystal field theory.
10. Explain very briefly the splitting of d-orbitals of the metal under octahedral field.
11. Calculate the crystal filed stabilization energy for the complex, $[MnF_6]^{3-}$.
12. Explain very briefly one of the consequences of crystal filed theory.
13. Although Ni^{2+} forms some tetrahedral complexes, Pd^{2+} and Pt^{2+} do not. Explain this difference.
14. What do you mean by John-Teller distortion?
15. Explain synergic bond in M-CO complexes.
16. What is the central-metal oxidation state and the overall charge on the complex ion in $NH_4[Cr(NH_3)_2(NCS)_4]$?
17. Write the structural formula for the isomers indicated for each of the following coordination compounds:
18. linkage isomer of $[Rh(NH_3)_5(ONO)]Cl$
19. ionization isomer of $[Fe(NH_3)_5I]Br_2$
20. Square planar d8 complexes are extremely rare. Account for this observation with a crystal field argument.
21. Using valence bond theory, explain why $[CoF_6]^{3-}$ is a high spin complex, whereas $[Co(NH_3)_6]^{3+}$ is low spin complex.
22. Explain Why the magnetic moment of an octahedral monothiocarbamate complex of Iron(III) drops from 5.8 μB at 300 K to 4 μB at 78 K?
23. Sketch the molecular orbital (MO) diagram for Mo(CO)6 complex
24. Chromium(II) fluoride and manganese (II) fluoride both have a central metal ion surrounded by six fluoride ligands. The Mn–F bond lengths are equidistant , but four of the Cr–F distances are long and two are short. Provide an explanation for this difference.

Exercise Questions: Type 2

25. Explain Werner theory of coordination complexes.
26. Name the following compounds.
    (i) $[Co(en)_2Br_2]Cl$
    (ii) $K_3[Cr(CN)_6]$
    (iii) $Na[Cr(en)C_2O_4Cl_2]$
    (iv) $[Co(en)_3][Cr(C_2O_4)_3]$
27. A nickel complex has the formula NiL_4^{2-}.
    (i) What are the possible structures for the complex?
    (ii) Explain how the magnetic moment could be used to determine the structure of the complex.
28. For each of the following complexes; give the hybrid orbital type and the number of unpaired electrons.
    (a) $[Co(H_2O)_6]^{2+}$; (b) $[FeCl_6]^{3-}$; (c) $[PdCl_4]^{2-}$; (d) $[Cr(H_2O)_6]^{3+}$
29. Would Jahn-Teller distortion be as significant for tetrahedral complexes as it is for octahedral complexes? For which of the electron configurations would Jahn-Teller distortion occur?
30. Calculate the crystal filed stabilization energy and decide whether the following complexes are weak filed or strong field
    i. $[Fe(CN)_6]^{3-}$
    ii. $Mo(CO)_6$
31. Explain the Jahn-Teller distortion in $[Cu(H_2O)_6]^{2+}$.
32. Write the molecular orbital diagrams to show that π ligands cause small Δ_o and also large Δ_o values in octahedral complexes.
33. Write the molecular orbital diagram for $[Co(NH_3)_6]^{3+}$ complex. Explain its salient features.
34. Determine the electron configuration (t2gx egy), number of unpaired electrons and crystal-field stabilization energy (in ΔO or ΔT) for the following complexes:
    $[FeCl_4]^{2-}$
    $[Ru(bipy)_3]^{2+}$
    $[Re(H_2O)_6]^{3+}$
35. What is meant by Jahn-Teller effect? How electronic spectra could be used to detect Jahn-Teller distortion in transition metal complexes?
36. $[Mn(H_2O)_6]^{3+}$ is a high spin complex, but $[Mn(NO_2)_6]^{3-}$ is a low-spin complex. Justify.
37. The compound Na_2IrCl_6 reacts with triphenylphosphine in diethylene glycol under an atmosphere of CO to give trans-$[IrCl(CO)(PPh_3)_2]$, which after reaction with $NaBH_4$ in ethanol gives $[IrH(CO)_2(PPh_3)_2]$. Draw the structures and name the two complexes.

OBJECTIVE TYPE QUESTIONS:

1. The oxidation state of the metal ion in the complex, $[Pd(NH_3)_2(ONO)_2I_2]$ is,
a) 2+ b) 4+ c) 5+ d) 6+
2. How many ions per mole will you expect to find in solution when a compound with the empirical formula $PtCl_4 \cdot 3NH_3$ is dissolved in water?
a) 2 b) 3 c) 4 d) 5
3. The IUPAC name of the complex compound, $[Ni(NH_3)_2(H_2O)_4](ClO_4)_2$ is,
a) diamminetetraaquonickel(II) perchlorate b) tetraaquodiamminenickel(II) pechlorate
c) diamminetetraaquonickel(II) bis(perchlorate) d) diamminetetraaquonickel(II) diperchlorate
4. The correct formula of the compound, hexaamminechromium(III) nitrate is,
a) $[Cr(NH_3)_6](NO_3)_3$ b) $[Cr(NO_3)_3](NH_3)_6$
c) $[Cr(NH_2)_6]NO_3$ d) $[Cr(NH_3)_6](NO_2)_3$
5. Which of the following complexes shows geometric isomerism?
a) $[Co(NH_3)_5Cl]SO_4$ b) $[Co(NH_3)_6]Cl_3$ c) $[Co(NH_3)_5Cl]Cl_2$ d) $K[Co(NH_3)_2Cl_4]$

6. According to valence bond theory $[Mn(H_2O))_6]^{3+}$ complex ion is,
a) outer orbital, high spin and paramagnetic b) inner orbital, low spin and diamagnetic
c) outer orbital, low spin and diamagnetic d) inner orbital, high spin and paramagnetic

7. According to valence bond theory, the hybridisation of the following complexes $[Zn(OH))_4]^{2+}$, $[Ni(CN))_5]^{3-}$ & $[Ti(H_2O))_6]^{3+}$ are
a) dsp^2, dsp^3 & d^2sp^3 b) sp^3, dsp^3 & d^2sp^3 c) sp^3, dsp^3 & sp^3d^2 d) sp^3, sp^3d & d^2sp^3

8. According to crystal field theory, ligands such as CO, CN^-:
(a) usually produce high spin complexes and small crystal field splittings.
(b) usually produce low spin complexes and small crystal field splittings.
(c) usually produce low spin complexes and high crystal field splittings.
(d) usually produce high spin complexes and high crystal field splittings

9. The electron configuration ($t_{2g}^x\ e_g^y$), number of unpaired electrons & crystal-field stabilization energy (in Δ_O) for the complex $[Fe(OH_2)_6]^{2+}$ are,
a) $t_{2g}^4\ e_g^2$, 4 electrons & CFSE = $-0.4\Delta_O$ b) $t_{2g}^6\ e_g^0$, 0 electrons & CFSE = $-2.4\Delta_O$
c) $t_{2g}^4\ e_g^2$, 4 electrons & CFSE = $-0.4\Delta_O + P$ d) $t_{2g}^6\ e_g^0$, 0 electrons & CFSE = $-2.4\Delta_O + 2P$

10. For which pair of complexes is the order of values of Δ_O correct?
a) $[Fe(CN)_6]^{4-} > [Fe(CN)_6]^{3-}$ b) $[Rh(NH_3)_6]^{3+} > [Co(NH_3)_6]^{3+}$
c) $[Cr(OH_2)_6]^{2+} > [Cr(OH_2)_6]^{3+}$ d) $[CrF_6]^{3-} > [Cr(CN)_6]^{3-}$

11. Using the Jahn-Teller theorem, chose the correct degree of deviation from idealized octahedral structure for complexes: $[Cr(CN)_6]^{4-}$, $[Cu(OH_2)_6]^{2+}$ and $[Cr(OH_2)_6]^{3+}$,
a) $[Cr(OH_2)_6]^{3+} < [Cu(OH_2)_6]^{2+} < [Cr(CN)_6]^{4-}$ b) $[Cr(OH_2)_6]^{3+} < [Cr(CN)_6]^{4-} < [Cu(OH_2)_6]^{2+}$
c) $[Cu(OH_2)_6]^{2+} < [Cr(OH_2)_6]^{3+} < [Cr(CN)_6]^{4-}$ d) $[Cr(CN)_6]^{4-} < [Cr(OH_2)_6]^{3+} < [Cu(OH_2)_6]^{2+}$

12. Based on molecular orbital theory, which of the following types of ligand causes lowest degree of Δ_O,
a) σ-donor ligand b) π – acceptor ligand c) Polydentate ligand d) π – donor ligand

13. The Russell-Saunders ground term (spectroscopic state) for d^3 configuration is,
a) ^2D b) ^3F c) ^4F d) ^5D

14. The number of microstates possible for a d^2 configuration, including both weak and strong field limits are a) 15 b) 45 c) 10 d) 90

15. The number of unpaired electrons and the magnetic moments at 25°C for $[Ru(NH_3)_6]^{3+}$ are,
a) 5 electrons and μ_B= 5.92 BM. b) 1 electron and μ_B= 1.73 BM
c) 0 electron and μ_B= 0 BM d) 4 electrons and μ_B= 4.83 BM

Chapter 2

ELECTRONIC SPECTRA OF COORDINATION COMPOUNDS

2.1 INTRODUCTION

The majority of complexes of transition metal ions are colored. The reason that these complexes absorb visible light is that there are electronic transitions of appropriate energy difference possible between the *d* orbitals. These orbitals are split according to the patterns that we have just described. The absorptions are not limited to the visible region, and some are seen in the ultraviolet and infrared regions of the spectrum. However, there are striking differences between the absorption of light by $[Ti(H_2O)_6]^{3+}$ and by a hydrogen atom. The absorption of light by $[Ti(H_2O)_6]^{3+}$ appears as a broad *band*, which has a maximum at 20,300 cm^{-1}.

Electronic spectroscopy is also known as UV-vis spectroscopy and covers the region from 200 to 800 nm of the electromagnetic spectrum. The region from 200 to 400 nm is called the ultraviolet (UV) and that from 400 to 800 nm is known as the visible region. A substance appeared colored when the emitted light falls in this visible region. This spectroscopy is used to find out the geometry of complexes because the number of bands seen in the visible region due to d-d transitions depends on the geometry of the complex.

In simple crystal field theory, the electronic transitions are considered to be occurring between the two groups of *d* orbitals of different energy. We have already alluded to the fact that when more than one electron is present in the *d* orbitals, it is necessary to take into account the spin-orbit coupling of the electrons. In ligand field theory, these effects are taken into account, as are the parameters that represent inter-electronic repulsion.

A very simplistic view of the reason that absorption bands are broad arises from considering what happens during the vibration of a complex having the formula ML_6. The ligand field undergoes instantaneous changes during a vibration, and for simplicity, we will consider the symmetric stretching vibration that preserves the O_h symmetry of the complex. As a result of the ligands moving in and out from the metal ion, there will be a slight difference produced in Δ_o as the ligands are moving the small distance that accompanies the vibration. Because the electronic transition takes place on a time scale that is much shorter than that corresponding to vibration, the transition actually occurs between ligand field states that are changing slightly during the vibration. As a result, a *range* of energies is absorbed, which results in a *band* rather than a *single line* as is the case for a gaseous atom or ion.

Another factor that causes broad absorption bands is that as the electronic energy is changed, so are vibrational energies, which causes a *range* of energy to be absorbed rather than a discrete line. Even though the bands are broad, they provide the basis for analysis of the spectrum of a transition metal complex to extract information about the ligand field and electronic interactions. The magnitudes of the energy gap Δ_o between the t_{2g} and e_g levels can be measured by recording UV – visible spectrum of the complex.

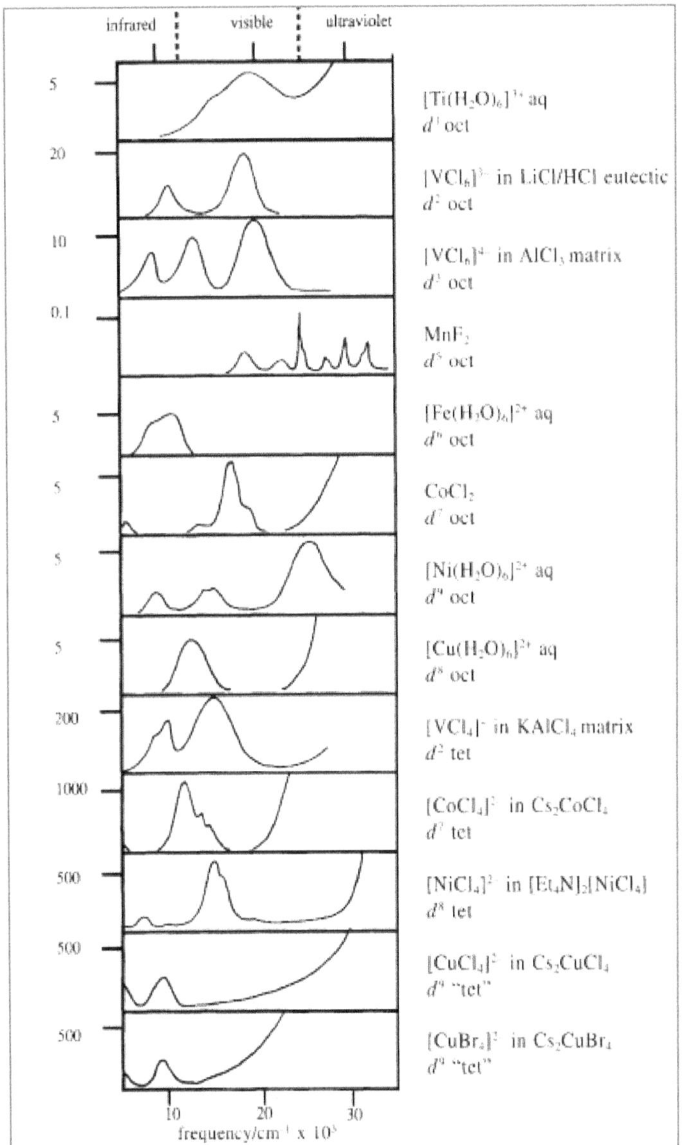

Typical absorption spectra of transition-metal complexes.

Electronic absorption spectra of a few typical transition-metal complexes are shown in the figure given below. The following features are to be noted.

a. All absorptions are broad, often up to 2000 cm^{-1} wide yet occasionally down to 100 cm^{-1}. Atomic line spectra are of the order 1 cm^{-1} in width.

b. Most bands in the near IR, visible and near UV are weak and about 10^2 to 10^4 times weaker than bands characterizing dyestuffs. These are called *'d-d'* bands.

c. Often, much more intense bands occur at higher energies, usually in the ultraviolet region. These comprise so-called 'charge-transfer' bands.

d. The spectra of most octahedral complexes of ions with the configurations d^1, d^4, d^6 and d^9 are characterized by a single absorption, while those for many corresponding d^2, d^3, d^7 and d^8 complexes have up to three main absorptions.

e. The spectra of d^5 complexes show a large number of very weak absorptions, some of which are relatively sharp.

f. The perceived colours of these complexes by transmission are those complementary to the absorptions. Suppose white light impinges upon a sample from a direction labelled z.

At appropriate absorption frequencies, the electronic arrangements in the molecules change as energy is absorbed. About 10 ~18 seconds later, the same energy (frequency) is re-emitted and the ground state electronic arrangement is recovered. However, the light is emitted equally in all directions normal to the incident electric displacement. When viewed along direction z, less of the absorbed frequencies are observed than if no resonance had occurred and we record a net absorption in our spectrum. Further, the colour that we *observe* with our eyes is, of course, determined by absorptions occurring only in the *visible* part of the electromagnetic spectrum.

2.2 SPECTROSCOPIC STATES

In the simplest way, electronic spectra may be discussed as though one electron is being excited into a higher energy level; ignoring the fact that other electrons do interact with the excited electron. For example, $[Ti(H_2O)_6]^{3+}$ contains one electron in the *d* orbital and hence no complication when this single electron absorbs energy. For most of the complex compounds containing more than one *d* electron, the interactions or couplings between the quantum numbers for the individual electrons must be considered in explaining the spectrum.

In a two-electron system (d^2), three types of electronic interactions are possible,

1. spin-spin coupling, in which the spin angular momenta of the two electrons are coupled.
2. orbit-orbit coupling, in which the orbital angular momenta of the two electrons are coupled.
3. spin-orbit coupling, in which coupling of spin and orbital angular momenta on the same electron are coupled.

The coupling between the spin of one electron with the orbital angular momentum of a different electron is negligible and hence ignored. These couplings are called **Russell Saunders coupling.**

Russell Saunders coupling

The ways in which the angular momenta associated with the orbital and spin motions in many-electron-atoms can be combined together are many and varied. In spite of this seeming complexity, the results are frequently readily determined for simple atom systems and are used to characterise the electronic states of atoms. An orbiting electronic charge produces magnetic field perpendicular to the plane of the orbit.

Hence the orbital angular momentum and spin angular momentum have corresponding magnetic vectors. As a result, both of these momenta couple magnetically to give rise to total orbital angular momentum.

There are two principal coupling schemes used:

1. Russell-Saunders (or L - S) coupling
2. j - j coupling.

In the Russell Saunders scheme it is assumed that

spin-spin coupling > orbit-orbit coupling > spin-orbit coupling.

This is found to give a good approximation for first row transition series where J coupling is ignored, however for elements with atomic number greater than thirty, spin-orbit coupling becomes more significant and the j-j coupling scheme is used.

Spin-Spin coupling

S - the resultant spin quantum number for a system of electrons. The overall spin S arises from adding the individual m_s together and is as a result of coupling of spin quantum numbers for the separate electrons.

$S = (s_1+s_2), (s_1+s_2-1), \ldots |s_1-s_2|$

For two electrons, S=0 when spins are coupled opposite (↑↓) or S=1 (↑↑) when spins are coupled parallel.

For three electrons: coupling of the type ↑↑↑ results in S=3/2

coupling of the type ↑↑↓ results in S=1/2

For four electrons: coupling of the type ↑↑↑↑ results in S=2

coupling of the type ↑↑↓↓ results in S=0

The spin multiplicity is given by (2S+1). Hence, If n is the number of unpaired electrons, spin multiplicity is given by n + 1.

Spin multiplicity value depends on the arrangement of electrons, that is, whether they are paired

or not. S= 0→"Singlet" S= ½→"Doublet" S= 1→"Triplet"

No of Electrons	Arrangement	S	2S+1	
1	↑	½	2(1/2)+1=2	doublet
2	↑↑	1	2(1)+1 =3	triplet
2	↑↓	0	2(0)+1 =1	singlet

Orbit-Orbit coupling

L - the total orbital angular momentum quantum number defines the energy state for a system of electrons. The allowed values of L are given below:

$L = (l_1 + l_2), (l_1 + l_2 - 1), (l_1 + l_2 - 2), \ldots\ldots, 0, \ldots\ldots, |l_1 - l_2|$, where l is the azimuthal quantum number.

The resultant L may be once again 0, 1, 2, 3, 4.... which are referred to as S, P, D, F G,...respectively in units of h/2π.The orbital multiplicity is given by (2L+1). These states or term letters are represented as follows:

Total Orbital Momentum

Total Orbital Momentum						
L	0	1	2	3	4	5
	S	P	D	F	G	H

Spin-Orbit coupling

Coupling occurs between the resultant spin and orbital momenta of an electron which gives rise to J the total angular momentum quantum number. The resultant S and L couple to give a total angular momentum, J. The possible values of J quantum number are given as

$J = (L+S), (L+S-1), (L+S-2), (L+S-3), \ldots\ldots L-S,$

The symbol | | indicates that the absolute value (L – S) is employed, i.e., no regard is paid to ± sign. Thus for L = 2 and S = 1, the possible J states are 3, 2 and 1 in units of h/2π. This scheme of coupling is known as spin-orbit coupling or j -j coupling.

Among elements beyond Z = 30, spin orbit coupling becomes significant and therefore with heavy elements j-j coupling is used.

The Russell Saunders term symbol that results from these considerations is given by:

$^{(2S+1)}L$ As an example, for a d^1 configuration:L=2 and S = ½, the Russell Saunders Ground Term is written as ^2D. The Russell Saunders term symbols for the other free ion configurations are given in the Table below.

Terms for $3d^n$ free ion configurations		
Configuration	Ground Term	Excited Terms
d^1, d^9	^2D	-
d^2, d^8	^3F	^3P, ^1G, ^1D, ^1S
d^3, d^7	^4F	^4P, ^2H, ^2G, ^2F, 2 x ^2D, ^2P
d^4, d^6	^5D	^3H, ^3G, 2 x ^3F, ^3D, 2 x ^3P, ^1I, 2 x ^1G, ^1F, 2 x ^1D, 2 x ^1S
d^5	^6S	^4G, ^4F, ^4D, ^4P, ^2I, ^2H, 2 x ^2G, 2 x ^2F, 3 x ^2D, ^2P, ^2S

Configuration	Ion	Term symbol
d^1	Ti^{3+}, V^{4+}	
d^9	Cu^{2+}	^2D
d^2	Ti^{2+}, V^{3+}, Cr^{4+}	
d^8	Ni^{2+}	^3F, ^3P, ^1G, ^1D, ^1S
d^3	Cr^{3+}, V^{2+}, Mn^{4+}	
d^7	Ni^{3+}, Co^{2+}	^4F, ^4P, ^{2}(H, G, F, D, D, P)
d^4	Cr^{2+}, Mn^{3+}	
d^6	Fe^{2+}, Co^{3+}	^5D, ^{3}(H, G, F, F, , D, P, P), ^{1}(I, ,G, G, F, D, D, S,S)
d^5	Mn^{2+}, Fe^{3+}	
d^{10}	Zn^{2+}	^6S, ^{4}(G, F, D, P), ^{2}(I, H, G, G, F, F), ^{2}(D, D, D, P, S)
		^6S

Number of microstates

The number of microstates possible for any electronic configuration may be calculated from the formula,

Number of microstates = n! / r! (n - r)!

Where n is the twice the number of orbitals, r is the number of electrons and ! is the factorial.

for a d^2 configuration, the number of microstates is given by

10! / 2! (10 – 2)!

$$\frac{10\times9\times8\times7\times6\times5\times4\times3\times2\times1}{2\times1(8\times7\times6\times5\times4\times3\times2\times1)}=45$$

Thus a d^2 configuration will have 45 microstates.

Microstates of different d^n configuration

d^n configuration	d^1,d^9	d^2,d^8	d^3,d^7	d^4,d^6	d^5	d^{10}
No.of microstates	10	45	120	210	252	1

2.3 HUND'S RULES

The Ground Terms are deduced by using Hund's Rules.

The two rules are:

1) The Ground Term will have the maximum multiplicity.

2) If there is more than 1 Term with maximum multipicity, then the Ground Term will have the largest value of L.

A simple graphical method for determining just the ground term alone for the free ions uses a "fill in the boxes" arrangement.

d^n	2	1	0	-1	-2	L	S	Ground Term
d^1	↑					2	1/2	^2D
d^2	↑	↑				3	1	^3F
d^3	↑	↑	↑			3	3/2	^4F
d^4	↑	↑	↑	↑		2	2	^5D
d^5	↑	↑	↑	↑	↑	0	5/2	^6S
d^6	↑↓	↑	↑	↑	↑	2	2	^5D
d^7	↑↓	↑↓	↑	↑	↑	3	3/2	^4F
d^8	↑↓	↑↓	↑↓	↑	↑	3	1	^3F
d^9	↑↓	↑↓	↑↓	↑↓	↑	2	1/2	^2D

To calculate **S**, simply sum the **unpaired** electrons using a value of ½ for each.

To calculate **L**, use the labels for each column to determine the value of **L** for that box, then add all the individual box values together.

For a d^7 configuration, then:

in the +2 box are 2 electrons, so **L** for that box is 2*2= 4

in the +1 box are 2 electrons, so **L** for that box is 1*2= 2

in the 0 box is 1 electron, **L** is 0

in the -1 box is 1 electron, **L** is -1*1= -1

in the -2 box is 1 electron, **L** is -2*1= -2 Total value of **L** is therefore +4 +2 +0 -1 -2 or **L**=3.

Note that for 5 electrons with 1 electron in each box then the total value of **L** is 0.

This is why **L** for a d^1 configuration is the same as for a d^6. The other thing to note is the idea of the "hole" approach.

A d^1 configuration can be treated as similar to a d^9 configuration. In the first case there is 1 electron and in the latter there is an absence of an electron ie a hole.

The overall result shown in the Table above is that:

4 configurations (d^1, d^4, d^6, d^9) give rise to D ground terms,

4 configurations (d^2, d^3, d^7, d^8) give rise to F ground terms and the d^5 configuration gives an S ground term.

For the d^2 case, the *l* value for each electron is 2 and so L can take the values 4, 3, 2, 1 or O, corresponding to the labels G, F, D, P, *S*. The s value for each electron is 1/2, so *S* can take values 1 or O and there may thus arise terms ^{3}G, ^3F, ^{3}D, ^{3}P, $^3S'$, ^{1}G, ^{1}F, ^{1}D, 1P, 1S. However, not all these terms do arise for the 'equivalent' *d* electrons of the d^2 configuration, because some electronic arrangements violate Pauli's exclusion principle while others ignore the fact that electrons are indistinguishable. For the d^2 configuration, the first electron may have any one of five m_l values and any one of two *ms* - or 10 possibilities altogether. The second electron is left with only 9 choices of the (m_l, m_s) combination because of the exclusion principle - hence yielding 90 arrangements for the pair. However, the indistinguishability of electrons means we cannot assign meaning to the adjectives 'first' and 'second' here and have thus counted each arrangement twice. Altogether, therefore, the d^2 configuration is 45-fold degenerate. It can be shown to give rise to the term set: ^1G, ^3F, ^1D, ^3P, ^1S. The (2L + 1)(2S + 1) - fold degeneracies of these terms are 9, 21, 5, 9 and 1 respectively, adding to 45, as required.

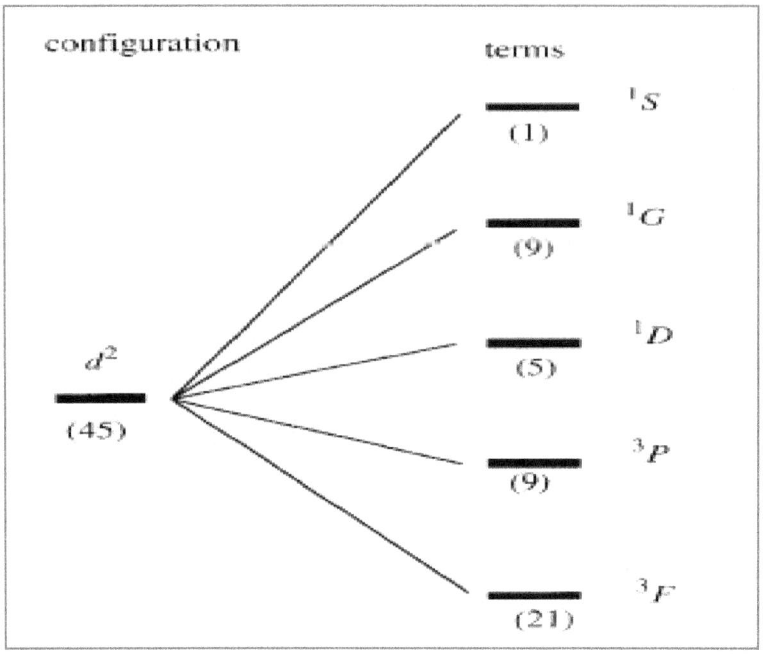

Hole formalism

A d^9 system can be considered as the inverted d^1 system as far as energy levels can be considered because d^9 system has an electron vacancy, which is called a ' hole'. Similarly, d^8 system is considered as inverted d^2 system as far as the energy levels are considered. This is called hole formalism. In short, an inverted energy level relationship exists between d^n and d^{10-n} systems. Hence, with the help of only these two diagrams, all the d^n energy level diagrams can be explained in terms of d^1 and d^2 systems as given below:

$d^9 = d^{10-1}$ = inverted d^1 system

$d^8 = d^{10-2}$ = inverted d^2 system

$d^7 = d^{5+2}$ = same as d^2 system

$d^6 = d^{5+1}$ = same as d^1 system

$d^4 = d^{5-1}$ = inverted d^1 system

d3 = d^{5-2} = inverted d^2 system

Russell-Saunders coupling scheme for a d^2 configuration

It is from the analysis of spectra that we obtain information about energies of spectroscopic states in metal ions and the effects produced by different ligands on the *d* orbitals. However, it is first necessary to know what spectroscopic states are appropriate for various metal ions. The analysis then progresses to how the spectroscopic states for the metal ions are affected by the presence of the ligands and how ligand field parameters are determined for spectral data. An understanding of spin-orbit coupling is necessary to determine the spectroscopic states that exist for various electron configurations, d^n ions that have degenerate d-orbitals.

Electronic configuration is split into terms due to interelectronic repulsion, which is split into states by spin-orbit coupling, which is further split into microstates by magnetic field (Fig. 65):

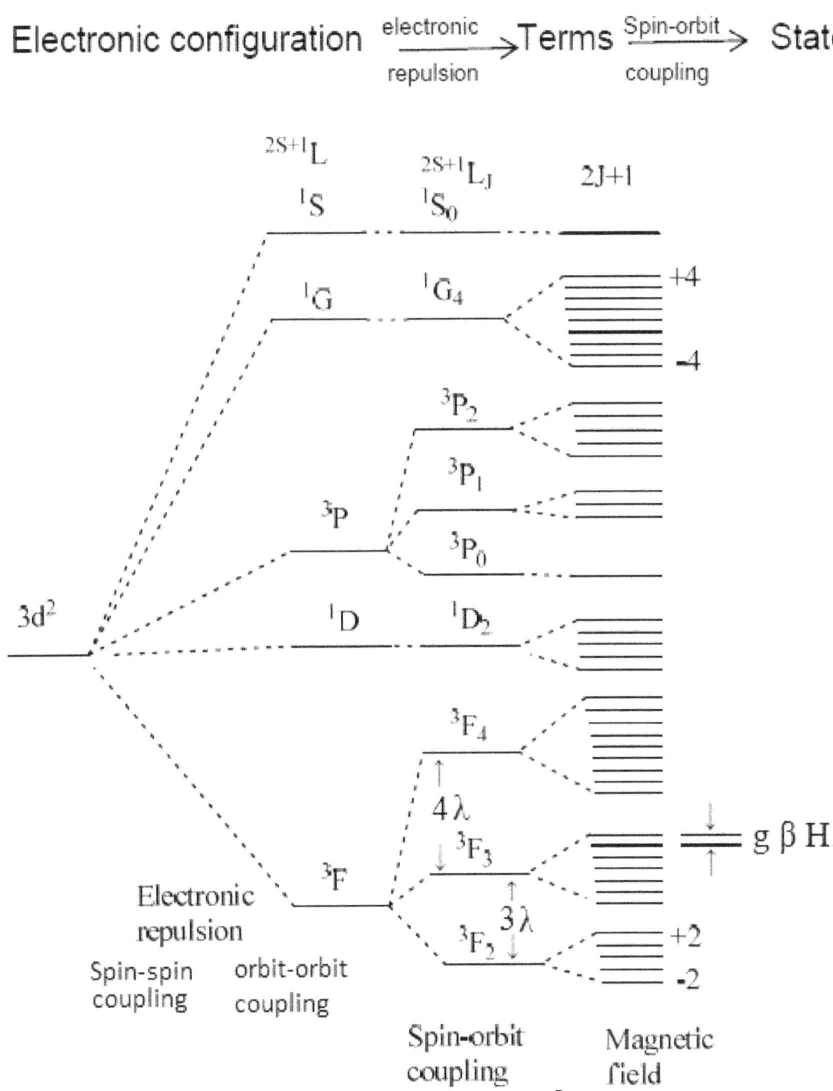

Fig. 65. Russell-Saunders coupling scheme for d^2 configuration

According to Russell- Saunders coupling scheme,

1. spin-orbit coupling constant (λ). It is the energy separations between levels produced by spin-orbit coupling.

2. The separations between the levels of J values, J and (J+1) = (J=1) λ

3. For the ^3F term, L = 3 and S = 1; therefore, J = (3 = 1), 3 and (3 -1); these corresponds to the three states $^3F_4$, $^3F_3$ and $^3F^2$. {J takes up all the values between (L + S) and (L – S)}.

4. The separation between $^3F_4$ and $^3F_3$ is 4 λ and that between $^3F_3$ and $^3F_2$ is 3 λ.

5. These energy levels ($^3F_4$, $^3F_3$ and $^3F^{2)}$ further split into microstates on application of magnetic field (Zeeman Effect). The J values are split into (2J +1); these split microstates are equally spaced and their number corresponds to the number of values that can be assumed by the magnetic quantum number m (-J,……0.,,,,,,+J.)

The separation between these energy levels is given as g β H; Where g is the Lande splitting factor, β is the Bohr magneton and H is applied magnetic field. The Lande factor is given by the equation,

$$g = \frac{3}{2} + \frac{S(S+1) - L(L+1)}{2J(J+1)}$$

2.4 SELECTION RULES

1. ΔS = 0 The Spin Rule- Transitions between states of different multiplicities are forbidden. This is called the *spin selection rule.* The rule says that allowed transitions involve the promotion of electrons without a change in their spin.

Thus transitions such as $^2S \rightarrow {}^2P$ and $^3D \rightarrow {}^3P$ are allowed, but transition such as $^1S \rightarrow {}^3P$ is forbidden. The same rule is also stated in the form of a statement, Electronic Transitions between the different states of spin multiplicity are forbidden. The selection Rule for total angular momentum, J, is $\Delta J = 0$ or ± 1
The transitions such as $^2P_{1/2} \rightarrow {}^2D_{3/2}$ and $^2P_{3/2} \rightarrow {}^2D_{3/2}$ are allowed, but transition such as $^2P_{1/2} \rightarrow {}^2D_{5/2}$ is forbidden since Δ J= 2.

There is no selection rule governing the change in the value of n, the principal quantum number. Thus in hydrogen, transitions such as 1s → 2p, 1s → 3p, 1s → 4p are allowed. Usually, electronic absorption is indicated by reverse arrow, ← , and emission is indicated by the forward arrow, → , though this rule is not strictly obeyed.

2. ΔL = +/- 1 The Orbital Rule (Laporte)-

The second rule says that transitions within a given set of p or d orbitals (i.e. those which only involve a redistribution of electrons within a given subshell) are forbidden if the molecule has a centre of symmetry.

In other words, the selection rule for total orbital angular momentum is $\Delta L = \pm 1$. This is La Porte allowed transitions. Thus transition such as $^1S \rightarrow {}^1P$ and $^2D \rightarrow {}^2P$ are allowed but transition such as $^3D \rightarrow {}^3S$ is forbidden since $\Delta L = -2$ .That is, transition should involve a change of one unit of angular momentum. Hence transitions from gerade to ungerade (g to u) or vice versa are allowed, i.e., u → g or g → u but not u → u or g → g. In the case of p sub shell, both ground and excited states are odd and in the case of d sub shell both ground and excited states are even. As a rule transition should be from even to odd or vice versa. The same rule is also stated in the form of a statement instead of an equation: *Electronic transitions within the same p or d sub-shell are forbidden, if the molecule has centre of symmetry.*

Relaxation of the Rules

a) Spin-Orbit coupling - this gives rise to weak spin forbidden bands. The spin rule can be relaxed very slightly for the first row transition metals by spin-orbit coupling. Typical molar absorbitivities are less than 1 L/mol-cm, with very pale color observed.

b) Vibronic coupling – The orbital rule is relaxed due to vibrations of the complex that cause a loss of the center of symmetry. As a result, molar absorbitivities of 10-50 L/mol-cm are observed.

c) π-acceptor and π-donor ligands can mix with the d-orbitals so transitions are no longer purely d-d.

Mechanism of breakdown of selection rules

Spin selection rule

For electronic transition to take place, $\Delta S = 0$ and $\Delta L = \pm 1$ in the absence of spin-orbit coupling. However, spin and orbital motions are coupled. Even, if they are coupled very weakly, a little of each spin state mixes with the other in the ground and excited states by an amount dependent. **upon the energy difference in the orbital states and magnitude of spin –orbit coupling constant.**

Therefore electronic transitions occur between different states of spin multiplicity and also between states in which ΔL is not equal to ± 1. For example, if the ground state were 99% singlet and 1% triplet (due to spin– orbit coupling) and the excited state were 1% singlet and 99 % triplet, then the intensity would derive from the triplet –triplet and singlet-singlet interactions. Spin-orbit coupling provides small energy differences between degenerate states.

This coupling is of two types. The single electron spin orbit coupling parameter ζ, gives the strength of the interaction between the spin and orbital angular momenta of a single electron for a particular configuration. The other parameter, λ, is the property of the term.

For high spin complexes,

$$\lambda = \pm \xi / 2S$$

Here positive sign holds for shells less than half field and negative sign holds for more than half filled shells. S is the same as the one given for the free ion. The λ values in crystals are close to their free ion values. Λ decreases in crystal with decreasing Racah parameters B and C. For high spin d5 configuration, there is no spin orbit coupling because 6S state is unaffected by the ligand fields. The λ and ζ values for 3d series are given below:

Ion	Ti(II)	V(II)	Cr(II)	Mn(II)	Fe(II)	Co(II)	Ni(II)
Ξ (cm^{-1})	121	167	230	347	410	533	649
λ(cm^{-1})	60	56	57	0	-102	-177	-325

λ and ζ values for 3d series

La Porte selection rule

Physically 3d (even) and 4p (odd) wave functions may be mixed, if centre of inversion (i) is removed. There are two processes by which i is removed.

a. The central metal ion is placed in a distorted field (tetrahedral field, Tetragonal distortions, etc.,) The most important case of distorted or asymmetric field is the case of

a tetrahedral complex. Tetrahedron has no inversion centre and so d-p mixing takes place. So electronic transitions in tetrahedral complexes are much more intense, often by a factor 100, than in a analogous octahedral complexes. Trans isomer of $[Co(en)_2Cl_2]^+$ in aqueous solution is three to four times less intense than the cis isomer because the former is centro-symmetric. Other types of distortion include Jahn –Teller distortions.

b. Odd vibrations of the surrounding ligands create the distorted field for a time that is long enough compared to the time necessary for the electronic transition to occur (Franck Condon Principle).Certain vibrations will remove the centre of symmetry. Mathematically this implies coupling of vibrational and electronic wave functions. Breaking down of La Porte rule by vibrionic coupling has been termed as "Intensity Stealing". If the forbidden excited term lies energetically nearby a fully allowed transition, it would produce a very intense band. Intensity

Stealing by this mechanism decreases in magnitude with increasing energy separation between the excited term and the allowed level.

2.5 CRYSTAL FIELD SPLITTING OF RUSSELL-SAUNDERS TERMS

The effect of a crystal field on the different orbitals (s, p, d, etc.) will result in splitting into subsets of different energies, depending on whether they are in an octahedral or tetrahedral environment. The magnitude of the d orbital splitting is generally represented as a fraction of Δ_{oct} or 10Dq.The ground term energies for free ions are also affected by the influence of a crystal field and an analogy is made between orbitals and ground terms that are related due to the angular parts of their electron distribution. The effect of a crystal field on different orbitals in an octahedral field environment will cause the d orbitals to split to give t_{2g} and e_g subsets and the D ground term states into T_{2g} and E_g, (where upper case is used to denote states and lower case orbitals). f orbitals are split to give subsets known as t_{1g}, t_{2g} and a_{2g}. By analogy, the F ground term when split by a crystal field will give states known as T_{1g}, T_{2g}, and A_{2g}. Note that it is important to recognise that the F ground term here refers to states arising from d orbitals and not f orbitals and depending on whether it is in an octahedral or tetrahedral environment the lowest term can be either A_{2g} or T_{1g}. The table below shows that the Mulliken symmetry labels, developed for atomic and molecular orbitals, have been applied to these states but for this purpose they are written in CAPITAL LETTERS.For splitting in a tetrahedral crystal field the components are similar, except that the symmetry label g (gerade) is absent. The ground term for first-row transition metal ions is either D, F or S, which in high spin octahedral fields gives rise to A, E or T states. This means that the states are either non-degenerate, doubly degenerate or triply degenerate.

The Crystal Field Splitting of Russell-Saunders terms in high spin octahedral crystal fields.

Russell-Saunders Terms	Crystal Field Components
S	A_{1g}
P	T_{1g}
D	E_g , T_{2g}
F	A_{2g} , T_{1g} , T_{2g}
G	A_{1g} , E_g , T_{1g} , T_{2g}
H	E_g , 2 x T_{1g} , T_{2g}
I	A_{1g} , A_{2g} , E_g , T_{1g} , T_{2g}

Mulliken Symbols	
Mulliken Symbol for atomic and molecular orbitals	**Explanation**
a	Non-degenerate orbital; symmetric to principal C_n
b	Non-degenerate orbital; unsymmetric to principal C_n
e	Doubly degenerate orbital
t	Triply degenerate orbital
(subscript) g	Symmetric with respect to center of inversion
(subscript) u	Unsymmetric with respect to center of inversion
(subscript) 1	Symmetric with respect to C_2 perp. to principal C_n
(subscript) 2	Unsymmetric with respect to C_2 perp. to principal C_n
(superscript) '	Symmetric with respect to s_h
(superscript) "	Unsymmetric with respect to s_h

Effect of weak and strong cubic- crystal fields (Octahedral and tetrahedral) on S, P, D and F terms

The correlation diagrams for multi-electron configurations can be quite complicated, but are extremely useful in explaining the spectra of transition metal complexes. For a d^2 configuration, the free ion has the following ground level energy state: ^3F. The other term states are ^3P, ^1S, ^1D, and ^1G.

A correlation diagram for this configuration is shown below. Note that the free ion terms will split slightly in a weak field, and resolve into term states which correspond to recognizable electron configuration in an infinitely strong octahedral field. If the ligands cause large splitting between the terms, it is called strong field case. If they cause small splitting between the terms, it is called weak field case. In strong field case, electrons will try to pair whenever possible and in weak field case, the electrons will try to remain unpaired. The free ion terms, weak field and strong field environments for a d^2 configuration are shown in the following figure,

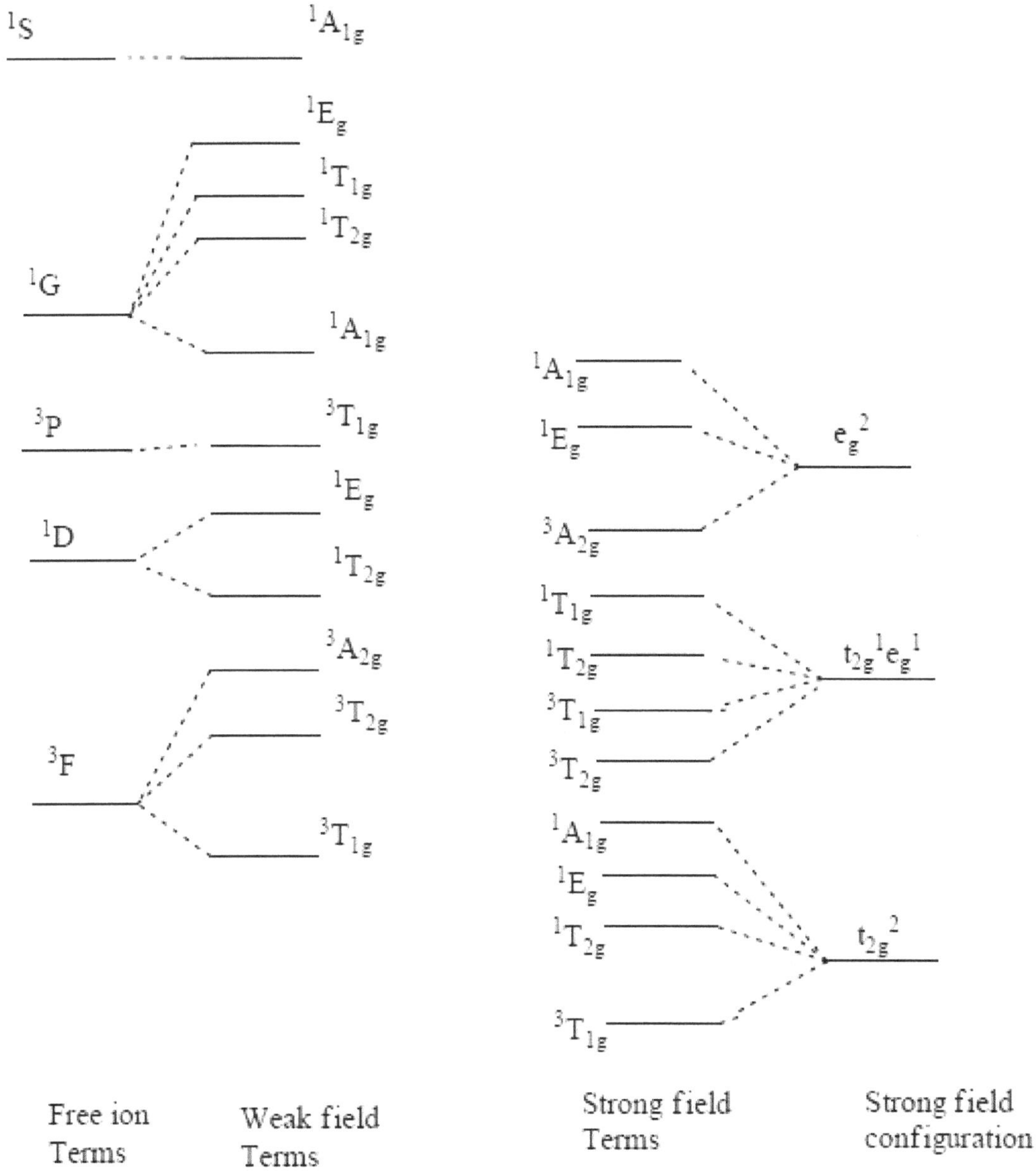

The notes at the far right side of the diagram show the electron configuration of the metal ion in an infinitely strong octahedral field. The ground energy state of the ion in an infinitely strong octahedral field will be t_{2g}^2, because this would have two electrons in the lowest set of orbitals with parallel spins. The next highest energy state would have the configuration t_{2g}^1 e_g^1, representing one electron promoted to the higher, e_g set of orbitals. the highest energy state would correspond to a configuration of e_g^2, with two electrons in the upper set of orbitals.

In actual coordination complexes, the environment of the metal is intermediate between the free ion case and the infinitely strong ligand field. As can be seen from the correlation diagram, each free-ion irreducible representation is matched with (correlates with) a strong-field irreducible representation having the same symmetry.

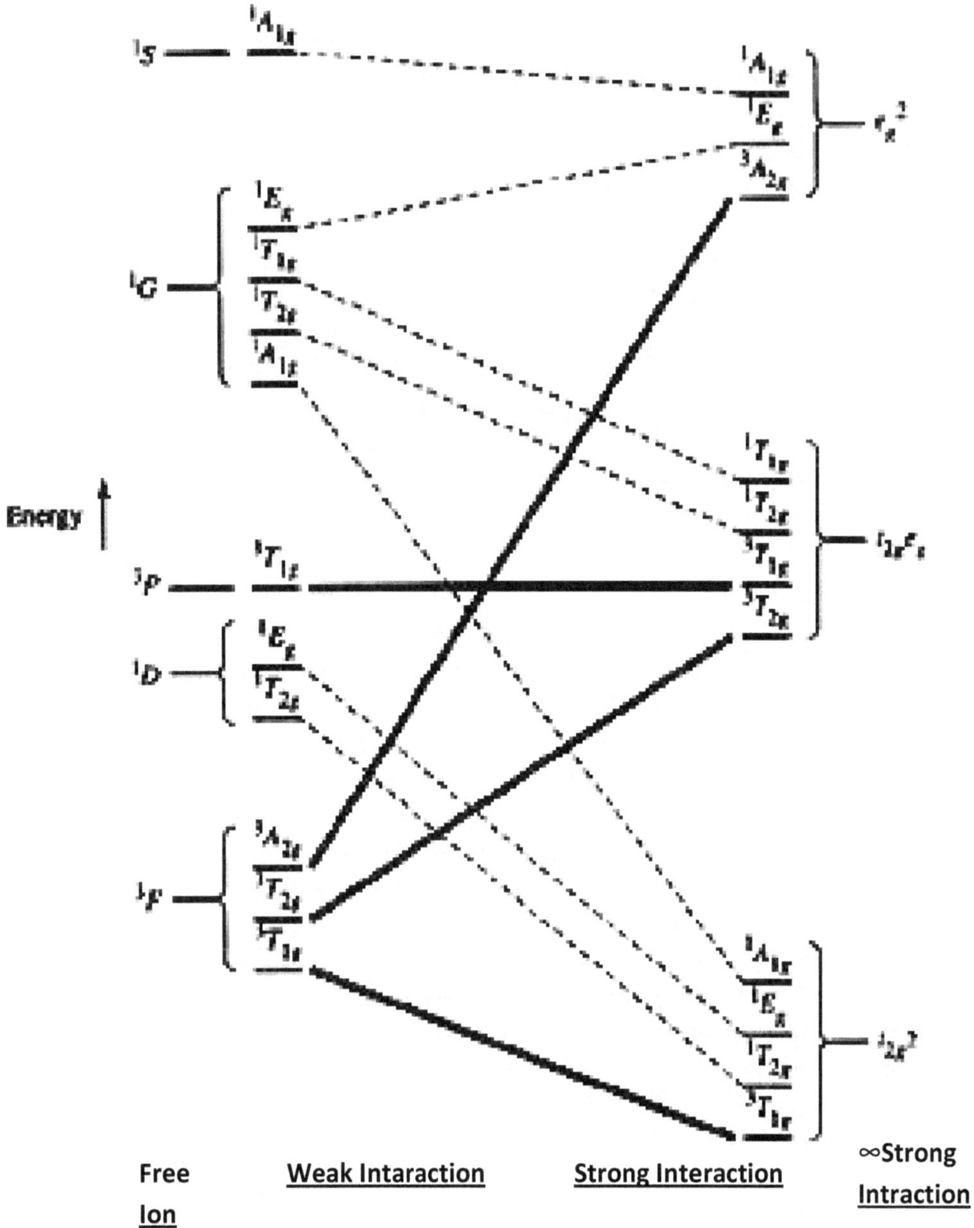

Correlation of spectroscopic ground terms with electron configurations for d^n ions in octahedral field

***d*^{1} ion: The ground configuration is t_{2g}^1.** The single electron can occupy any of the three degenerate t_{2g} orbitals; therefore, it has an orbital degeneracy of 3 corresponding to a ground term. When this electron is excited to e_g orbital, there it has degeneracy of only 2; this excited eg1 configuration corresponds to an E term at an energy Δ above T.

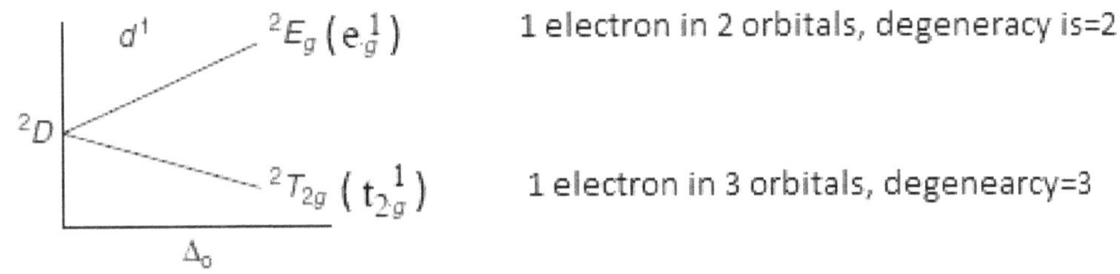

Applying the same considerations on the other d- ions, the correlation of spectroscopic ground terms with electron configurations can be derived as follows,

***d*^{2} ion: The ground configuration is t_{2g}^2.**

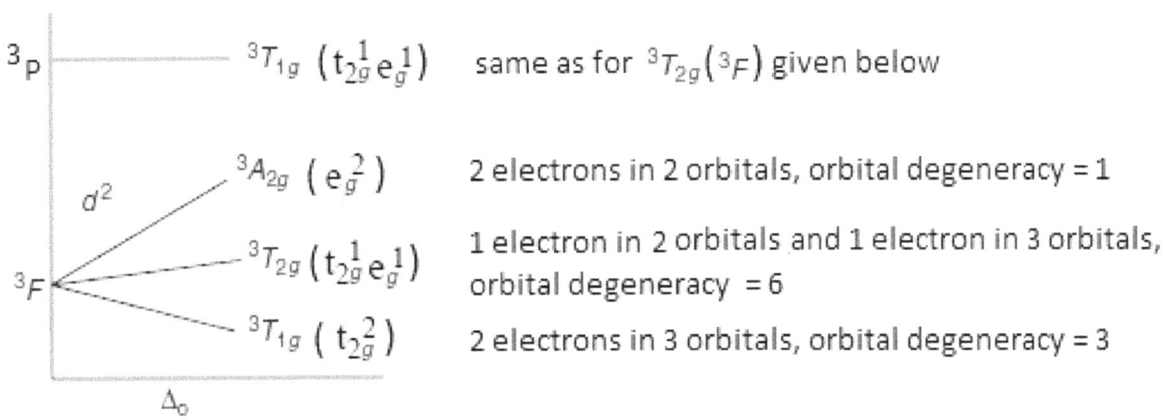

Term	Configuration	Specific Degenerate Configurations
$^3T_{1g}(F)$	$t_{2g}^{\ 2}$	$(d_{xy})^1\,(d_{xz})^1\,(d_{yz})^0$
		$(d_{xy})^1\,(d_{xz})^0\,(d_{yz})^1$
		$(d_{xy})^0\,(d_{xz})^1\,(d_{yz})^1$
$^3T_{2g}$	$t_{2g}^{\ 1}e_g^{\ 1}$	$(d_{xy})^1\,(d_{z^2})^1$
		$(d_{xz})^1\,(d_{x^2-y^2})^1$
		$(d_{yz})^1\,(d_{x^2-y^2})^1$
$^3T_{1g}(P)$	$t_{2g}^{\ 1}e_g^{\ 1}$	$(d_{xy})^1\,(d_{x^2-y^2})^1$
		$(d_{xz})^1\,(d_{z^2})^1$
		$(d_{yz})^1\,(d_{z^2})^1$
$^3A_{2g}$	$e_g^{\ 2}$	$(d_{x^2-y^2})^1\,(d_{z^2})^1$

Why Are There Two Different-Energy Triply Degenerate States for the Configuration $t_{2g}^{\ 1}e_g^{\ 1}$?

- In all three microstates of $^3T_{2g}$ the two electrons are placed in d orbitals that are geometrically orthogonal to each other:

$$(d_{xy})^1\,(d_{z^2})^1 \qquad (d_{xz})^1\,(d_{x^2-y^2})^1 \qquad (d_{yz})^1\,(d_{x^2-y^2})^1$$

- In all three microstates of $^3T_{1g}(P)$ the two electrons are placed in d orbitals that lie in the same plane:

$$(d_{xy})^1\,(d_{x^2-y^2})^1 \qquad (d_{xz})^1\,(d_{z^2})^1 \qquad (d_{xz})^1\,(d_{z^2})^1$$

- Electron-electron repulsions are less in the microstates of $^3T_{2g}$ than they are in the microstates of $^3T_{1g}(P)$, so the energy of the $^3T_{2g}$ state is lower.

d^3 ion: The ground configuration is t_{2g}^3

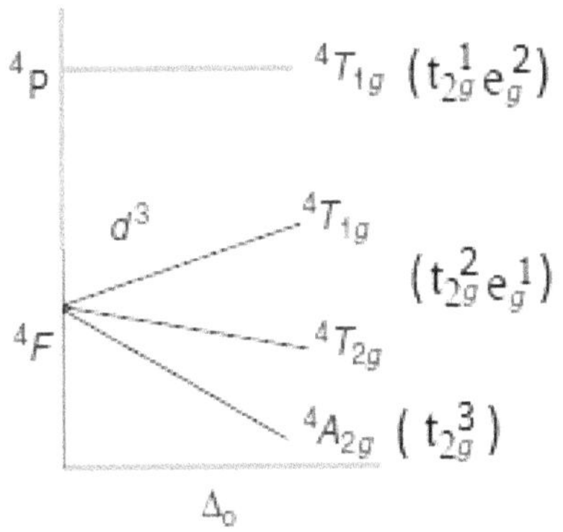

1 electron in 3 orbitals and 2 electrons in 2 orbitals, orbital degeneracy = 3

2 electrons in 3 orbitals and 1 electron in 2 orbitals, orbital degeneracy = 6

3 electrons in 3 orbitals, orbital degeneracy = 1

d^4 ion: The ground configuration is $t_{2g}^3 e_g^1$

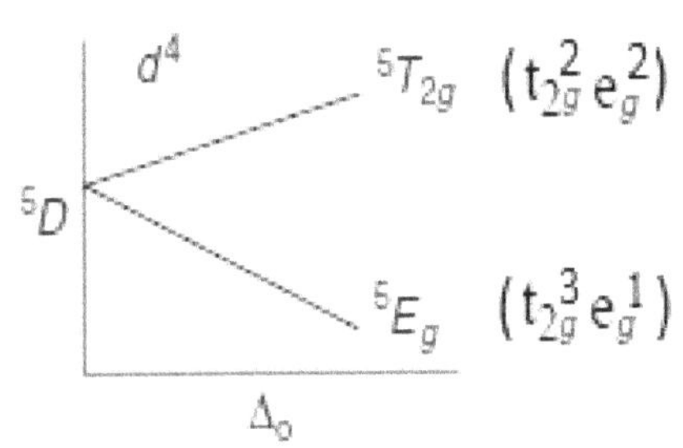

2 electrons in 3 orbitals, and 2 electrons in 2 orbitals orbital degeneracy = 3

3 electrons in 3 orbitals and 1 electron in 2 orbitals, orbital degeneracy = 6

d^5 ion: The ground configuration is $t_{2g}^3 e_g^2$

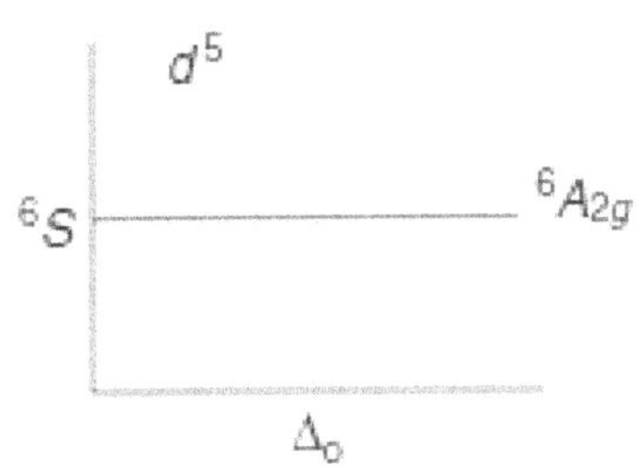

5 electrons in 5 orbitals orbital degeneracy = 1

d^6 ion: The ground configuration is $t_{2g}^4e_g^2$

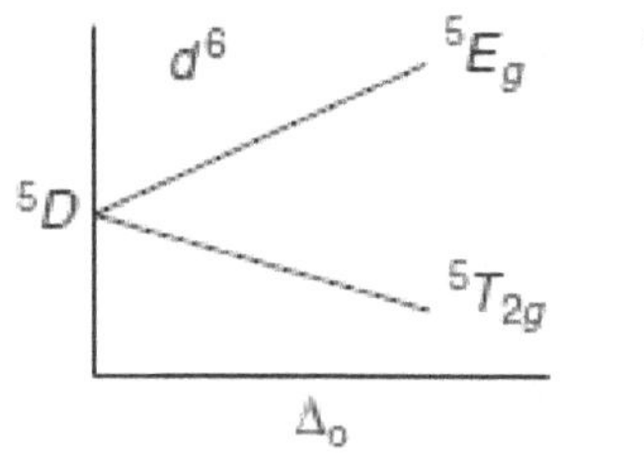

with 4 positive holes; d^6 becomes the exact inverse of d^4

d^7 ion: The ground configuration is $t_{2g}^5e_g^2$

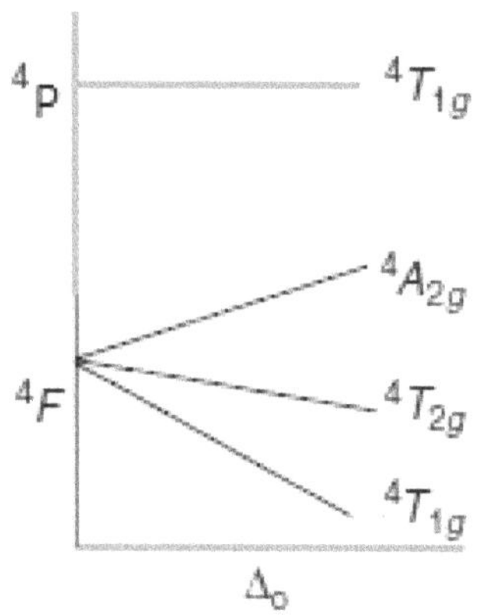

with three positive holes,
d^7 is the exact inverse of d^3

d^8 ion: The ground configuration is $t_{2g}^6e_g^2$

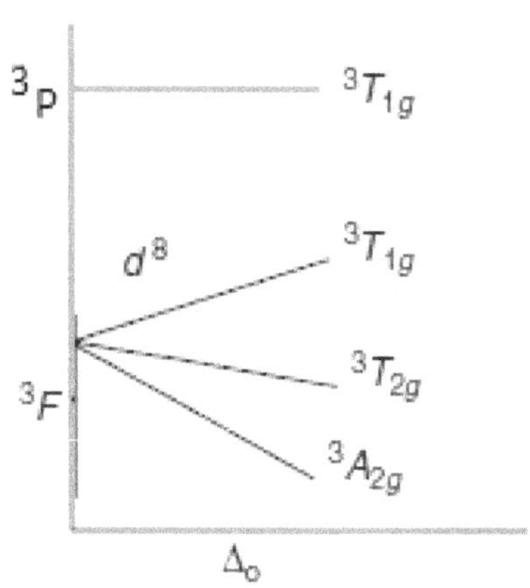

with 2 positive holes, d^8 is the exact inverse of d^2

d^9 ion: The ground configuration is $t_{2g}^6e_g^3$

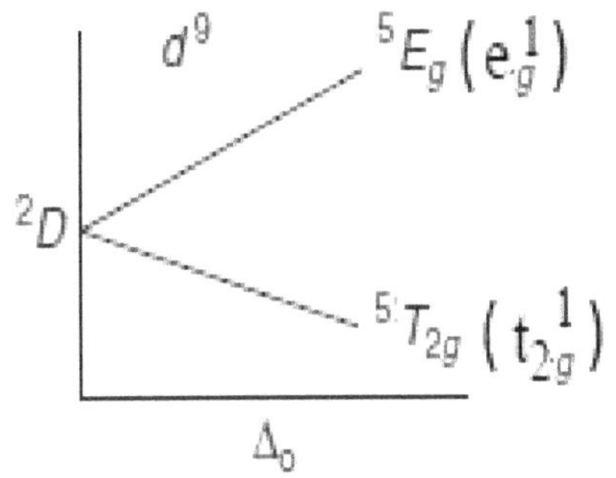

with 1 positive hole, d^9 is the exact inverse of d^1

Comparison of the d^n octahedral and d^n tetrahedral energy level diagrams

- The energy level diagrams of a d^n system in octahedral field is the inverse of the diagram of the same ion in tetrahedral field.
- Ligands in a tetrahedral field produces 4/9 the splittings of the octahedral field.
- When a d^n ion in octahedral field is switched to a d^n ion in tetrahedral field, all splittings of the Russell-Saunders states are reversed.
- Similar inversions are caused on changing from the d^n configuration from an octahedral field to the d^{10-n} configuration in an octahedral field; same is true for the change d^n Tetrahedral to d^{n-1} Tetrahedral.
- Various pairs of configuration- geometry combinations have qualitatively identical energy level diagrams in crystal fields: the difference lies only in splitting of the individual free ion states.

  d^1_{oct} and d^9_{tet} : *Reverse of* d^9_{oct} and d^1_{tet}

  d^2_{oct} and d^8_{tet} : Reverse of d^8_{oct} and d^2_{tet}

  d^3_{oct} and d^7_{tet} : Reverse of d^7_{oct} and d^3_{tet}

  d^4_{oct} and d^6_{tet} : Reverse of d^6_{oct} and d^4_{tet}

2.6 SPLITTING PATTERNS FOR GROUND STATE D AND F TERMS

When a metal ion is surrounded by ligands in a coordination compound, those ligands generate an electrostatic field that removes the degeneracy of the *d* orbitals. The result is that e_g and t_{2g} subsets of orbitals are produced. Because the *d* orbitals are no longer degenerate, spin-orbit coupling is altered so that the states, no longer apply to a metal ion *in a complex*. However, just as the *d orbitals* are split in terms of their energies, the *spectroscopic states* are split in the ligand field. The spectroscopic states are split into components that have the same multiplicity as the free ion states from which they arise.

A single electron in a *d* orbital gives rise to a ^2D term for the gaseous ion, but in an octahedral field the electron will reside in a t_{2g} orbital, and the spectroscopic *state* for the t_{2g} *configuration* is $^2T_{2\ g}$. If the electron were excited to an e_g orbital, the spectroscopic state would be ^2E_g. Thus, transition between $^2T_{2g}$ and ^2E_g states would not be spin forbidden because both states are doublets. Note that lowercase letters are used to describe *orbitals*, whereas capital letters describe spectroscopic *states*.

A gaseous ion having a d^2 configuration gives rise to a ^3F ground state as a result of spin-orbit coupling. Although they will not be derived, in an octahedral ligand field the $t_{2g}{}^2$ configuration gives three different spectroscopic states that are designated as $^3A_{2g}$, $^3T_{1g}$, and $^3T_{2g}$. These states are often referred to as *ligand field states*. The energies of the three states depend on the strength of the ligand field, but the relationship is not a simple one. The larger the ligand field splitting, the greater the difference between the ligand field states of the metal. The table given below, shows a summary of the states that result from splitting the gaseous state terms of metal ions in an octahedral field produced by six ligands.

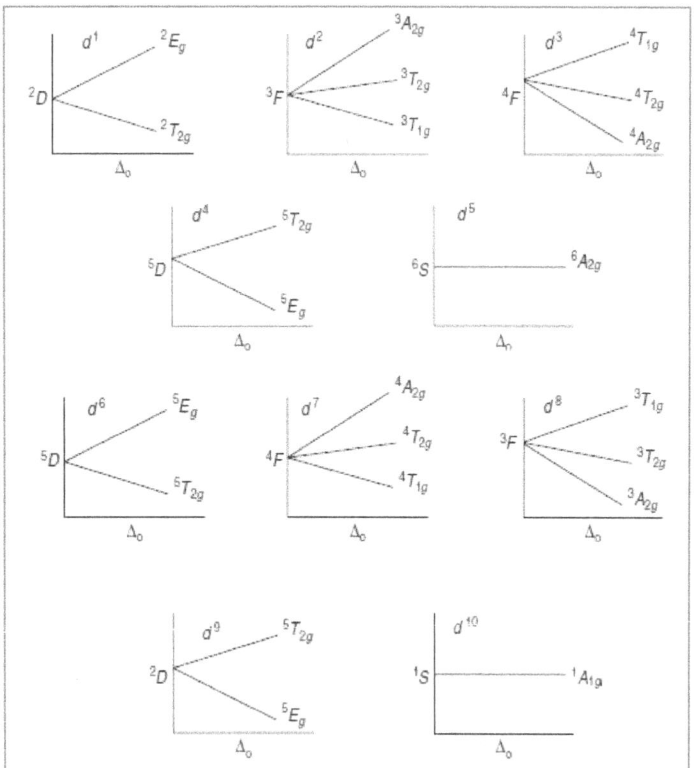

The splitting patterns for ground-state *D* and *F* terms in an octahedral field.

Note that the ground-state terms for d^n ions (except for d^5) are all either *D* or *F* terms and that the state splitting occurs so that the center of energy is maintained. For the ligand field states that are produced by splitting the ^3F term (which results from a d^2 configuration), the center of energy is also preserved even though there are three states in the ligand field.

For a specific d^1 electron configuration, there are usually several spectroscopic states that correspond to energies above the ground state term. However, they may not have the same *multiplicity* as the ground state. When the spectroscopic state for the free ion becomes split in an octahedral field, each ligand field component has the same multiplicity as the ground state *Transitions between spectroscopic states having different multiplicities are spin forbidden.*

Because the T_{2g} and E_g spectroscopic states in a ligand field have the same multiplicity as the ground states from which they arise, it can be sees that the $T_{2g} \rightarrow E_g$ transition is the only spin-allowed transition for ions that have *D* ground states.

In cases where the ground state is an *F* term, there is a *P* state of higher energy that has the same multiplicity. That state gives a T_{1g} state (designated as T_{1g} (*P*)) having the same multiplicity as the ground state T_{1g} term. Accordingly, spectroscopic transitions are possible from the ground state in the ligand field to the *T* state that arises from the *P* term. Therefore, for ions having *T* and *A* ground states, the spin-allowed transitions in an octahedral field are as follows:

Octahedral Field			
For *T* ground states:		For *A* ground states:	
ν_1	$T_{1g} \rightarrow T_{2g}$	ν_1	$A_{2g} \rightarrow T_{2g}$
ν_2	$T_{1g} \rightarrow A_{2g}$	ν_2	$A_{2g} \rightarrow T_{1g}$
ν_3	$T_{1g} \rightarrow T_{1g}(P)$	ν_3	$A_{2g} \rightarrow T_{1g}(P)$

The spin-allowed transitions for ions having *T* and *A* ground states in tetrahedral fields are as follows:

Tetrahedral Field			
For *T* ground states:		For *A* ground states:	
ν_1	$T_1 \rightarrow T_2$	ν_1	$A_2 \rightarrow T_2$
ν_2	$T_1 \rightarrow A_2$	ν_2	$A_2 \rightarrow T_1$
ν_3	$T_1 \rightarrow T_1(P)$	ν_3	$A_2 \rightarrow T_1(P)$

As shown, for both octahedral and tetrahedral complexes, *three* absorption bands are expected. Many complexes do, in fact, have absorption spectra that show three bands. However, charge transfer absorption makes it impossible in some cases to see all three bands, so spectral analysis must often be based on only one or two observed bands.

2.7 ORGEL DIAGRAMS

Energy Level Diagrams are described by two independent schemes - Orgel Diagrams which are applicable to weak field complexes and Tanabe –Sugano (or simply T-S) Diagrams which are applicable to both weak field and strong field complexes.

Orgel diagrams

The splitting patterns of the spectroscopic states can be reduced to graphical presentation in two diagrams. These diagrams were developed by L. E. Orgel, and they have since become known as *Orgel diagrams*. Fig. 66 shows the Orgel diagram that is applicable for ions that give *D* ground states.

In an Orgel diagram, energy is represented as the vertical dimension, and the vertical line in the center of the diagram represents the gaseous ion where there is no ligand field ($\Delta = 0$). Note that the right-hand side of the diagram applies to d^1 and d^6 ions in *octahedral* fields or d^4 and d^9 ions in *tetrahedral* fields. This situation arises because the ligand field states are inverted for the two cases, and the electron-hole formalism also causes the orbitals to be inverted. As a result, the ligand field states are the same for a d^1 ion in an octahedral field as they are for a d^4 ion in a tetrahedral field. . Fig. 67 shows the Orgel diagram that is applicable for ions that give *F* ground states.

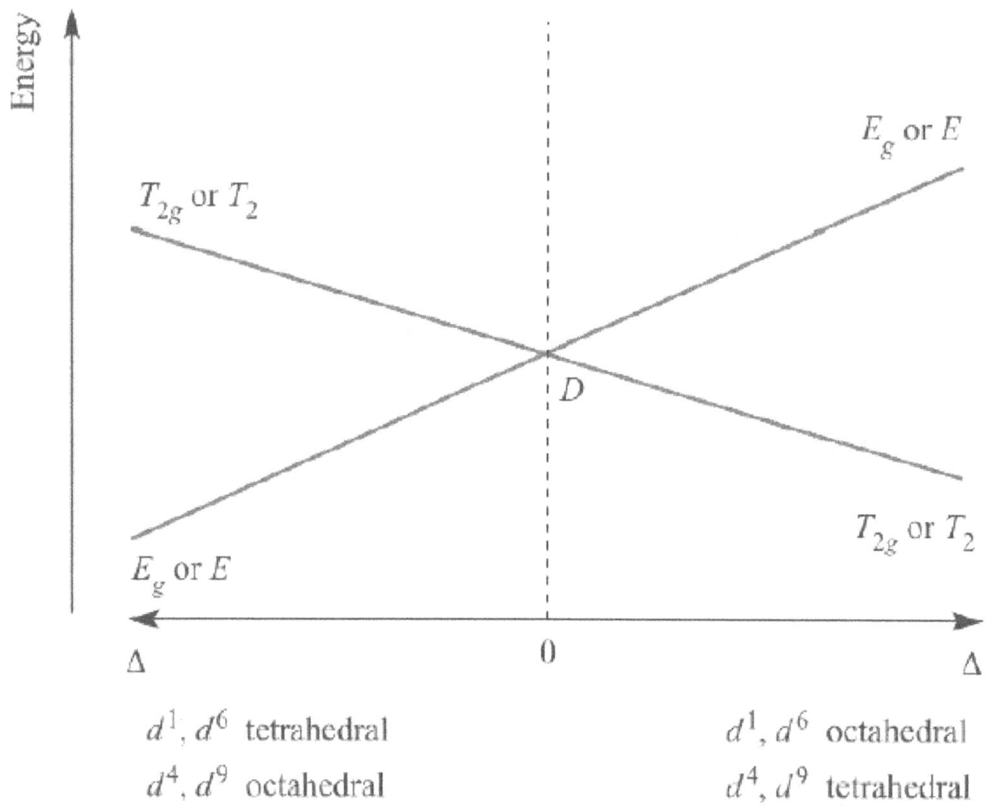

Fig. 66. Orgel diagram for D terms

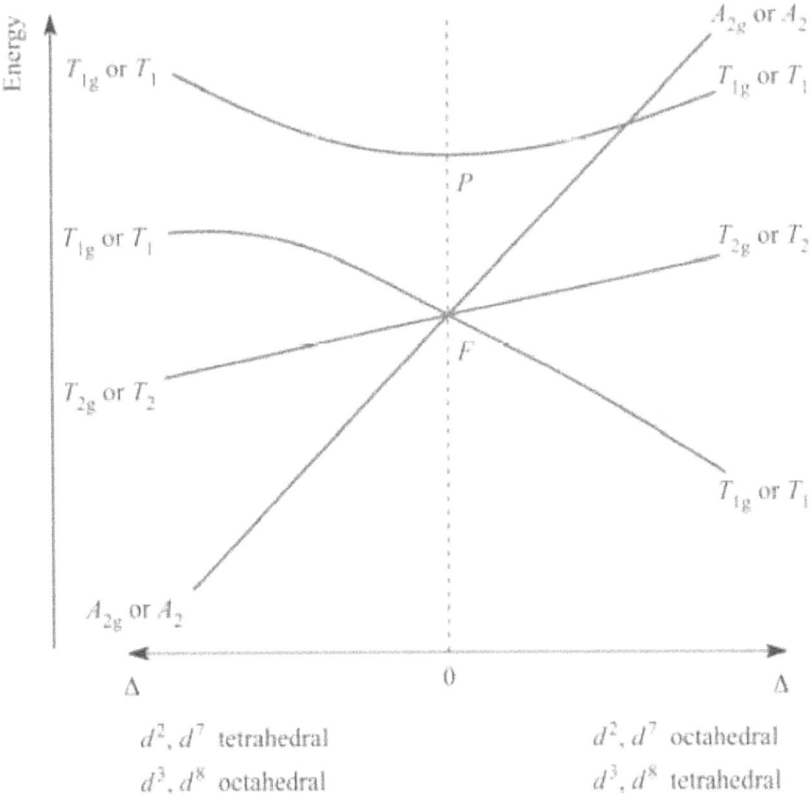

Fig. 67. Orgel diagram for F terms

2.8 ELECTRONIC SPECTRA OF FIRST ROW TRANSITION METAL IONS

d^1 – configuration: Electronic spectrum of $[Ti(H_2O)_6]^{3+}$

The electronic configuration of Ti^{3+} is [Ar] $3d^1$ $4s^2$. It has five-fold degeneracy and its ground state term symbol is ^2D. In an octahedral crystal field, the five-fold degeneracy is split into $^2T_{2g}$ and ^2E_g states. Thus only one single electron transition, $^2T_{2g} \rightarrow {}^2E_g$, is expected in an octahedral crystal field (Fig. 69).

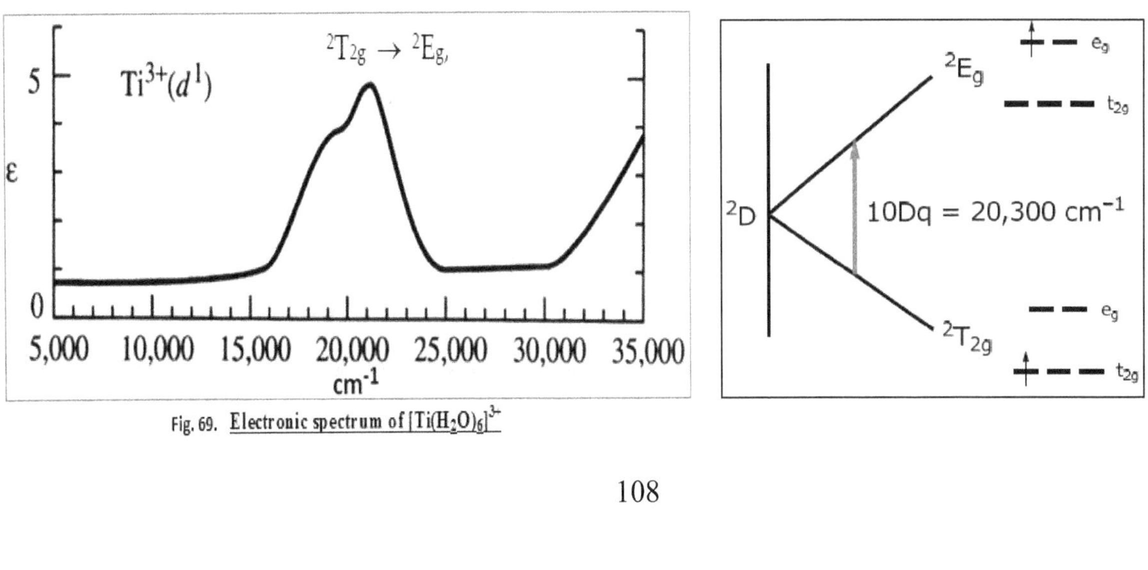

Fig. 69. Electronic spectrum of $[Ti(H_2O)_6]^{3+}$

Normally, the ground $^2T_{2g}$ state is split due to Jahn-Teller effect and hence lowering of symmetry is expected for Ti (III) ion. This state splits into $^2B_{2g}$ and ^2E_g states in tetragonal symmetry and the excited term ^2E_g also splits into $^2B_{1g}$ and $^2A_{1g}$ levels. Thus, three bands are expected for tetragonal (C_{4v}) symmetry.

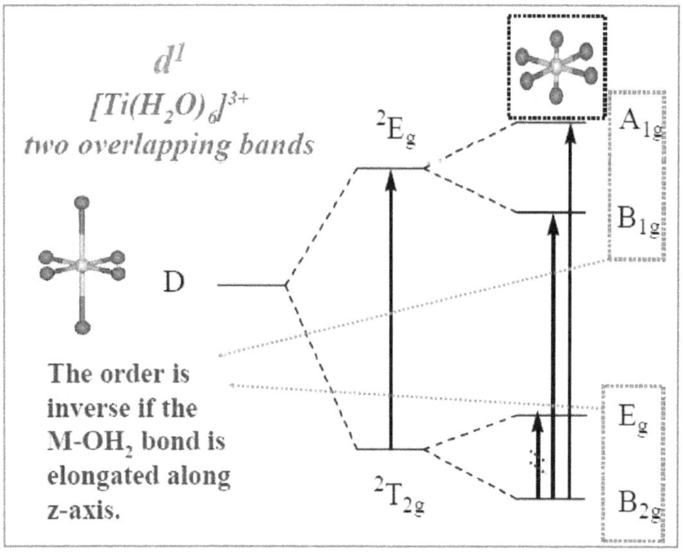

d^2 – configuration **Electronic spectrm of $[V(H_2O)_6]^{3+}$**

1. In $[V(H_2O)_6]^{3+}$ (d^2), ground state is $^3T_{1g}(F)$; under ordinary conditions this is the only electronic state that is appreciably occupied.

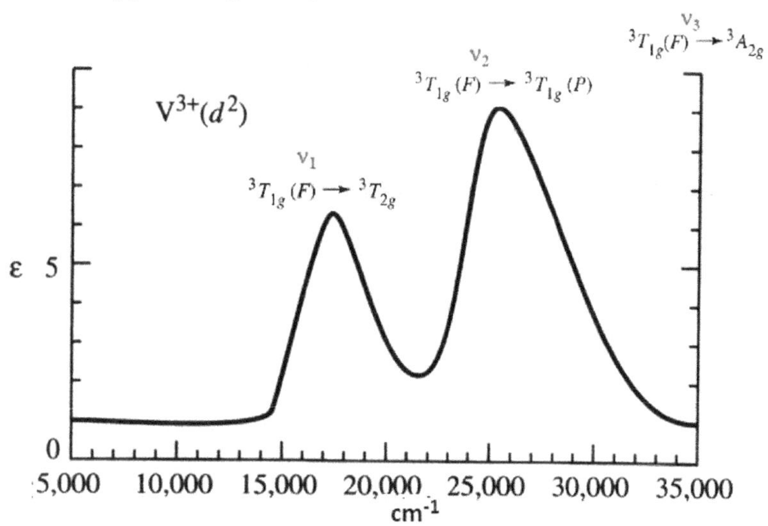

Fig. 70. Electronic spectrm of $[V(H_2O)_6]^{3+}$

2. *There are three transitions from $^3T_{1g}$ (F) to $^3T_{2g}$, $^3T_{1g}$ (P), & $^3A_{2g}$ (spin conservation).*

3. *Three allowed transitions are expected (Fig. 70). However, two bands are readily observed at 17,800 **(562 nm)** &25,700 cm^{-1} **(389 nm)**.*

4. *A third band, at approximately 38,000 cm^{-1} (263 nm), is apparently obscured in aqueous solution by charge transfer band.*

Orgel diagram showing the transitions among different states in case of V^{2+} (d^2) ion is as given below:

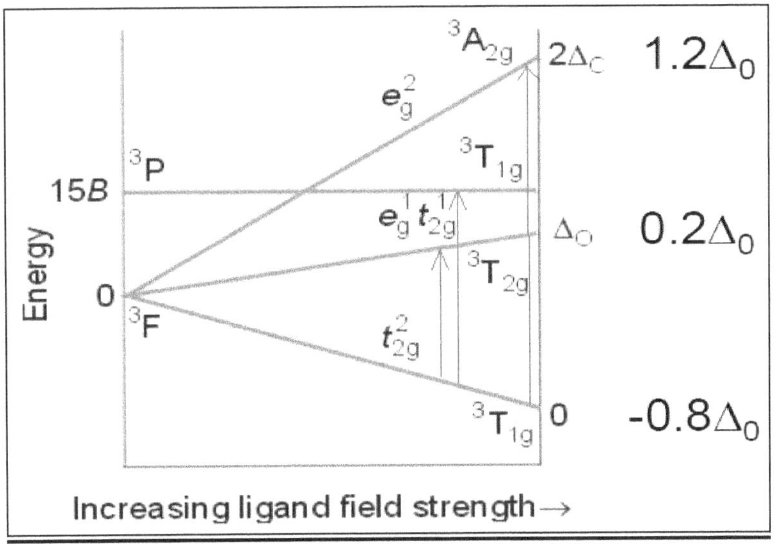

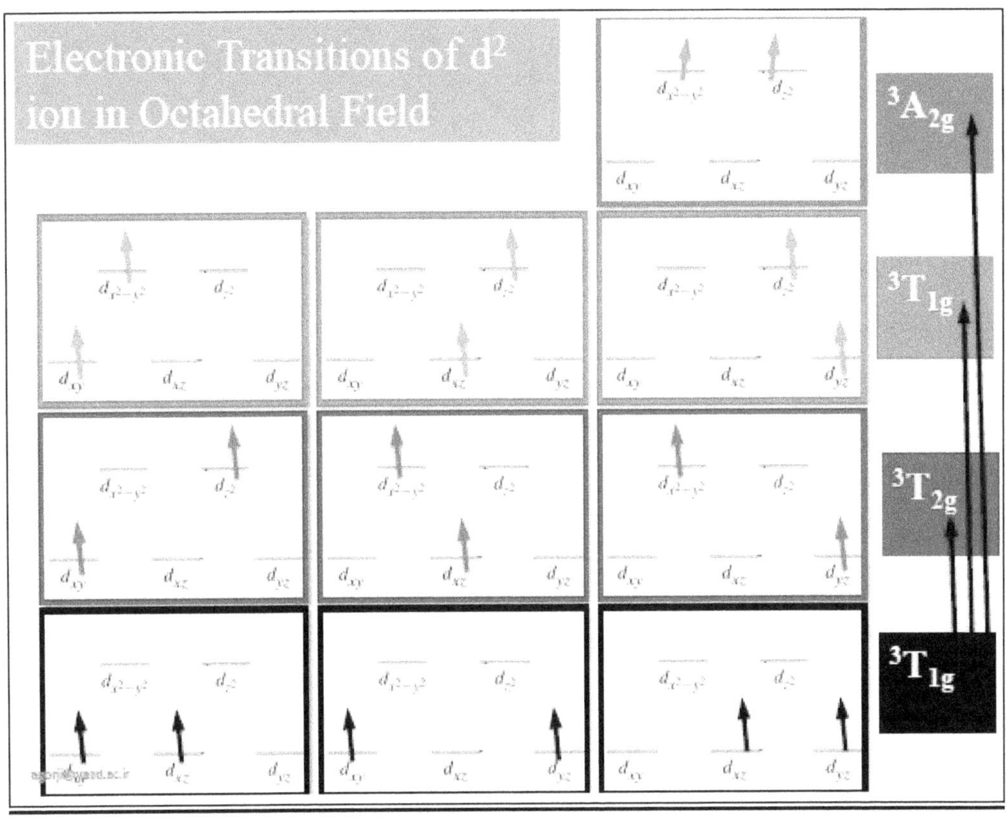

d^3 – configuration: Electronic spectrum of $[Cr(H_2O)_6]^{3+}$

In octahedral symmetry, the three unpaired electrons are in t_{2g}^3 orbitals which give rise to $^4A_{2g}$, ^4E_g, $^4T_{1g}$ and $^2T_{2g}$ states. Of these $^4A_{2g}$ is the ground state. Hence, three spin allowed transitions, observed in high spin state, are shown in the spectrum (Fig.71) given below,

$^4A_{2g}$ (F) → $^4T_{2g}$ (F) (υ1),

$^4A_{2g}$ (F)→ $^4T_{1g}$ (F) (υ2) and

$^4A_{2g}$ (F) → $^4T_{1g}$ (P) (υ3).

These spin allowed bands split into two components when the symmetry of Cr^{3+} ion is lowered from octahedral to C_{4V} or C_{3V}. Generally, $^4A_{2g}$ (F) → $^4T_{1g}$ (P) occurs in the UV-Vis region.

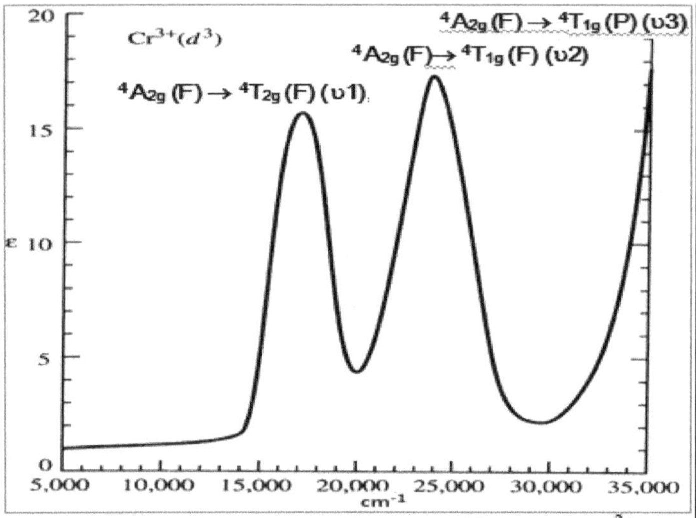

Fig. 71. Electronic spectrum of $[Cr(H_2O)_6]^{3+}$

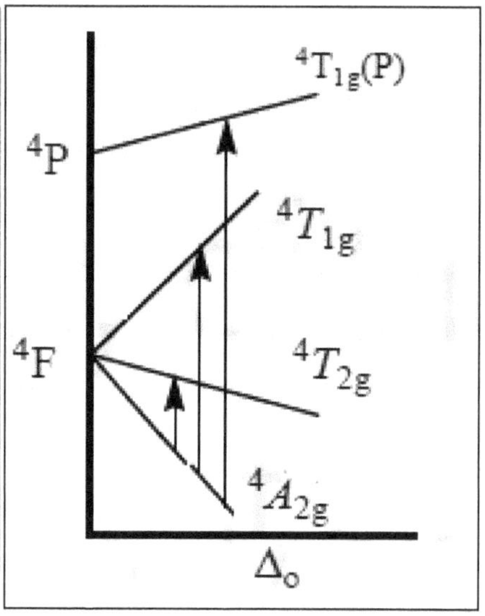

d^4 – configuration: Electronic spectrum of $[Cr(H_2O)_6]^{2+}$

Cr^{2+} has a d^4 configuration and forms high spin complexes only for crystal fields less than 2000 cm^{-1}. The electronic spectra of $[Cr(H_2O)_6]^{2+}$ is shown in the Figure 72 given below. The ground state term in an octahedral crystal field is ^5Eg belongs to the t_{eg}^3 e_g^1 configuration. The excited state $^5T_{2g}$ corresponds to promotion of one single electron to give t_{eg}^2 e_g^2 configuration. One single electron transition, ^5E_g→ $^{25}T_{2g}$ is expected in an octahedral crystal field.

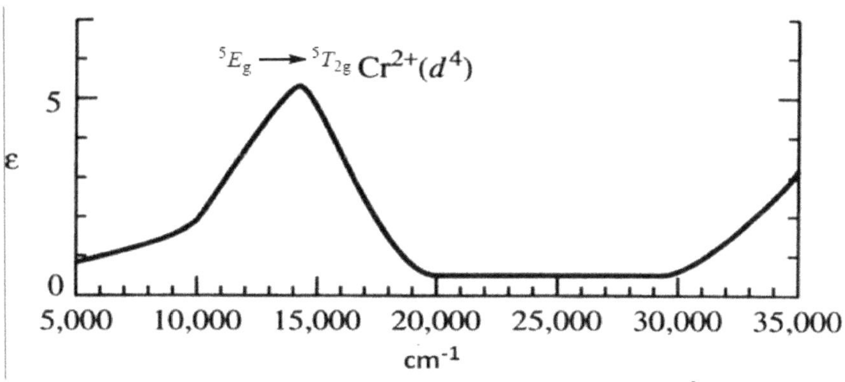

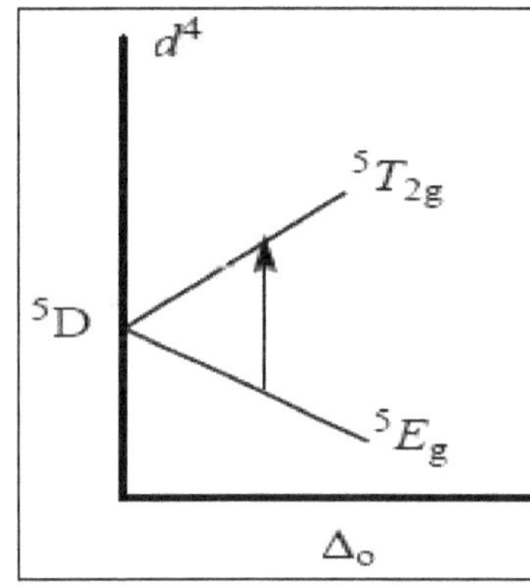

Fig. 72. Electronic spectrum of $[Cr(H_2O)_6]^{2+}$

d^5 – configuration: Electronic spectrum of $[Mn(H_2O)_6]^{2+}$

$[Mn(H_2O)_6]^{2+}$ is a high-spin complex, because H_2O is a rather weak-field ligand. The ground state for weak-field d^5 is the $^6A_{1g}$. There are no excited states of the same spin multiplicity, & consequently there can be no spin-allowed absorptions. The spectrum of $[Mn(H_2O)_6]^{2+}$, as shown below (Fig. 73), consist of weak spin forbidden bands. These bands are due to possible pairing of electrons. These bands are extremely weak; but few are sharper.

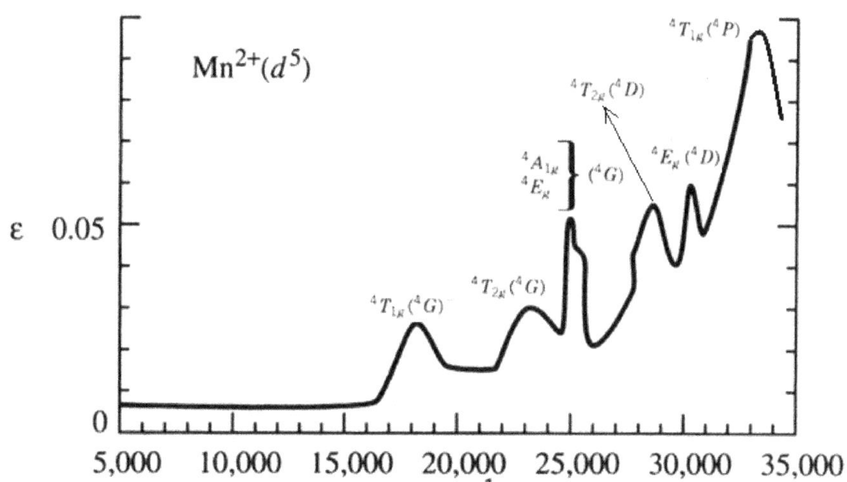

Fig. 73. Electronic spectrum of $[Mn(H_2O)_6]^{2+}$

d^6 – configuration: Electronic spectrum of $[Fe(H_2O)_6]^{2+}$

In divalent iron (d^6), the free ion ground term is ^5D and the excited terms are triplet states (^3H, ^3P, ^3F, ^3G, ^3D) and singlet states (^1I, ^1D). In an octahedral field, the ^5D term splits into an upper ^5E_g level and a lower $^5T_{2g}$ level of which the latter forms the ground state. The only allowed transition is $^5T_{2g} \rightarrow {}^5E_g$ which gives an intense broad absorption band. This band splits into two bands due to Jahn-Teller effect (Fig. 74

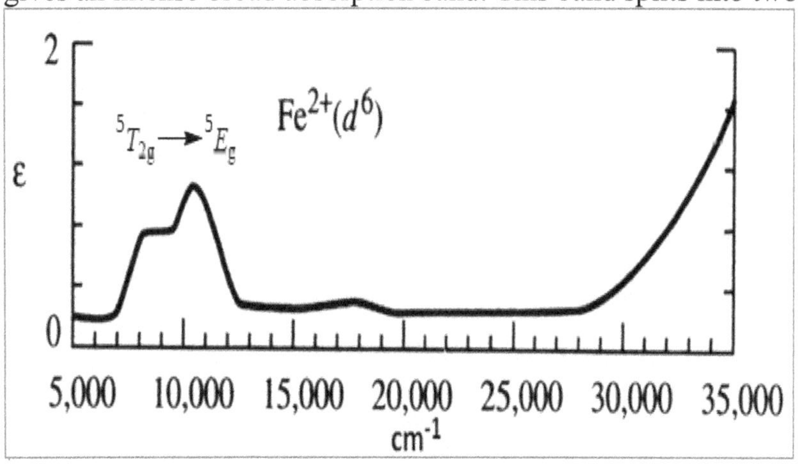

Fig. 74. Electronic spectrum of $[Fe(H_2O)_6]^{2-}$

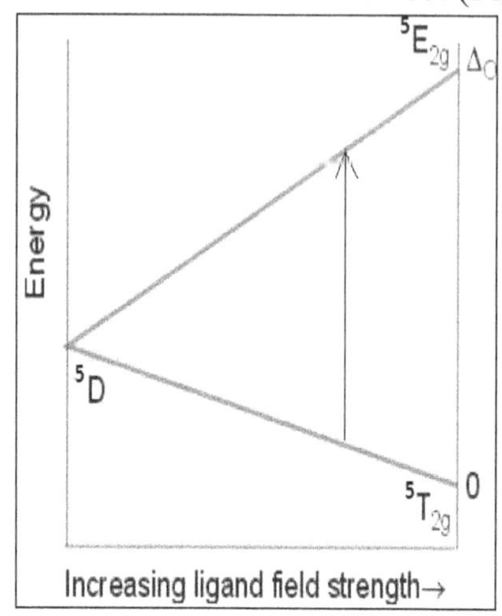

d^7 – configuration: Electronic spectrum of $[Co(H_2O)_6]^{2+}$

The colour of Co(II) complexes has interested chemists for many years and the pale pink, octahedral to bright-blue, tetrahedral colour change is seen in such devices as weather guides and in the dye in silica gel in the laboratory. Assignment of the bands for these spectra can present some problems however. For the aqua ion, a band is observed at around 8000 cm^{-1} and a broad band centred around 20,000 cm^{-1} (ε for these bands are less than 1 m^2 mol^{-1}).

The lowest energy band must correspond to: $^4T_{1g} \rightarrow {}^4T_{2g}$ which leaves the bands at 16,000. A tentative assignment puts the $^4T_{1g} \rightarrow {}^4T_{1g}$ (P) transition at 19,400 and hence the 21,600 band is due to $^4T_{1g} \rightarrow {}^4A_{2g}$. The band at 21,600 cm^{-1} is believed to come from spin-orbit effects (Fig. 75).

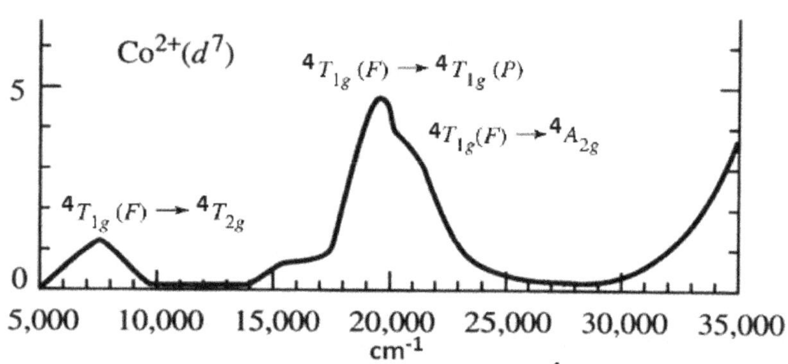

Fig. 75. Electronic spectrum of $[Co(H_2O)_6]^{2-}$

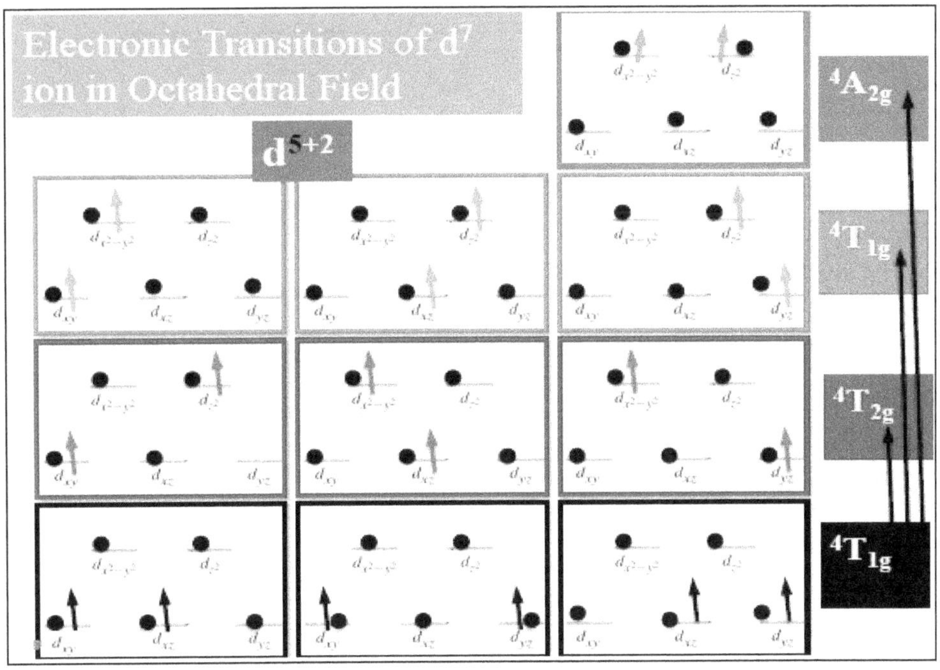

d [8] – configuration: **Electronic spectrum of $[Ni(H_2O)_6]^{2+}$**

The electronic distribution of Ni (II) ion (d^8) is $t_{2g}^6e_g^2$ which gives rise to ^3F, ^3P, ^1D and ^1S terms of which ^3F is the ground state. The ground state is the $^3A_{2g}$ (F). Hence three spin allowed transitions are possible and the others are spin forbidden. The three spin allowed transitions (Fig. 76) are: $^3A_{2g}(F) \rightarrow {}^3T_{1g}(P)$, $^3A_{2g}(F) \rightarrow {}^3T_{1g}(F)$ and $^3A_{2g}(F) \rightarrow {}^3T_{2g}(F)$.

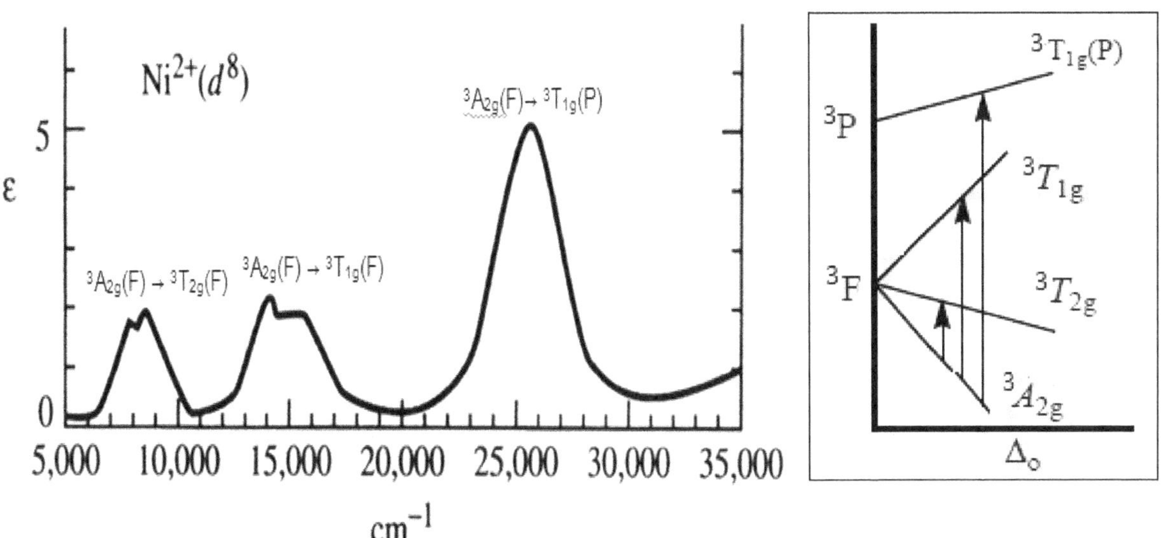

Fig. 76. Electronic spectrum of $[Ni(H_2O)_6]^{2-}$

Some absorption bands (most notably the highest energy ${}^3A_{2g} \rightarrow {}^3T_{1g}({}^3P)$ bands) have only a single maximum for octahedral complexes of nickel(II), others, especially the "middle" $3A_{2g} \rightarrow {}^3T_{1g}({}^3F)$ band in spectra of nickel(II) complexes with halide or oxygen ligands, show a more complicated band shape with two maxima. This can be clearly seen in the spectrum of $[Ni(H_2O)_6]^{2+}$. Shoulders are sometimes observed in the lowest energy ${}^3A_{2g} \rightarrow {}^3T_{2g}$ band of complexes with nitrogen donor ligands. A comparison of the experimental spectra with the Tanabe-Sugano diagram reveals that double maxima are observed for situations in the neighborhood of a crossing of the lowest-energy singlet excited state with the first or second triplet excited states.

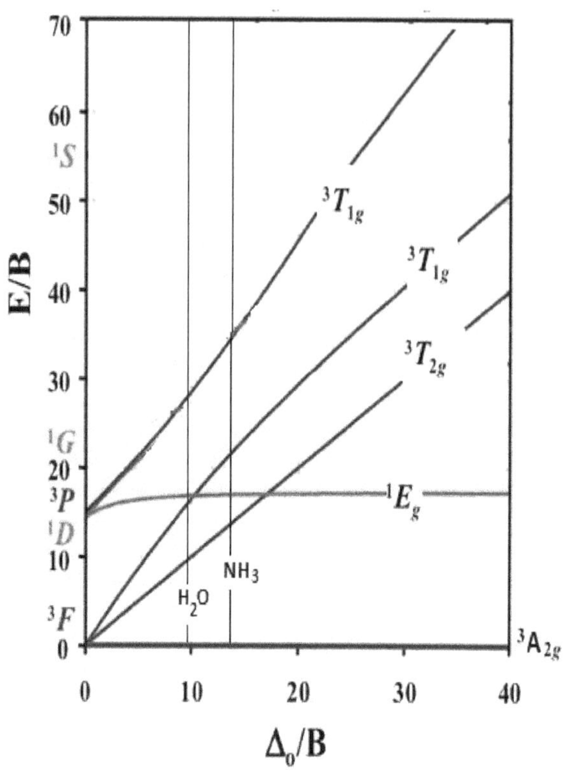

d^9 – configuration: **Electronic spectrum of $[Cu(H_2O)_6]^{2+}$**

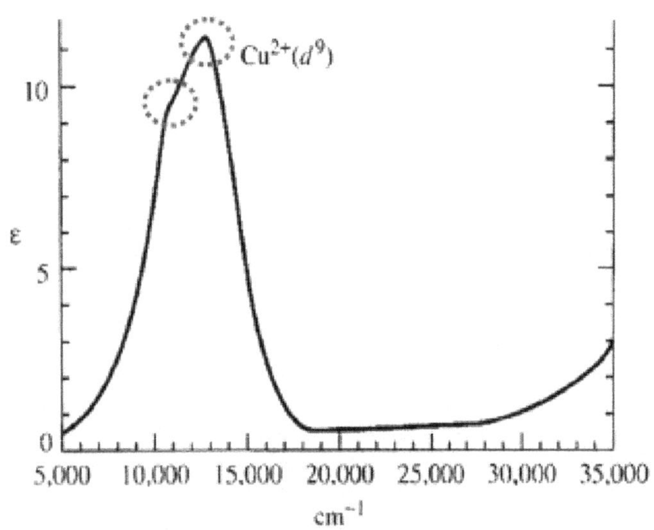

Fig. 77. **Electronic spectrum of $[Cu(H_2O)_6]^{2+}$**

In octahedral crystal field, the ground state electronic distribution of Cu^{2+} is $t_{2g}^{6} e_g^{3}$ which yields ${}^2Eg$ term. The excited electronic state is $t_{2g}^5 e_g^4$ which corresponds to ${}^2T^{2g}$ term. Thus only one single electron transition, i.e., ${}^2Eg \rightarrow {}^2T^{2g}$, is expected in an octahedral crystal field. The difference is 10Dq. Octahedral coordination is distorted either by elongation or compression of octahedron leading to tetragonal symmetry. Normally, the ground ${}^2Eg$ state is split due to Jahn-Teller effect and hence lowering of symmetry is expected for Cu(II) ion. This state splits into ${}^2B_{1g}(dx^2\text{-}y^2)$ and ${}^2A_{1g}(dz2)$ states in tetragonal symmetry and the excited term ${}^2T_{2g}$ also splits into ${}^2B_{2g}(dxy)$ and ${}^2E_g(dxz,dyz)$ levels. Thus, three bands are expected for tetragonal (C4v) symmetry (Fig. 77).

2.9 INTER ELECTRONIC REPULSION PARAMETERS: RACAH PARAMETERS

The inter-electronic repulsions within a configuration are linear combinations of Coulombic and exchange integrals above the ground term. They are expressed by either of the two ways: Condon - Shortley parameters, F_0, F_2 and F_4 and Racah parameters, A, B and C. The magnitude of these parameters varies with the nature of metal ion.

Coloumbic repulsions exist among electrons in the molecular orbitals of the complex ion and hence different terms of a configuration have different energies. The repulsion energy of each term of a configuration is expressed as sums of three quantities. The three different sums of integrals are called the Racah parameters denoted as A B and C.

Racah parameters A,B and C represent interelectronic repulsions.

- Racah parameter A corresponds to the partial shift of all terms of a given electronic configuration. Hence in the optical transition considerations, it is not taken into account.
- The B and C values are a measure of spatial arrangement of the orbitals of the ligand and the metal ion.
- The parameter, B measures the interelectronic repulsion among the electrons in the d-orbitals. The decrease in the value of the interelectronic repulsion parameter, B leads to formation of partially covalent bonding.

The ratio between the crystal B^1 parameter and the free ion B parameter is known as Nephelauxetic ratio and it is denoted by β. The value of β is a measure of covalency. The smaller the value, the greater is the covalency between the metal ion and the ligands.

Racah redefined the empirical Condon –Shortley parameters so that the separation between states having the maximum multiplicity (for example, difference between is a function of ^3F and ^3P or ^4F and ^4P) is a function of a single parameter, B. However, separations between terms of different multiplicity involve both B and C.

The energy of a term originating from a given configuration is expressed as linear combination of the three Racah parameters. Accurate evaluation of these parameters is not possible. These parameters are calculated from spectral data of free ions. For example, for a d^2 system, the energies of the terms are as given below,

$E(^1S) = A + 14B + 7C$

$E(^1G) = A + 4B + 2C$

$E(^1D) = A - 3B + 2C$

E (^{3}P) = A + 7B

E (^3F) = A - 8B

The observed energies of the terms obtained from spectral data, can be substituted in these expressions to evaluate A, B and C.

- Racah parameters have positive values.
- Racah parameter A is common to all terms, so it can be ignored.
- Racah parameter B is used to express the difference in energy between ground term and excited term. The energy difference between the ground term ^3F and excited term ^{3}P of d^2 system is 15B.
- For most of the transition metal ions, B values are about 1000 cm^{-1} and C ≈ 4B, as evaluated from spectral data.

The actual values of B and C for some of the 3d metal ions are given in the table below,

Table 21 Values of Racah parameters B and C (in cm^{-1}) for some transition metal ions.

Configuration	Metal ion	B	C
$3d^2$	Ti^{3+}	718	2629
	V^{3+}	861	4165
	Cr^{3+}	1039	4238
$3d^3$	Sc^{+}	480	---
	V^{2+}	766	2855
	Cr^{3+}	918	3850
	Mn^{4+}	1064	---
$3d^4$	Cr^{2+}	830	3430
	Mn^{3+}	1140	3675
$3d^5$	Mn^{2+}	960	3325
$3d^6$	Fe^{2+}	1058	3901
	Co^{3+}	1100	---
$3d^7$	Co^{2+}	971	4366
$3d^8$	Ni^{2+}	1041	4831

(Reproduced from 'concise coordination chemistry, R Gopalan and V Ramalingam, Vikas publishing house, India).

Nephelauxetic effect (Electron cloud expansion effect)

When a metal ion is surrounded by ligands in a complex, the ligand orbitals being directed toward the metal ion produce changes in the total electron environment of the metal ion. The interelectronic repulsions in the case of free metal ions is higher than that in case of complex ions. This is due to the effect of delocalization of metal electrons over molecular orbitals that encompass both the metal and ligands. This phenomenon of electron cloud expansion in complex ion is called Nephelauxetic effect. The B value for a complex (B') is always smaller than that for the corresponding free ion (the value of B' is always less than that of B). This is due to Nephelauxetic effect.

The ratio B'/B = β is called the Nephelauxetic ratio and is always less than 1 (varies between 0.6 and 0.9). It decreases with increasing delocalisation. The value of β depends up on both the metal ion and the ligand. For a series of complexes of a particular metal ion with different ligands, the β values are in the order,

$I^- > Br^- > CN^- > Cl^- > OH^- > NH_3 > H_2O > F^-$

Similarly the following is the order of β values metal for ions,

$Pt^{4+} > Pd^{4+} > Rh^{3+} > Co^{3+} > V^{3+} > Cr^{3+} > Fe^{3+} > Fe^{2+} > Co^{3+} > Ni^{2+} > Mn^{2+}$

2.10 TANABE-SUGANO DIAGRAMS

Tanabe-Sugano diagrams are special correlation diagrams that are especially useful in the interpretation of the electronic spectra of coordination compound, proposed by Japanese chemists Tanabe and Sugano. Quantitative interpretation of electronic absorption spectra is possible by using Tanabe – Sugano diagrams or simply T-S diagrams.

These diagrams are widely employed to correlate and interpret spectra for ions of all types, from d^2 to d^8. Orgel diagrams are useful only qualitatively for high spin complexes whereas T-S diagrams are useful both for high spin and low spin complexes. Exact solutions for the excited sate energy levels in terms of Dq, B and C are obtained from Tanabe-Sugano matrices. The T-S diagrams are valid only if the value of B, C and Dq ae lower for a complex than for the free ion value.

1. *In T-S diagrams, the lowest (ground state) energy state is plotted along the horizontal axis. Regardless of the ligand field strength, the horizontal axis represents the energy of the ground state. The ligand field strength increases from left to right on the horizontal axis in units of $\Delta o/B$.*

2. *The energies of the various electronic(Term) states are given on the vertical axis is in units of E/B.*

3. *The unit of energy in T-S diagram is B, Racah Parameter.*

4. *The quantities plotted in the Tanabe-Sugano diagrams are as follows:*

   Along the x-axis, Δo/B: where Δo is the octahedral ligand field splitting.

   Along the y – axis, E/B: where E is the energy (of excited states) above the ground state.

   B is the Racah parameter.

The symbols in the diagram omit the subscript, g, with the understanding that all states are gerade states. The energy of the lowest term is taken as zero for drawing this diagram. Some lines in these diagrams are non linear because of the mixing of terms of the same type. If the increasing ligand field causes two weak field terms of the same symmetry to approach, they do not cross but bend apart from each other. This is called non-crossing rule.

For some configurations, there is an abrupt change of slope and the ground state term also changes. This happens when there is a transition from a High –Spin to a Low – Spin state. The values of Δo and B for a complex ion can be evaluated by fitting the observed spectral data in the Tanabe-Sugano diagram of the complex ion.

d^1, d^9 and d^{10} – T S diagrams are not necessary

d^1

There is no electron repulsion in a d^1 complex, and the single electron resides in the t_{2g} orbital ground state. A d^1 octahedral metal complex, such as $[Ti(H_2O)_6]^{3+}$, shows a single absorption band in a UV-vis experiment.The term symbol for d^1 is ^2D, which splits into the $^2T_{2g}$ and ^2E_g states. The t_{2g} orbital set holds the single electron and has a $^2T_{2g}$ state energy of -4Dq. When that electron is promoted to an e_g orbital, it is excited to the ^2E_g state energy, +6Dq. This is in accordance with the single absorption band in a UV-vis experiment. Thus, this simple transition from $^2T_2$ to ^2E_g does not require a Tanabe-Sugano diagram.

d^9

Similar to d^1 metal complexes, d^9 octahedral metal complexes have ^2D spectral term. The transition is from the $(t_{2g})^6(e_g)^3$ configuration (^2E_g state) to the $(t_{2g})^5(e_g)^4$ configuration ($^2T_{2g}$ state). This could also be described as a positive "hole" that moves from the e_g to the t_{2g} orbital set. The sign of Dq is opposite that for d^1, with a ^2E_g ground state and a $^2T_{2g}$ excited state. Like the d^1 case, d^9 octahedral complexes do not require the Tanabe-Sugano diagram to predict their absorption spectra.

d^{10}

There are no d-d electron transitions in d^{10} metal complexes because the d orbitals are completely filled. Thus, UV-vis absorption bands are not observed and a Tanabe-Sugano diagram does not exist.

d^1, d^4 (high spin), d^6 (high spin), d^9

In each of these cases, the absorption spectra will result from the excitation of an electron from a t_{2g} orbital to an e_g orbital, with the excited electron configuration having the same spin multiplicity as the initial configuration. In each case, there is a single excited state of the same spin multiplicity as the ground state. As a result, there is a single spin-allowed absorption, with the energy of absorbed light equal to Δo. Examples of these complexes include $[Ti(H_2O)_6]^{3+}$, $[Cr(H_2O)_6]^{3+}$, $[Fe(H_2O)_6]^{2+}$, and $[Cr(H_2O)_6]^{2+}$. The spectra of these complexes show a single absorption band, with occasional splitting due to a Jahn-Teller distortion.

The Tanabe –Sugano diagrams for configurations from d^2 to d^8 are as given below:

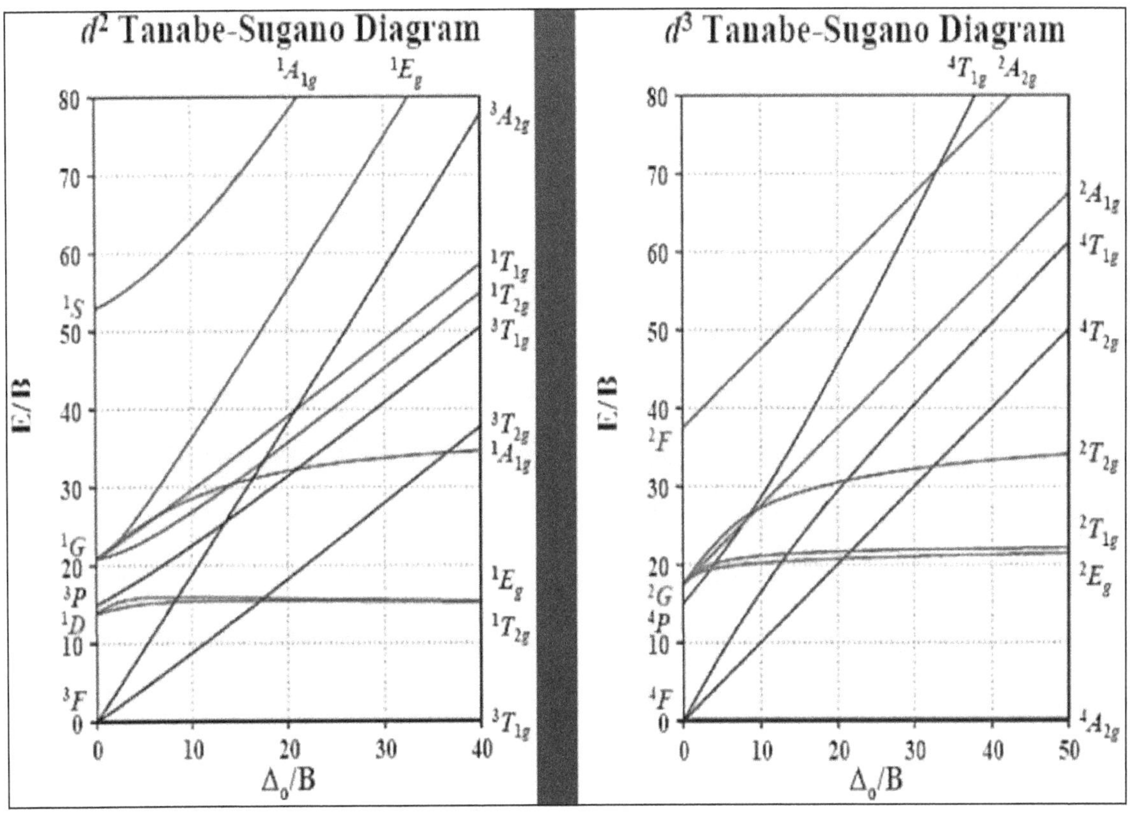

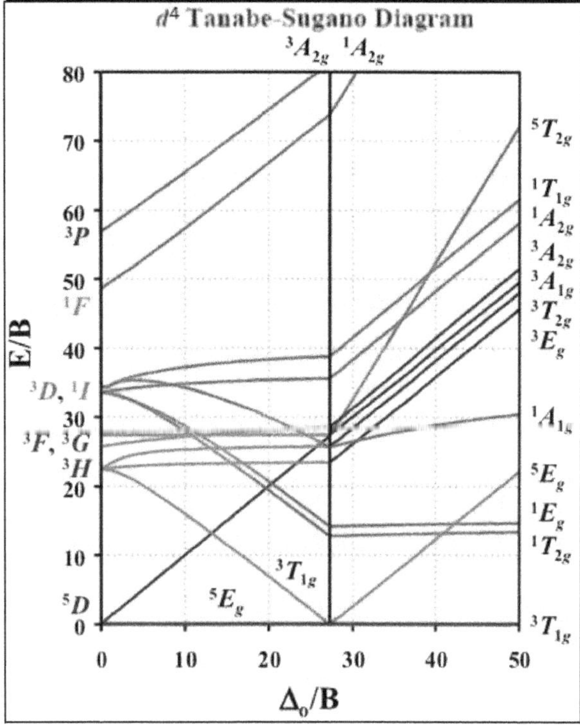
d^4 Tanabe-Sugano Diagram
$^3A_{2g}$ $^1A_{2g}$
$^5T_{2g}$
$^1T_{1g}$
$^1A_{2g}$
$^3A_{2g}$
$^3A_{1g}$
$^3T_{2g}$
^3E_g
$^1A_{1g}$
^5E_g
^1E_g
$^1T_{2g}$
$^3T_{1g}$
^3P
^1F
^3D, ^1I
^3F, ^3G
^3H
^5D
^5E_g
E/B
Δ_o/B

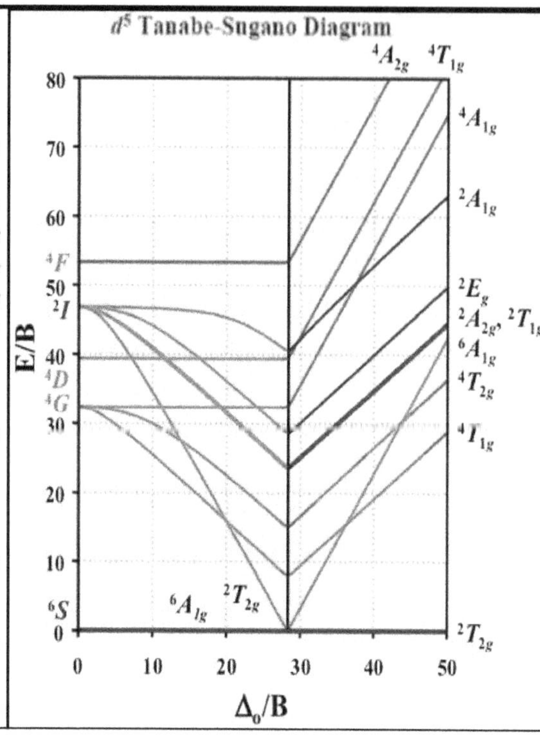
d^5 Tanabe-Sugano Diagram
$^4A_{2g}$ $^4T_{1g}$
$^4A_{1g}$
$^2A_{1g}$
^2E_g
$^2A_{2g}$, $^2T_{1g}$
$^6A_{1g}$
$^4T_{2g}$
$^4T_{1g}$
$^2T_{2g}$
^4F
^2I
^4D
^4G
^6S
$^6A_{1g}$ $^2T_{2g}$
E/B
Δ_o/B

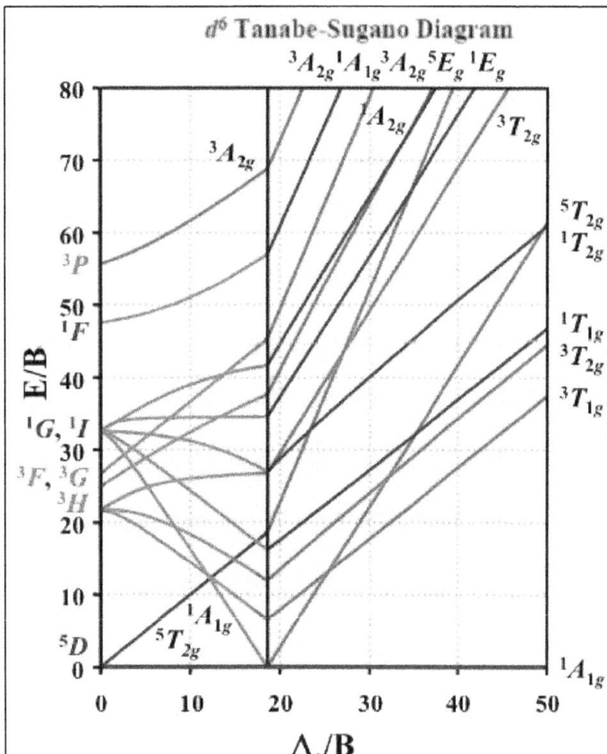
d^6 Tanabe-Sugano Diagram
$^3A_{2g}$ $^1A_{1g}$ $^3A_{2g}$ ^5E_g ^1E_g
$^1A_{2g}$
$^3T_{2g}$
$^3A_{2g}$
$^5T_{2g}$
$^1T_{2g}$
$^1T_{1g}$
$^3T_{2g}$
$^3T_{1g}$
$^1A_{1g}$
^3P
^1F
^1G, ^1I
^3F, ^3G
^3H
^5D
$^1A_{1g}$
$^5T_{2g}$
E/B
Δ_o/B

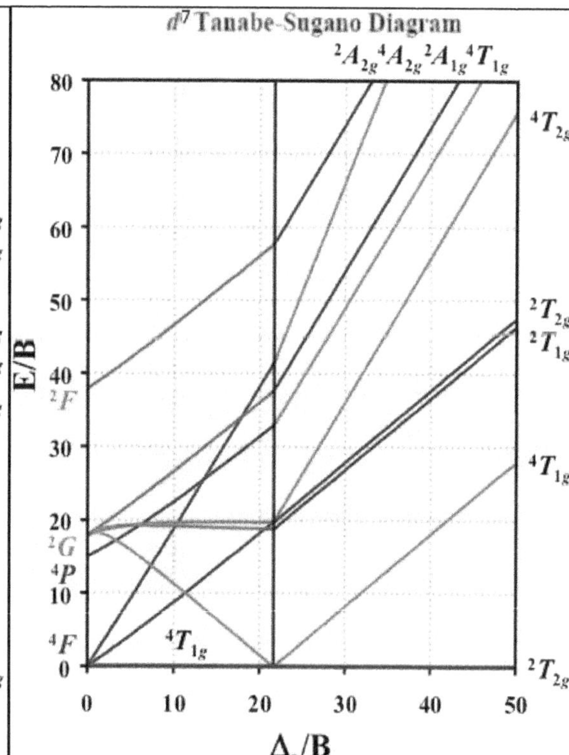
d^7 Tanabe-Sugano Diagram
$^2A_{2g}$ $^4A_{2g}$ $^2A_{1g}$ $^4T_{1g}$
$^4T_{2g}$
$^2T_{2g}$
$^2T_{1g}$
$^4T_{1g}$
$^2T_{2g}$
^2F
^2G
^4P
^4F
$^4T_{1g}$
E/B
Δ_o/B

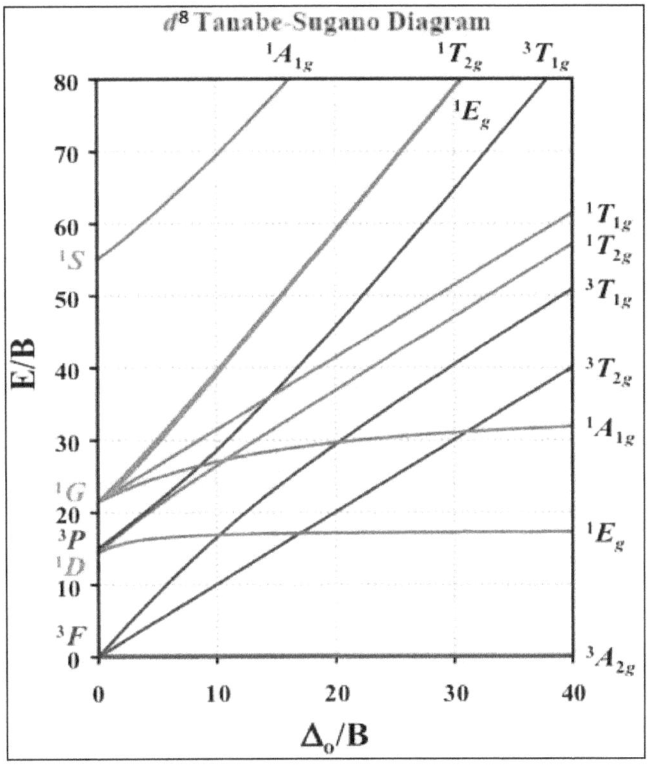

2.11 INTERPRETATION OF ELECTRONIC SPECTRA USING T-S DIAGRAMS

Calculation of Racah parameter B: **The electronic spectrum of the d^2 $[V(H_2O)_6]^{3+}$**

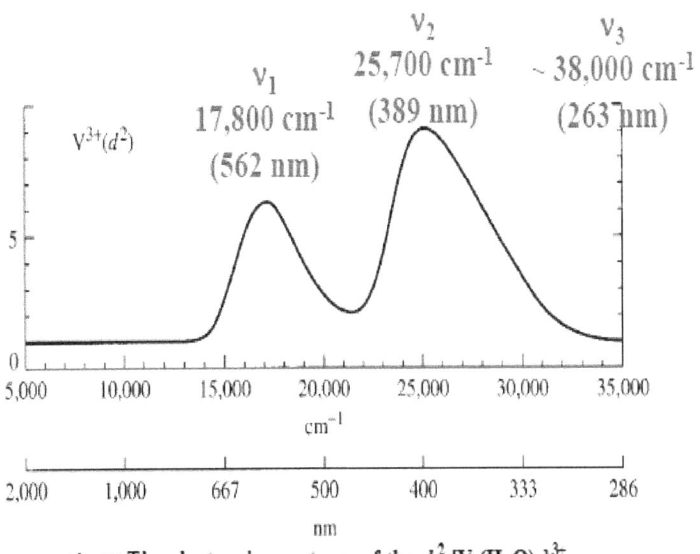

Fig. 78. The electronic spectrum of the d^2 $[V(H_2O)_6]^{3+}$

The electronic spectrum of the V^{3+} ion (Fig. 78), where V(III) is doped into alumina (Al_2O_3), shows three major peaks with frequencies of: ν_1=17400 cm^{-1}, ν_2=25400 cm^{-1} and ν_3=34500 cm^{-1}.

These have been assigned to the following spin-allowed transitions.

$^3T_{1g}$ (F) → $^3T_{2g}$ (F) ν_1=17800 cm^{-1}

$^3T_{1g}$ (F) → $^3T_{1g}$ (P) ν_2=25700 cm^{-1}

$^3T_{1g}$ (F) → $^3A_{2g}$ (F) ν_3=38000 cm^{-1}

The ratio between the first two transitions is calculated as $\mathbf{v_2 / v_1}$ which is equal to 25700 / 17800 = 1.444. In order to calculate the Racah parameter, B, the position on the horizontal axis where the ratio between the lines representing v_2 and v_1 is equal to 1.448, has to be determined. On the diagram (Fig. 79) below, this occurs at Δ/B=30.9. Having found this value, a vertical line is drawn at this position.

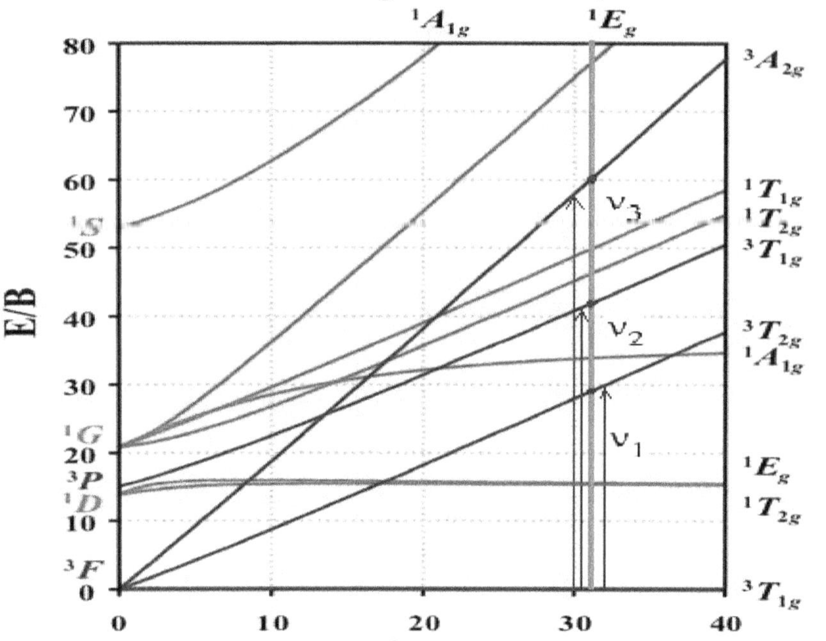

Fig. 79. T-S diagram for d^2 configuration

On moving up the line from the ground term to where lines from the other terms cross it, we are able to identify both the spin-forbidden and spin-allowed transition and hence the total number of transitions that are possible in the electronic spectrum.

Next, find the values on the vertical axis that correspond to the spin-allowed transitions so as to determine the values of v_1/B, v_2/B and v_3/B. From the diagram above these are 28.78, 41.67 and 59.68 respectively.

Knowing the values of v_1, v_2 and v_3, we can now calculate the value of B.

Since v1/B=28.78 and v_1 is equal to 17,800 cm-1, then B=v1/28.78 = 17800/28.78 or B=618.5cm^{-1}. The calculated B (618.5 cm^{-1}) is less than the free ion B value (1030 cm^{-1}). This is due to Nephelauxetic effect as a consequence to the decreased interelectronic repulsion.

Then it is possible to calculate the value of Δ. Since Δ/B=30.9, then: Δ=B x 30.9 and hence: Δ = 618.5 x 30.9 = 19111 cm^{-1}.

The Nephelauxetic ratio (β) for this d^2 system is calculated as follows,
$\beta = B'/B = 618.5 / 1030 = 0.6$.

The electronic spectrum of the d^3 $[Cr(H_2O)_6]^{3+}$

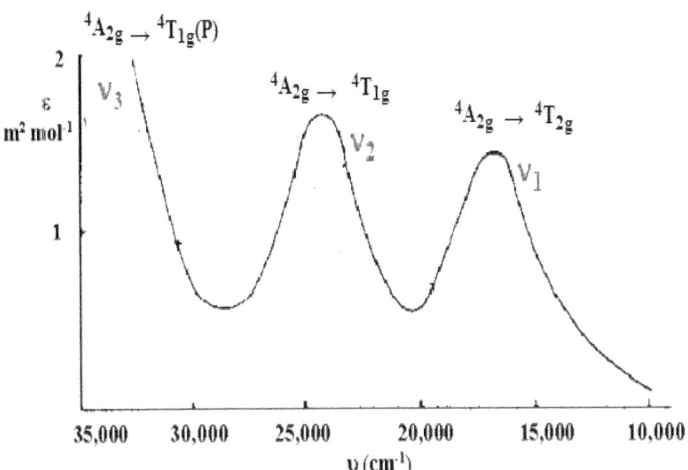

Fig. 80. The electronic spectrum of the d^3 $[Cr(H_2O)_6]^{3+}$

The electronic spectrum of the $[Cr(H_2O)_6]^{3+}$ shows three major peaks with frequencies of: ν_1=17000 cm^{-1}, ν_2=24000 cm^{-1} and ν_3=37000 cm^{-1}(Fig. 80).

These have been assigned to the following spin-allowed transitions.

$^4A_{2g} \rightarrow {}^4T_{2g}$ **ν_1=17000 cm^{-1}**

$^4A_{2g} \rightarrow {}^4T_{1g}$ **ν_2=24000 cm^{-1}**

$^4A_{2g} \rightarrow {}^4T_{1g}(P)$ **ν_3=37000cm^{-1}**

From the information given, the ratio ν_2 / ν_1 = 24000 / 17000 = 1.412. Using a Tanabe-Sugano diagram for a d^3 system this ratio is found at Δ/B=24.00.

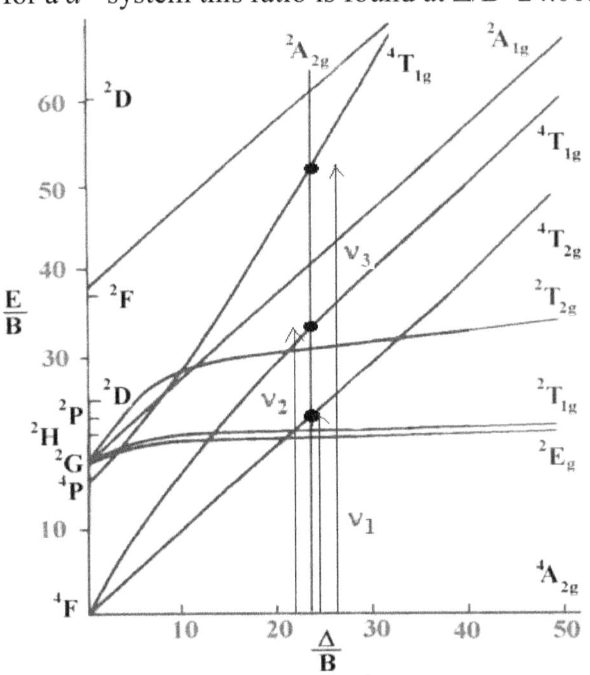

Fig. 81. T-S diagram for d^3 configuration

Interpolation of the graph to find the Y-axis values for the spin-allowed transitions gives:ν_1/B =24.00, ν_2/B =33.90, ν_3/B =53.11.

Recall that ν_1=17000 cm-1. Therefore for the first spin-allowed transition,
17000 /B =24.00 from which B can be obtained, B=17000 / 24.00 or B=708.3 cm^{-1}.

This information is then used to calculate Δ (Fig. 81). Since Δ / B=24.00 then Δ = B x 24.00 = 708.3 x 24.00 = 17000 cm^{-1}.

It is observed that the value of Racah parameter B in the complex is 708.3 cm^{-1}, while the value of B in the free Cr^{3+} ion is 1030cm^{-1}. This shows a 31% reduction in the Racah parameter indicating a strong Nephelauxetic effect.

The electronic spectrum of the d^5 $[Mn(H_2O)_6]^{2+}$

$[Mn(H_2O)_6]^{2+}$ is high-spin octahedral complex with a very pale pink color, owing to a series of weak spin-forbidden transitions. The free ion ground term is ^6S. No transitions are spin allowed. The spectrum of $[Mn(H_2O)_6]^{2+}$ Consist of weak spin forbidden bands (Fig. 82). These bands are due to possible pairing of electrons. These bands are extremely weak; but few are sharper.

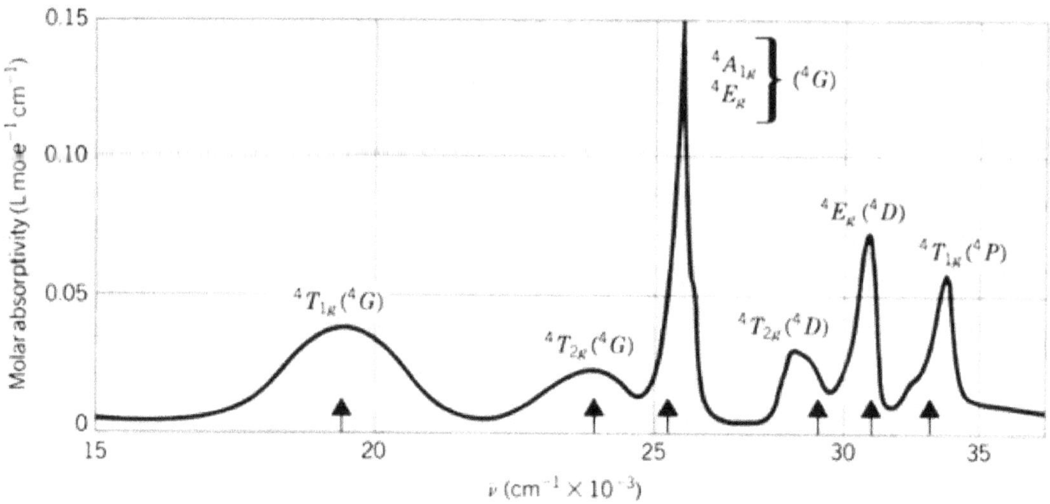

Fig. 82. The electronic spectrum of the d^5 $[Mn(H_2O)_6]^{2+}$

The electronic spectrum of the d^6 $[Fe(H_2O)_6]^{2+}$

The free ion ground term for the d^6 ion is ^5D; therefore only one absorption is expected. This occurs at 10500 cm^{-1}.

The electronic spectrum of the d^7 $[Co(H_2O)_6]^{2+}$

The free ion ground term for this complex ion is ^4F. Three transitions are expected in this case. In an octahedral field, the splitting of d^7 is the same as for the octahedral d^2 ion and therefore the spectrum can be interpreted by using the same T-S diagram used for d^2 (V^{3+}) system.

The electronic spectrum of the d^8 $[Ni(H_2O)_6]^{2+}$

The free ion ground term for this complex ion is ^3F. Three transitions are expected in this case. In an octahedral field, the splitting of d^8 is the same as for the octahedral d^3 ion and therefore the spectrum can be interpreted by using the same T-S diagram used for d^3 system. Hence three spin allowed transitions are possible and the others are spin forbidden The three spin allowed transitions are: $^3A_{2g}(F) \rightarrow {}^3T_{2g}(F)$, $^3A_{2g}(F) \rightarrow {}^3T_{1g}(F)$ and $^3A_{2g}(F) \rightarrow {}^3T_{1g}(P)$.

For an octahedral Ni (II) complex, three peaks were observed at 8000, 13200 and 22800 cm^{-1}. From the ratio v_2/v_1 of 1.65 this gives a value of Δ/B' of 10.0.

This can be shown in the following diagram (Fig.83),

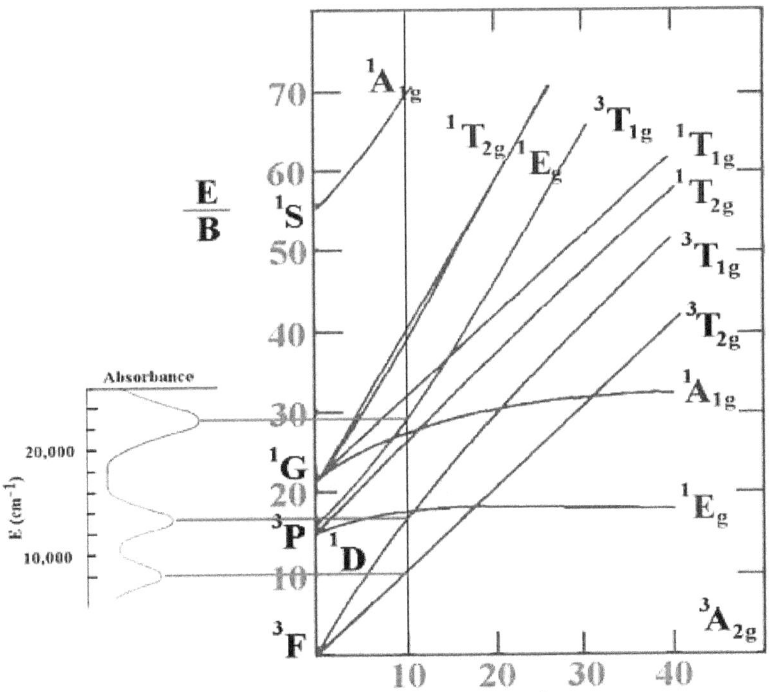

Fig. 83. T-S diagram for d^8 $[Ni(H_2O)_6]^{2+}$

Since $v_1 = \Delta$ in this case (and equals 8000 cm^{-1}) then B' can be evaluated to be 800 cm^{-1}. The spin-forbidden lines that would be between v_2 and v_3 in energy are not observed in the spectrum nor are any lines seen at higher energy.

2.12 MAGNETIC PROPERTIES OF COORDINATION COMPOUNDS

There are two main types of magnetic compounds, those that are diamagnetic (compounds that are repelled by a magnetic field) and those that are paramagnetic (compounds that are attracted by a magnetic field). All substances possess the property of diamagnetism due to the presence of closed shells of electrons within the substance. Note that diamagnetism is a weak effect while paramagnetism is a much stronger effect.

Paramagnetism derives from the spin and orbital angular momenta of electrons. This type of magnetism occurs only in compounds containing unpaired electrons, as the spin and orbital angular momenta is cancelled out when the electrons exist in pairs.

Compounds in which the paramagnetic centres are separated by diamagnetic atoms within the sample are said to be magnetically dilute.

If the diamagnetic atoms are removed from the system then the paramagnetic centres interact with each other. This interaction leads to ferromagnetism (in the case where the neighbouring magnetic dipoles are aligned in the same direction) and antiferromagnetism (where the neighbouring magnetic dipoles are aligned in alternate directions).

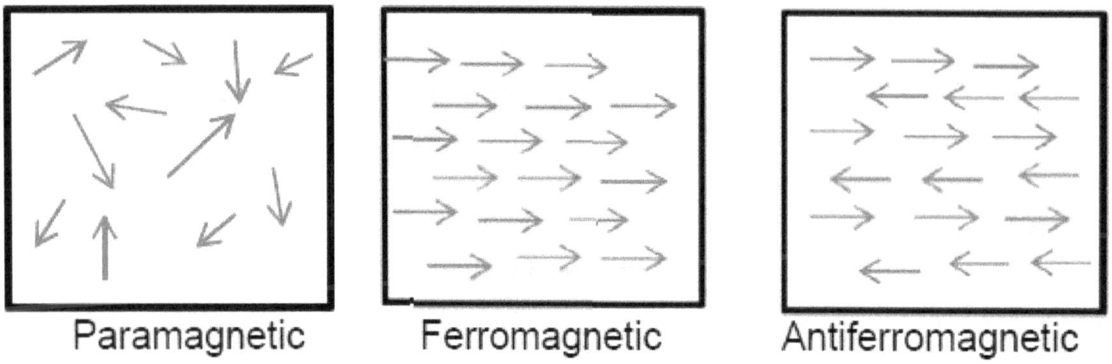

Fig. 84. Types of Alignment of magnetic moments in magnetic materials

These two forms of paramagnetism show characteristic variations of the magnetic susceptibility with temperature (Fig. 84).

Curie temperature (T_C): At this temperature, ferromagnetism changes to paramagnetism.

Neel temperature (T_N): Antiferromagnetism changes to paramagnetism at this temperature.

In the case of ferromagnetism, above the Curie point the material displays "normal" paramagnetic behaviour. Below the Curie point the material displays strong magnetic properties.

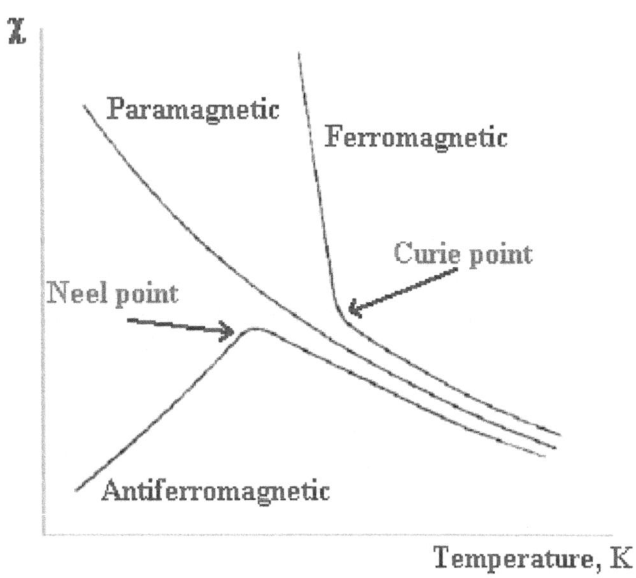

Fig.85. Effect of temperature on the magnetic properties of materials

Ferromagnetism is commonly found in compounds containing iron and in alloys.

For antiferromagnetism, above the Neel point the material displays "normal" paramagnetic behaviour. Below the Neel point the material displays weak magnetic properties which at lower and lower temperatures can become essentially diamagnetic (Fig. 85).

Antiferromagnetism is more common and is found to occur in transition metal halides and oxides, such as $TiCl_3$ and VCl_2.

Curie's law

In a paramagnetic material the magnetization of the material is (approximately) directly proportional to an applied magnetic field. However, if the material is heated, this proportionality is reduced: for a fixed value of the field, the magnetization is (approximately) inversely proportional to temperature. This fact is encapsulated by **Curie's law**:

$$\mathbf{M} = C \cdot \frac{\mathbf{B}}{T},$$

Where $\mathbf{M}$ is the resulting magnetisation

$\mathbf{B}$ is the magnetic field, measured in tesla's

T is absolute temperature, measured in kelvins

C is a material-specific Curie constant.

Curie–Weiss law

The **Curie–Weiss law** describes the magnetic susceptibility χ of a ferromagnet in the paramagnetic region above the Curie point:

$$\chi = \frac{C}{T - T_c}$$

where C is a material-specific Curie constant, T is absolute temperature, measured in kelvins, and T_c is the Curie temperature, measured in kelvin. The law predicts a singularity in the susceptibility at $T = T_c$. Below this temperature the ferromagnet has a spontaneous magnetization.

Zeeman Effect

Zeeman effect is the splitting of a single spectral line (see spectrum) into a group of closely spaced lines when the substance producing the single line is subjected to a uniform magnetic field. The effect was discovered in 1896 by the Dutch physicist Pieter Zeeman. In the so-called normal Zeeman effect, the spectral line corresponding to the original frequency of the light (in the absence of the magnetic field) appears with two other lines arranged symmetrically on either side of the original line. In the anomalous Zeeman effect (which is actually more common than the normal effect), several lines appear, forming a complex pattern.

The normal Zeeman effect was successfully explained by H. A. Lorentz using the laws of classical physics (Zeeman and Lorentz shared the 1902 Nobel Prize in Physics). The anomalous Zeeman effect could not be explained using classical physics; the development of the quantum theory and the discovery of the electron's intrinsic spin led to a satisfactory explanation. According to the quantum theory all spectral lines arise from transitions of electrons between different allowed energy levels within the atom, the frequency of the spectral line being proportional to the energy difference between the initial and final levels. Because of its intrinsic spin, the electron has a magnetic field associated with it. When an external magnetic field is applied, the electron's magnetic field may assume only certain alignments. Slight differences in energy are associated with these different orientations, so that what was once a single energy level becomes three or more.

The first order Zeeman effect causes splitting of energy levels on application of magnetic field and second order Zeeman effect involves effect of magnetic field in mixing different levels together. The first order Zeeman effect is proportional to the magnetic field, H and the second order Zeeman effect is proportional to the square of the magnetic field, H^2. The following energy level diagram (Fig. 86) is showing splitting of ground state derived from a t_{2g}^1 configuration.

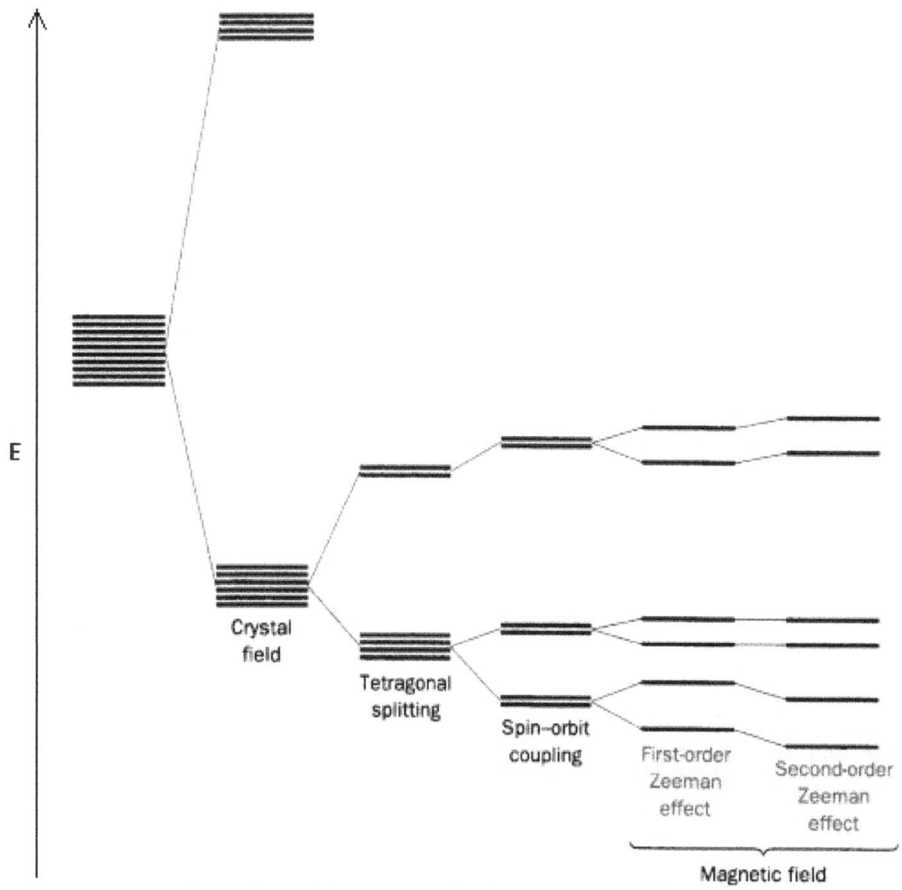

Fig. 86. Splitting of spectroscopic states due to Zeeman effect

Magnetic properties of the complexes

Magnetic moments are often used in conjunction with electronic spectra to gain information about the oxidation state and stereochemistry of the central metal ion in coordination complexes. Most of the compounds of transition elements are paramagnetic in nature and are attracted by the magnetic field. The transition elements involve the partial filling of d-sub shells. Most of the transition metal ions or their compounds have unpaired electrons in d-sub shell (configuration from d^1 to d^9) and therefore, they give rise to paramagnetic character. The magnetic character is expressed in *Bohr magnetons* abbreviated as B.M.

A common laboratory procedure for the determination of the magnetic moment for a complex is the Gouy method (Fig. 87) which involves weighing a sample of the complex in the presence and absence of a magnetic field and observing the difference in weight.

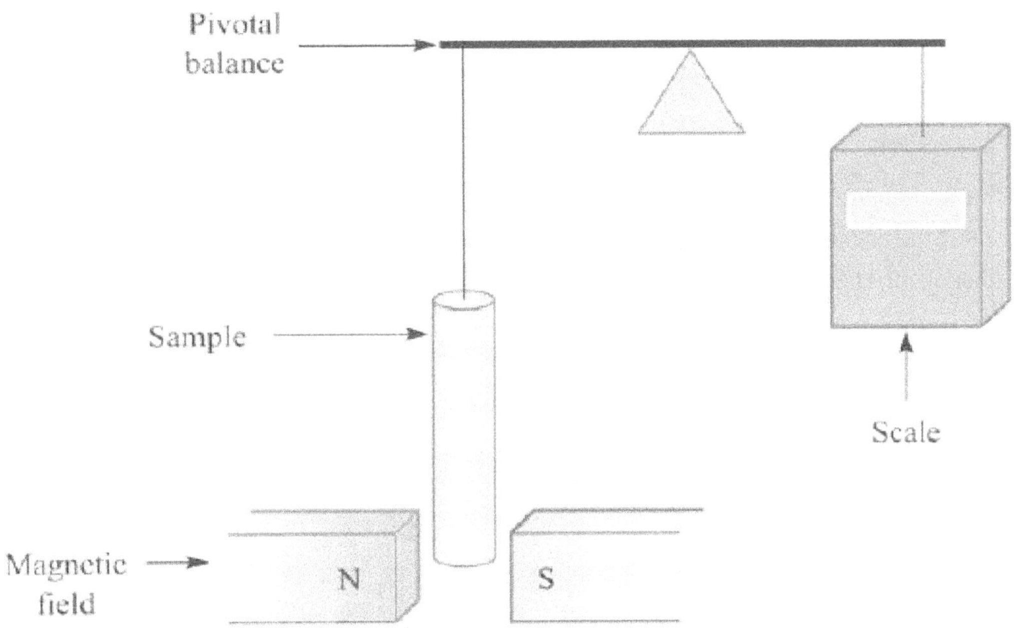

Fig. 87. Experimental setup for the determination of magnetic moment of complex by Guoy method

For first row transition metal ions in the free ion state, i.e. isolated ions in a vacuum, all 5 of the 3d orbitals are degenerate. A simple crystal field theory approach to the bonding in these ions assumes that when they form octahedral complexes, the d orbitals are no longer degenerate but are split such that two orbitals, the d_{x2-y2} and the d_{z2} (e_g subset) are at higher energy than the d_{xy}, d_{xz}, d_{yz} orbitals (the t_{2g} subset). For ions with between 4 and 7 d electrons, this gives rise to 2 possible arrangements called either high spin/weak field or low spin/strong field respectively. The energy gap is dependent on the position of the coordinated ligands in the SPECTROCHEMICAL SERIES.

In tetrahedral complexes, the energy levels of the orbitals are again split, such that two orbitals, the d_{x2-y2} and the d_{z2} (e subset) are now at lower energy (more favoured) than the remaining three d_{xy}, d_{xz}, d_{yz} (the t_2 subset) which are destabilised. **Tetrahedral complexes are ALL high spin since the difference between the 2 subsets of orbitals is much smaller than is found in octahedral complexes.** The usual relationship quoted between them is: $\Delta_{tet} = 4/9 \Delta_{oct}$. Square planar complexes are less common than tetrahedral and **few ions forming square planar complexes are d^8 e.g. Ni(II), Pd(II) and Pt(II).** As with octahedral complexes, the energy gap between the d_{xy} and d_{x2-y2} is Δ_{oct} and these are considered strong field / low spin hence they are all diamagnetic, μ=0 Bohr Magneton (B.M.). The presence of unpaired electrons makes transition metal atom or ion to behave as tiny magnet giving rise to magnetic moment. The magnetic moment is expressed in Bohr magnetrons μB.

$$\mu_B = \frac{eh}{\mu\pi\ mC}$$

The general equation for the magnetic moments of 3d-transition metal ions is

$$\mu_{(s+L)} = \sqrt{4s(s+1) + L(L+1)}\ .\ \mu_B$$

where s → total of the spin quantum numbers

L → Resultant of the orbital angular momentum quantum number.

In case transition is quenched. Therefore the spin only magnetic moment μ_s may be written as

$$\mu_s = \sqrt{4s(s+1)}\ \mu_B$$

The magnetic moment μ is related to the number of unpaired spins n by the equation.

$$\mu_s = \sqrt{n(n+2)}$$

This is the formula, called the spin only formula used to calculate the magnetic moment of the complexes (Table 22 and 23).

Table 22. Comparison of calculated spin-only magnetic moments with experimental data for some octahedral complexes

Ion	Config	μ_{so} / B.M.	μ_{obs} / B.M.
Ti(III)	d1 (t_{2g}1)	√3 = 1.73	1.6-1.7
V(III)	d2 (t_{2g}2)	√8 = 2.83	2.7-2.9
Cr(III)	d3 (t_{2g}3)	√15 = 3.88	3.7-3.9
Cr(II)	d4 high spin (t_{2g}3 e_g1)	√24 = 4.90	4.7-4.9
Cr(II)	d4 low spin (t_{2g}4)	√8 = 2.83	3.2-3.3
Mn(II)/ Fe(III)	d5 high spin (t_{2g}3 e_g2)	√35 = 5.92	5.6-6.1
Mn(II)/ Fe(III)	d5 low spin (t_{2g}5)	√3 = 1.73	1.8-2.1
Fe(II)	d6 high spin (t_{2g}4 e_g2)	√24 = 4.90	5.1-5.7
Co(III)	d6 low spin (t_{2g}6)	0	0
Co(II)	d7 high spin (t_{2g}5 e_g2)	√15 = 3.88	4.3-5.2
Co(II)	d7 low spin (t_{2g}6 e_g1)	√3 = 1.73	1.8
Ni(II)	d8 (t_{2g}6 e_g2)	√8 = 2.83	2.9-3.3
Cu(II)	d9 (t_{2g}6 e_g3)	√3 = 1.73	1.7-2.2

Table 23. Comparison of calculated spin-only magnetic moments with experimental data for some tetahedral complexes

Ion	Config	μ_{so} / B.M.	μ_{obs} / B.M.
Cr(V)	d1 (e^1)	√3 = 1.73	1.7-1.8
Cr(IV) / Mn(V)	d2 (e^2)	√8 = 2.83	2.6 - 2.8
Fe(V)	d3 (e^2 t_2^1)	√15 = 3.88	3.6-3.7
-	d4 (e^2 t_2^2)	√24 = 4.90	-
Mn(II)	d5 (e^2 t_2^3)	√35 = 5.92	5.9-6.2
Fe(II)	d6 (e^3 t_2^3)	√24 = 4.90	5.3-5.5
Co(II)	d7 (e^4 t_2^3)	√15 = 3.88	4.2-4.8
Ni(II)	d8 (e^4 t_2^4)	√8 = 2.83	3.7-4.0
Cu(II)	d9 (e^4 t_2^5)	√3 = 1.73	-

Orbital contribution to magnetic moments

From a quantum mechanics viewpoint, the magnetic moment is dependent on both spin and orbital angular momentum contributions. The spin-only formula used earlier was given as:
$\mu_{s.o.}=\sqrt{4S(S+1)}$
and this can be modified to include the orbital angular momentum
$\mu_{S+L}=\sqrt{4S(S+1)+L(L+1)}$

An orbital angular momentum contribution is expected when the ground term is triply degenerate i.e. a T state. These show temperature dependence as well. In order for an electron to contribute to the orbital angular momentum the orbital in which it resides must be able to transform into an exactly identical and degenerate orbital by a simple rotation (it is the rotation of the electrons that induces the orbital contribution). For example, in an octahedral complex the degenerate t_{2g} set of orbitals (d_{xz},d_{yx},d_{yz}) can be interconverted by a 90° rotation. However the orbitals in the e_g subset (d_{z2},d_{x2-y2}) cannot be interconverted by rotation about any axis as the orbital shapes are different; therefore an electron in the e_g set does not contribute to the orbital angular momentum and is said to be quenched. In the free ion case the electrons can be transformed between any of the orbitals as they are all degenerate, but there will still be partial orbital quenching as the orbitals are not identical.

Electrons in the t_{2g} set do not always contribute to the orbital angular moment. For example in the d^3, $t_{2g}{}^3$ case, an electron in the d_{xz} orbital cannot by rotation be placed in the d_{yz} orbital as the orbital already has an electron of the same spin. This process is also called quenching.
Tetrahedral complexes can be treated in a similar way with the exception that we fill the e orbitals first, and the electrons in these do not contribute to the orbital angular momentum.
The tables in the links below give a list of all d^1 to d^9 configurations including high and low spin complexes and a statement of whether or not a direct orbital contribution is expected.

The orbital angular momentum is high in the free metal ion and it is reduced when ligands are attached to it. When the orbital contribution to the magnetic moment is zero, we say that the orbital contribution to the magnetic moment is quenched. When the ground state of a complex is either A or E, the orbital contribution will be quenched. In other words, there is no orbital angular momentum for these states.

'A' state is non-degenerate because it has only one level. Hence, rotation cannot change one orbital into another equivalent orbital.

'E' state is doubly degenerate, which means that there are two energy levels with the same energy. The orbitals giving rise to this are d_{x2-y2} and d_{z2}. One orbital cannot be changed into another by rotation because their shapes are different. Hence, A and E terms do not have orbital angular momentum. In other words, orbital angular momentum is quenched in these cases.

Thus, complexes having A or E ground states will have no orbital angular momentum contribution to the magnetic moment and hence, they will have lower values than expected.

T state

This is a triply degenerate state caused by the t_{2g} orbitals, d_{xy}, d_{yz}, and d_{zx}. They have similar shape and hence, one orbital can be transformed into another by simple rotation. Thus, if an electron is present in a d_{xy} orbital, it can occupy the d_{yz} or d_{zx} by simple rotation about the proper axis and the electron can rotate in an orbit producing magnetic moment. Hence, the orbital angular momentum will not be quenched and the orbital angular momentum contribution will add to the total magnetic moment of the complex.

Configurations	Ground-state term	Orbital contribution
d^1	$^2T_{2g}$	Yes
d^2	$^3T_{1g}$	Yes
d^3	$^4A_{2g}$	No
d^4	^5E_g (high-spin)	No
	$^3T_{1g}$ (low-spin)	Yes
d^5	$^6A_{1g}$ (high-spin)	No
	$^2T_{2g}$ (low-spin)	Yes
d^6	$^5T_{2g}$ (high-spin)	Yes
	$^1A_{1g}$ (low-spin)	No
d^7	$^4T_{1g}$ (high-spin)	Yes
	^2E_g (low-spin)	No
d^8	$^3A_{2g}$	No
d^9	^2E_g	No

Consider a Ni(II) complex, electronic configuration is d^8

For a free metal ion,

S = 1, L = 3 and $\mu = \sqrt{L(L+1) + 4S(S+1)} = 4.47$ B.M.

For an octahedral complex

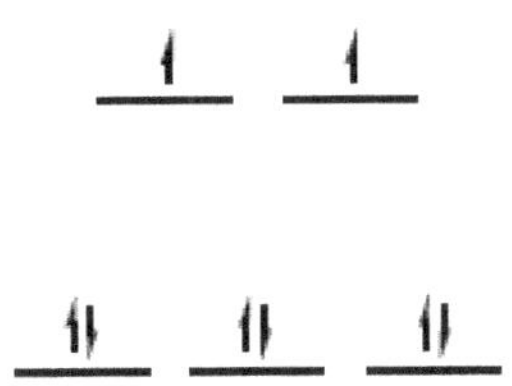

Orbital contribution is zero and magnetic moment is close to the spin only value.

For a tetrahedral complex

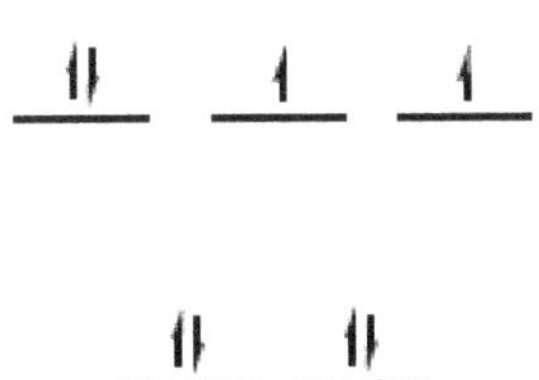

Magnetic moment is higher than the spin only value as there is positive orbital contribution

The change in manetic moment in case of cobalt complexes due to the influence of moisture and heat is shown below (Fig. 88),

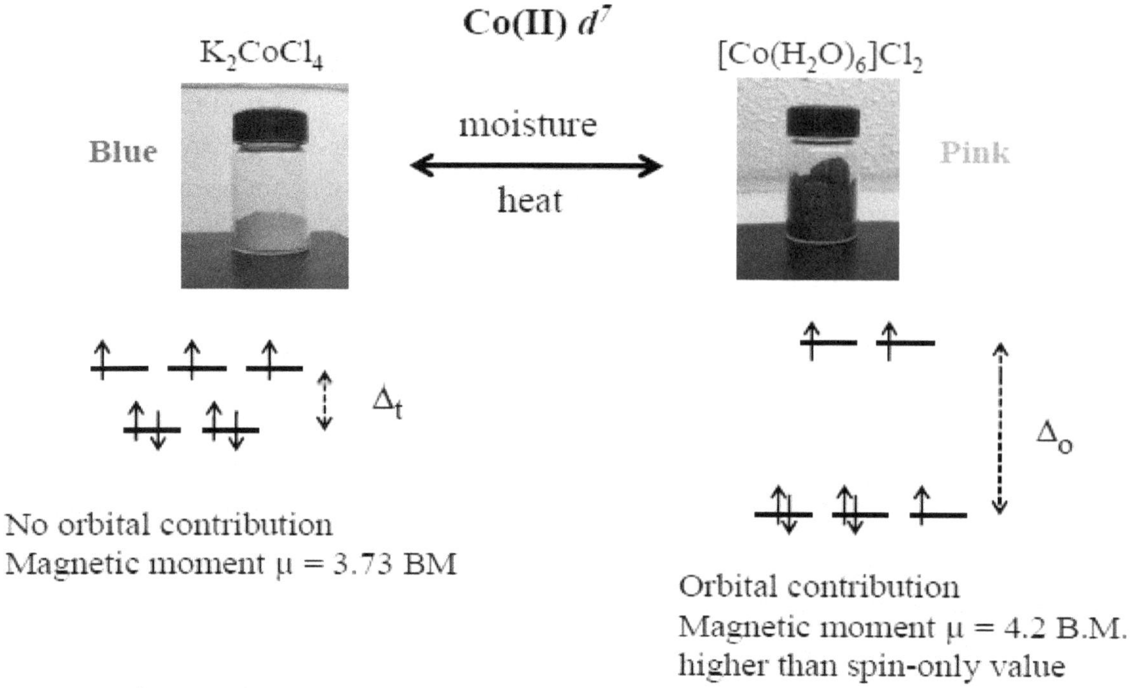

Fig. 88. Inluence of moisture and heat on the magnetic moment of cobalt complexes

A and E ground terms

The configurations corresponding to the A_1 (free ion S term), E (free ion D term), or A_2 (from F term) do not have a direct contribute to the orbital angular momentum. For the A_2 and E terms there is always a higher T term of the same multiplicity as the ground term which can affect the magnetic moment (usually by a only small amount).

$$\mu_{eff} = \mu(\text{spin-only})(1 - \alpha\lambda/\Delta_{oct})$$

where α is a constant (2 for an E term, 4 for an A_2 term)
λ is the spin-orbit coupling constant which is generally only available for the free ion but this does give important information since the sign of the value varies depending on the orbital occupancy.

some spin-orbit coupling constants for 1st row TM ions

metal ion	Ti(III)	V(III)	Cr(III)	Mn(III)	Fe(II)	Co(II)	Ni(II)	Cu(II)
d configuration	1	2	3	4	6	7	8	9
λ / cm^{-1}	155	105	90	88	-102	-172	-315	-830

for d^1 to d^4 the value is positive hence μ_{eff} is less than $\mu_{s.o.}$
for d^6 to d^9 the value is negative hence μ_{eff} is greater than $\mu_{s.o.}$
Δ is the crystal field splitting factor which again is often not available for complexes.

For the tetrahedral Co(II) ion, $CoCl_4^{2-}$, the observed experimental magnetic moment, μ_{obs} = 4.59 Bohr Magneton(B.M.)

The spin-only magnetic moment, $\mu_{s.o.}$ = 3.88 B.M. which is not in good agreement. How can we improve the analysis?

Since the ground term in the tetrahedral field is split from a ^4F to a $^4A_2$ term then we can apply the formula above.

For an A term the constant $\alpha = 4$. The spin-orbit coupling constant, λ for the free ion is -172 cm^{-1} which we can use as an approximation and Δ= 3100 cm^{-1}.

Hence μ_{eff} = 3.88 x (1 - (4* -172) / 3100)
which comes out at μ_{eff} = 4.73 B.M.

This gives a much better fit than the spin-only formula.

In the case of the series;

CoI_4^{2-}, $CoBr_4^{2-}$, $CoCl_4^{2-}$, $Co(NCS)_4^{2-}$
the magnetic moments have been recorded as
4.77, 4.65, 4.59, 4.40 BM
assuming that λ is roughly a constant, then this variation shows the inverse effect of the spectrochemical series on the magnetic moment, since Δ is expected to increase from I- to NCS-.

T ground terms

The configurations corresponding to the T_2 term (from D) or a T_1 term (from an F term) are those where there is a direct contribution to orbital angular momentum expected. The magnetic moments of complexes with T terms are often found to show considerable temperature dependence. This is as a result of spin-orbit coupling that produces levels whose energy differences are frequently of the order *k*T, so as a result, temperature will have a direct effect on the population of the levels arising in the magneticfield.

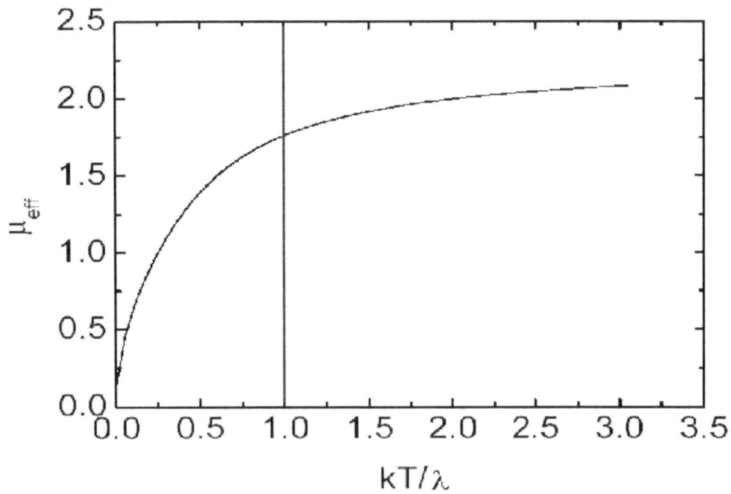

Fig. 89. kotani plot for octahedral Ti (III)

In a Kotani plot (Fig. 89) μ_{eff} is plotted against kT/λ and when this corresponds to a value of 1 then μ equals the "spin-only" value. If this is extrapolated to infinity then the value corresponds to μ_{S+L}

High-spin / Low-spin crossover equilibria

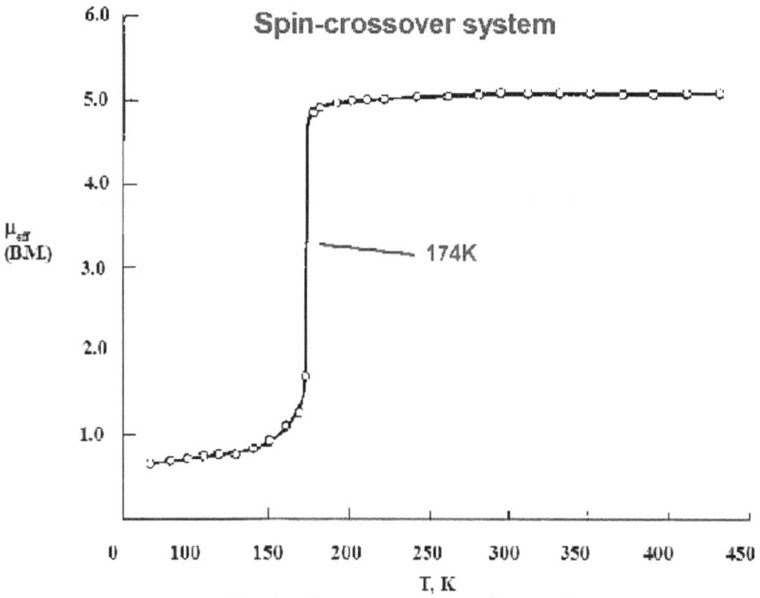

Fig. 90. Spin cross over observed in complexes

Octahedral complexes with between 4 and 7 d electrons can be either high-spin or low-spin depending on the size of Δ. When the ligand field splitting has an intermediate value such that the two states have similar energies, then the two states can coexist in measurable amounts at equilibrium. Many "crossover" systems of this type have been studied, particularly for iron complexes. In the d^6 case of $Fe(phen)_2(NCS)_2$, the crossover involves going from S=2 to S=0. At the higher temperature the ground state is $^5T_{2g}$ while at low temperatures it changes to $^1A_{1g}$. The changeover is found at about 174K as shown in figure 90.

Spin-cross over region and effect of temperature

Magnetic measurements will tell us whether the complex is a high-spin or lowspin complex as explained earlier. Many transition metal ions are able to form high-spin and low-spin complexes depending up on the strength of the ligand field. When the ligand is of intermediate field strength, both high-spin and lowspin complexes can coexist in equilibrium.

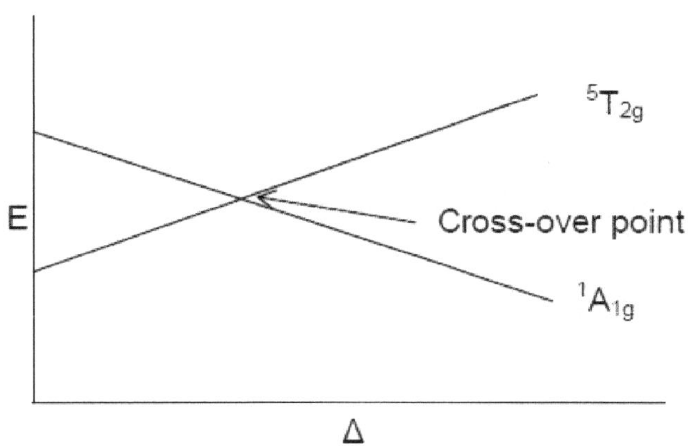

Fig. 91. Spin cross over between spectroscopic states

Example:

Fe^{2+} can form both high-spin, $[Fe(H_2O)_6]^{2+}$, and low-spin, $[Fe(CN)_6]^{2-}$, complexes. The high-spin complex has electronic configuration, $t_{2g}^4\ e_g^{\ 2}$ and has 4 unpaired electrons. Hence, S=4/2 =2 and is paramagnetic. The electronic configuration of low-spin complex is $t_{2g}^6\ e_g^0$. S = 0 and is diamagnetic as there are no unpaired electrons.

In octahedral complexes, the d^6 system will be having $^5T_{2g}$ as the ground state in weak-field cases and $^1A_{1g}$ as the ground state in strong-field cases.

This is shown below (Fig. 91),
Near the cross-over point, the difference in energy between the two spin states is very small and hence, they exist in equilibrium near this point. The actual population of the states depends on the temperature.

For the complexes of iron(II) mentioned above, the high-spin complex is predominant at high temperature and the low-spin complex is predominant at low temperature.

2.13 EXERCISE QUESTIONS

1. Answer the following questions about the electronic states of ions.

(a) Explain in words, what the Russell-Saunders terms represent.

(b) Determine the Russell-Saunders ground state term of an f^6 ion.

(c) Determine the states arising when a ^4D term is placed in a linear ligand field. Assume the ligands lie along the z axis.

(d) What is the total degeneracy of a ^5G term?

2. What do you mean by spin-spin coupling and spin –orbit coupling?

3. Calculate the number of microstates for d^3 configuration.

4. Explain Hund's rule for Term symbols.

5. Explain, with a neat diagram, Russell-Saunders coupling scheme for d^2 configuration.

6. Write short notes on Selection rules for electronic transitions in complexes.

7. What are Orgel diagrams? Draw the combined Orgel diagrams for D and F terms in octahedral and tetrahedral fields.

8. The absorption bands in electronic spectroscopy are broad. Why?

9. What is hole formalism in electronic spectroscopy?

10. Which complex will be more coloured either $[Co(NH_3)_6]^{3+}$ or $[Co(NH_3)_5\ Cl]^{2+}$.

Justify your answer.

11. Predict how many absorption bands there will be (assuming no overlap) in the UV/Vis spectrum of $[V(H_2O)_6]^{3+}$ with the spectrum and Orgel diagram.

12. Predict how many absorption bands there will be (assuming no overlap) in the UV/Vis spectrum of $[Co(H_2O)_6]^{2+}$ with the spectrum and Orgel diagram.
13. Answer the following questions about electronic spectroscopy.

(a)Explain why the Laporte symmetry rule is often relaxed (partially violated) in real complexes.

(b)Explain why the molar absorptivity of $[Co(H_2O)_6]^{2+}$ is 10 L mol^{-1} cm^{-1} while that for $[CoCl_4]^{2-}$ is 600 L mol^{-1}cm^{-1}

(c)Predict how many absorption bands there will be (assuming no overlap) in the UV/Vis spectrum of $[Cr(NH_3)_6]^{2+}$.

(d)How does this change if two of the NH_3 are replaced by Cl^- in trans- $Cr(NH_3)_4(Cl)_2$?

(e)Explain why, in terms of electronic transitions, substituting two of the NH_3 in $[Cr(NH_3)_6]^{2+}$ by bipy to form $[Cr(NH_3)_4(bipy)]^{2+}$ leads to much more intensely coloured ($\varepsilon > 45{,}000$ L mol^{-1}cm^{-1}) complex.

14. Circle the less intense electronic spectral band among each of the following pairs, and explain the reason for your choice:

$[Cr(NH_3)_6]^{3+}$: $^4A_{2g}\rightarrow{}^4T_{2g}$ or $[Cr(en)_3]^{3+}$: $^4A_{2g}\rightarrow{}^4T_{2g}$

cis -$[CoCl_2(NH_3)_4]^+$: $^1A_{1g}\rightarrow{}^1T_{1g}$ or trans-$[CoCl_2(NH_3)_4]^+$: $^1A_{1g}\rightarrow{}^1T_{1g}$

$[Cr(NH_2O)_6]^{2+}$: $^4A_{2g}\rightarrow{}^4T_{1g}$ or $[CoCl_4]^{2-}$; $^4A_2$ ---> $^4T_1$

15. For the complex $[Co(CO)_4]^-$, sketch the expected electronic spectrum due to d-d transitions. Be sure to label and include scales for the axes in your drawing.

16. Give the appropriate term symbols for the states with the following values of L and S. Also work out the possible values of J.

i) L = 3 S = 2

ii) L = 2 S = 3/2.

17. Using Hund's rules, arrange the following terms in the increasing order of energy: ^4H, ^4G, ^2I, ^2M, ^2P.

18. From the following spectral data and using Tanabe–Sugano diagram, calculate Δo for the complex $[VF_6]^{3-}$, which has absorption bands at 14,800 and 23,250 cm^{-1} and a third band in the ultraviolet region. Also calculate Racah parameter B for this complex.

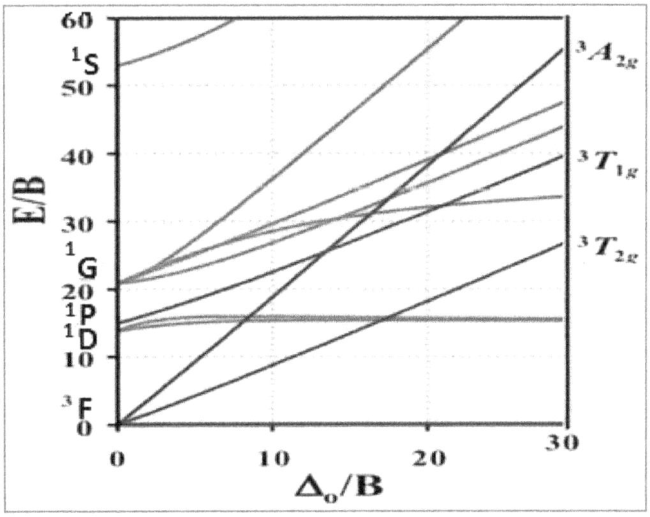

19. What are Tanabe-Sugano diagrams? Distinguish between Tanabe- Sugano diagrams and Orgel diagrams

20. Calculate the effective magnetic moment μ_{eff} for $[Ni(NH_3)_6]^{2+}$ complex with Δo =8500 cm^{-1}. Given that the value of the spin- orbit coupling constant λ, is -316 cm^{-1} for Ni^{2+} and α =4.

21. $[Mn(NCS)_4]^{2-}$ is tetrahedral complex whereas $[Mn(NCS)_6]^{4-}$ is octahedral; however, their molar magnetic susceptibilities at room temperature are the same. Justify your answer.

22. A complex $[Cr(H_2O)_6]Cl_3$ records three spin allowed transitions at 17400cm^{-1}, 24600 cm^{-1} and 37800 cm^{-1}. Calculate Nephelauxetic ratio β using the given T –S diagram and comment on the nature of M – L bond.

(B = 918 cm^{-1}).

Chapter 3

ORGANOMETALLIC CHEMISTRY

3.1 INTRODUCTION

Organometallic chemistry plays a significant role in organic, inorganic, biochemistry, materials science and chemical engineering. The origin of organometallic chemistry dates back to the mid of 1800, when Frankland discovered the methyl and ethyl derivatives of mercury, tin and zinc. However it was the successful introduction of the organomagnesium reagents by Victor Grignard around 1900 that provides a real breakthrough in the utilization of organometallics in organic synthesis. The discovery that an alkyl halide could be readily converted into an organomagnesium compound in a reaction with metallic magnesium Mg(0) and that in turn leading to a transformation of electrophilic carbon into nucleophilic carbon made it possible to develop unprecedented chemistry. The Grignard reaction provided an extremely powerful tool in organic synthesis, overcoming some of the limitations of classical organic synthesis.

Since then, an ever increasing number of organic transformations have been based on organometallic compounds. The reactivity of a wide range of organometallic starting materials intermediates and catalysts have been investigated and a huge number of new and efficient synthetic transformations involving their utilization have been developed. The way organic chemists design and realize synthetic processes has been deeply and irreversibly influenced by the use of organometallics. An important part of many synthetic processes both in academic and industrial laboratories are presently based on organometallic compounds.

Organometallic compounds had been known since the discovery of Zeise's salt, $K[Pt(C_2H_4)Cl_3]$, in 1827 and the preparation of metal alkyls by Sir Edward Frankland in 1849, but with the discovery of ferrocene and Ziegler-Natta polymerization in the early 1950s, organometallic chemistry took on a different level of importance. These discoveries were followed in 1955 by the preparation of dibenzenechromium. Although there are numerous organometallic compounds of all of the elements in any specific group, those of one element in the group are generally more important. For example, in group IA, the organic compounds of lithium are more numerous than those of sodium or potassium.

Organometallic compounds

Organometallic compounds (metal organyls, organometallics) are defined as materials which possess direct, more or less polar bonds $M^{\delta+}—C^{\delta-}$ between metal and carbon atoms. In addition to the traditional metals, lanthanides, actinides, and semimetals, elements such as boron, silicon, arsenic, and selenium are considered to form organometallic compounds, e.g. organoborane compounds such as triethylborane

(Et_3B). Compounds where metal binds to ligands via, N, O, S, other M-carboxylates, ethylenediamine and water are not considered as organometallic compounds.

Organometallic chemistry is the study of organometallic compounds. Since many compounds without such bonds are chemically similar, an alternative may be compounds containing metal element bonds of a largely covalent character. Organometallic chemistry combines aspects of inorganic chemistry and organic chemistry. By way of generalization, it may be stated that the chemistry of main-group organometallics is governed by the group the metal belongs to, whereas for organotransition-metal compounds, the nature of the ligand dominates.

- Organometallic chemistry is the chemistry of compounds that contain M-C bonds $M^{\delta+}—C^{\delta-}$, where M is either a main group metal or a transition metal.
- One common feature of all organometallic compounds is that the metal is in a very low oxidation state.
- Because of the low oxidation state on the metal, there are many *d*-electrons in the complex, so π-back bonding between the metal and ligands is very important. One of the most common ligands in organometallic chemistry is carbonyl, CO.
- Organometallic reagents are very important in synthesis because of their versatility. For example, metals will often activate the C-H bond, which lowers the activation energy for many organic transformations.
- Organometallic catalysts are used to make many important pharmaceuticals and are critical in the petrochemical industry.

Stability of organometallic compounds

- There is one problem with many organometallic reagents: because the metal is in a low oxidation state, if the complex is exposed to atmospheric oxygen the metal is readily oxidized. Therefore, organometallic reactions are usually carried out under an inert atmosphere of nitrogen or argon.
- All organometallic compounds are thermodynamically unstable to oxidation – Due to large negative free energies of formation of metal oxide, carbon dioxide and water.
- Many organometallic compounds are kinetically unstable to oxidation at or below room temperature– Associated with the presence of empty low lying orbitals on the metal atom, or a non bonding pair of electrons.
- Hydrolysis
    - Involves nucleophilic attack by water.
    - Facilitated by the presence of empty low lying orbitals on the metal atom which can accept electrons. Organometallic compounds of groups 1, 2, Zn, Cd, Al, Ga, In- readily hydrolysed.
    - Rate of hydrolysis is dependent on M-C bond polarity – greater polarity, faster rate

- Alkyls and Aryls of groups 14 and 15 are kinetically inert to hydrolysis by water because the metal atoms are surrounded by a filled shell of 8 electrons. Nucleophilic attack is no longer favoured.

In dealing with the stability of a compound, one should be clear about the type of stability under consideration. In general terms, the stability of an organometallic compound may refer to either its thermal stability or resistance to chemical attack. Obviously, these different types of stabilities would depend both on thermodynamic as well as kinetic factors. Most metal alkyls and aryls are thermodynamically unstable to oxidation and hydrolysis and therefore need protection from the atmosphere during their synthesis, isolation and storage. There are few other examples of organometallic compounds like tetramethyl lead which are thermodynamically unstable with respect to oxidation and hydrolysis, yet they survive exposure to air / moisture possibly because of kinetic factors.

a. Thermal stability of organometallic compounds

The thermodynamic stability of a compound relative to its constituent elements would depend upon the standard free energies of formation, ΔG^0_f values of the different species concerned.

The decomposition of the metal methyls into the component elements, i.e., metal, carbon and hydrogen could be expected to be an exothermic process in the case of heavier metal alkyls, e.g., $MgMe_2$, $CdMe_2$, $InMe_3$, $PbMe_4$ and $BiMe_3$, for which ΔH^0_f i.e., enthalpy of formation is positive. Conversely for methyls of lighter elements such as B, Al, Si, Ge and P, it should be endothermic process.

In an actual decomposition process, however, the metal alkyls yield the metal, hydrogen and mixtures of hydrocarbons rather than simple elementary carbon. In such cases, the enthalpy of formation of hydrocarbons is taken into consideration. For example, consider the decomposition of tetramethyllead into lead, methane and ethane only,

$$\begin{array}{lccccccc} & Pb(CH_3)_4(g) & \longrightarrow & Pb(s) & + & 2CH_4(g) & + & C_2H_4(g) \\ \Delta H_f^{\circ}\ (kJ\ mol^{-1}) & +137 & & 0 & & 2\times(-73) & & +52 \end{array}$$

The standard enthalpy change ΔH^0_m for this reaction is,

$$\Delta H^{\circ}_m = -150 + 52 - 137 = -235\ kJ\ mol^{-1}$$

The above reaction, therefore, is highly exothermic. This clearly indicates a higher stability for the products than the reactant molecules.

Similarly for the reaction,

$$\begin{array}{lcccc} & Si(CH_3)_4(g) \longrightarrow & Si(s) + & 2CH_4(g) + & C_2H_4(g) \\ \Delta H_f^{\circ}\ (kJ\ mol^{-1}) & -238 & 0 & 2\times(-75) & 52 \end{array}$$

The standard enthalpy change $\Delta H^0{}_m$ for this reaction is found to be + 140 kJ mol^{-1}, which indicates that tetramethylsilicon is more stable thermodynamically.

For resolving such anomalies, one would have to take into consideration whether a suitable reaction path is available for the decomposition to occur. The kinetic stability, therefore, appears to play a more predominant role.

b. Stability to oxidation

The thermodynamic instability of organometallic compounds may be ascribed to the large value of free energies of formation of the metal oxides, carbon dioxide and water. Actually the combustion of organometallic compounds is a highly exothermic process, e.g.,

$$Zn(CH_3)_2(g) + 4O_2(s) \longrightarrow ZnO(s) + 2CO_2(g) + 3H_2O(g)$$
$$\Delta H_m^{\circ} = -1918\ kJ\ mol^{-1}$$
$$Sn(CH_3)_4(g) + 8O_2(g) \longrightarrow SnO_2(s) + 4CO_2(g) + 6H_2O(g)$$
$$\Delta H_m^{\circ} = -3591\ kJ\ mol^{-1}$$

In fact dimethylzinc is spontaneously inflammable compound whereas tetramethyltin is air stable. The greater reactivity of dimethylzinc can be easily understood on the basis of its strong Lewis acid character. In addition to $ZnMe_2$, the lighter alkyls of Li, Na, Be, Mg, Zn, B, Al, Ga, In, Tl, As, Sb and Bi are also spontaneously inflammable. Thus the presence of empty low lying orbitals or a lone pair of electrons appears to be a source for kinetic reactivity to oxidation by air. The group IVB tetraalkyls such as $SnMe_4$ possess neither of these characteristics and therefore, they resist oxidation at ambient temperatures.

c. Stability to hydrolysis

The organometallic compounds are generally hydrolysed via nucleophilic attack by water, which is facilitated by:

i) The presence of empty low -lying orbitals on the metal, and
ii) The polarity of metal –carbon bonds.

As a consequence of these factors, the organoderivatives of alkali and alkaline earth metals as well as of zn, Cd, Al, Ga and In are readily hydrolysed. By contrast, the tryalkylboranes although coordinatively unsaturated are unaffected by water at ambient temperatures, possibly due to comparatively very low polarity of boron- carbon bonds. The alkyls and aryls of the group IV and V elements are also kinetically inert to hydrolysis by water.

3.2 ORGANOMETALLICS OF GROUP IA ELEMENTS: Organolithium compounds

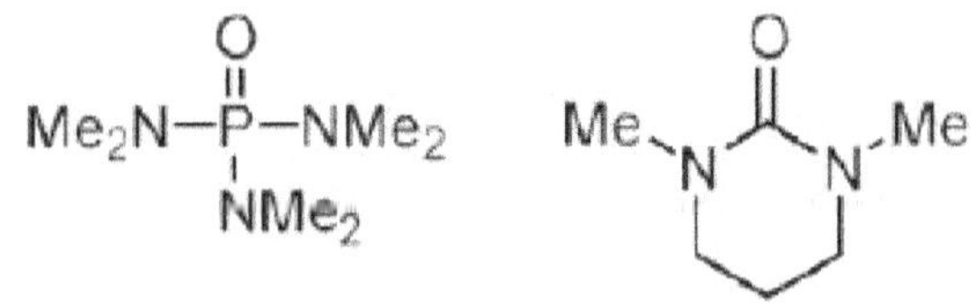

HMPA DMPU

Some donor molecules used to modify the degree of aggregation as well as reactivity of organolithium compounds

Lithium is strongly electropositive element and the carbon – lithium bond is polarized so as to place high electron density on the carbon. Therefore, Organolithium compounds are strong bases and nucleophiles. They are usually designated as RLi, which is useful but incomplete representation of their composition in solution where they exist as more complex structures. For example, simple Organolithium reagents exist mainly as tetramers in ethers and as hexameric structures in hydrocarbons. Their degree of chelating ligands such as tetramethylenediamine (TMEDA) or strong donaor molecules such as N,N-dimethylpropyleneurea (DMPU) and hexamethylphosphorictriamide (HMPA)

Preparation of Organolithium compounds

1. Oxidative additive reaction: Oxidation of the metal by the addition of a group RX. Simple Organolithium compounds are usually prepared by reaction of lithium metal with an appropriate organic halide. With alkyl halides, loss of stereochemical integrity at the site of reaction is usually observed. With vinyl halides, the configuration of the double bond is generally preserved.

$$RX + 2\,Li \longrightarrow RLi + LiX$$

$$BuCl + 2Li \xrightarrow[Ar]{Et_2O} LiCl(s) + BuLi$$

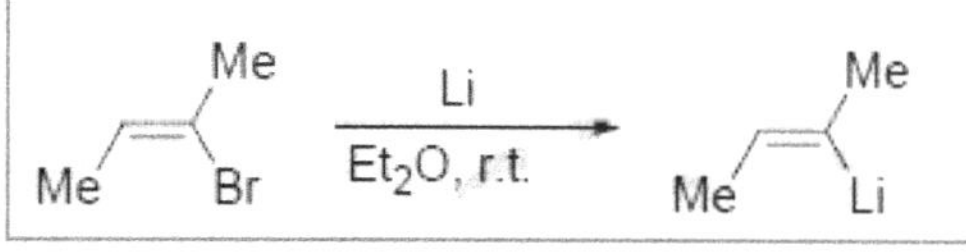

2. Metal – halogen exchange: Organolithium reagents can also be prepared by via a metal – halogen exhange process involving an organo halide and an alkyl lithium. The reaction proceeds in the direction favouring the formation of the Organolithium reagent derived from the more acidic compound. Thus, the greater stability of sp^2 cabanions in comparison with that of sp^3 carbanions is the driving force of the conversion of vinyl or aryl halides into the corresponding lithium derivatives in the presence of alkyl lithium reagents.

$$RX \xrightarrow{R^1Li} RLi$$

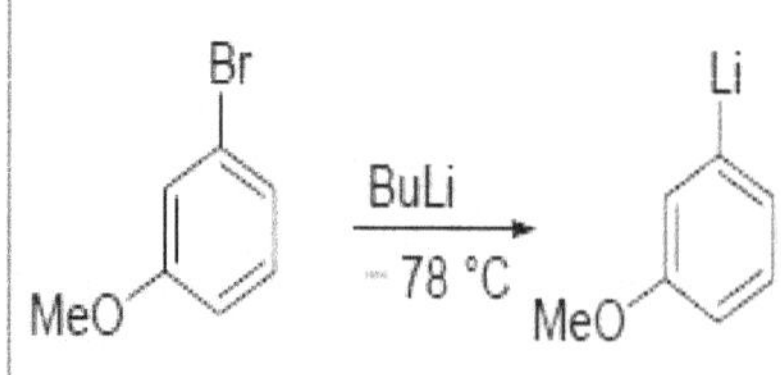

3. Hydrogen-metal exchange or metalation: Another general method which is very useful for the preparation of organolithium compounds is the hydrogen-metal exchange or metalation.

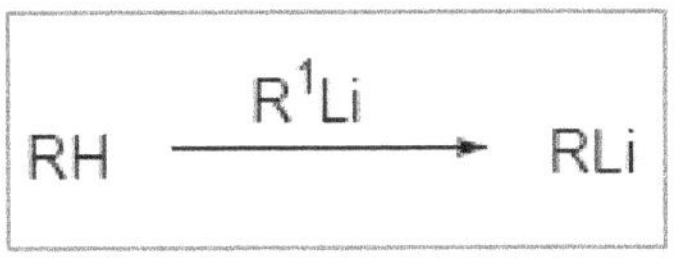

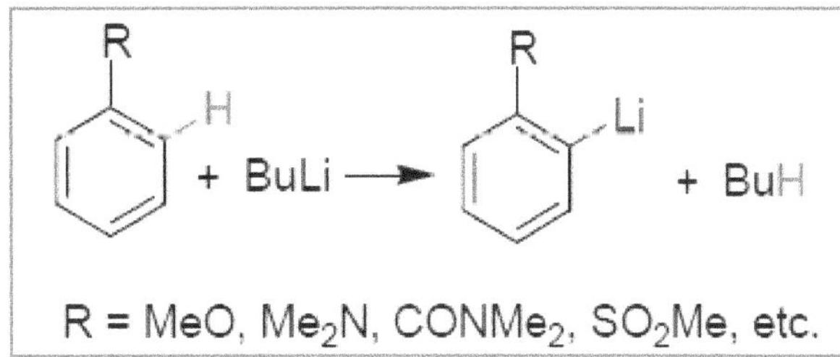

The directing effect of the substituents in the substrate and the acidity of the available hydrogens influence the position of lithiation. The latter factor, of course, plays an important role even in determining the basic strength of the Organolithium reagent required to perform the exchange reaction.

4. Transmetallation: Organolithium reagents can also be prepared through a metal-metal exchange process involving an alkyl lithium and an organometallic compound. The reaction proceeds so as to favor the formation of the organometallic derivative with the more stable carbanion bound to the more electropositive metal. An example of vinyltin compound and an alkyl lithium is given below,

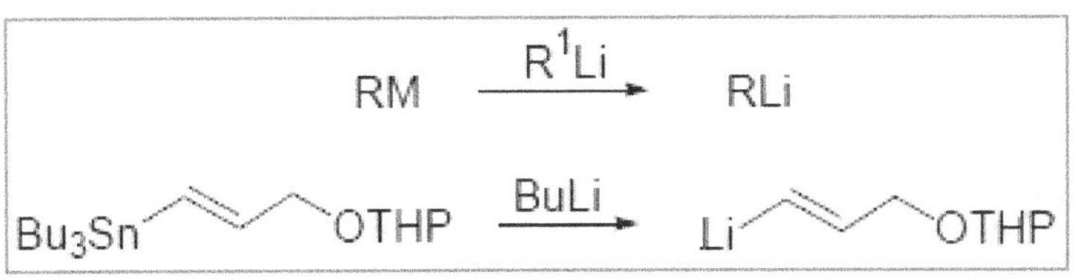

Structure and bonding of the tetrameric cluster of methyl lithium-_($(LiCH_3)_4$)-Structure

Methyl lithium, readily available from chemical suppliers as a solution in ether, has been shown to exist in the solid state using X-ray diffraction and in less-coordinating solvents as the tetrahedral Me_4Li_4 cage shown below. There is a central Li_4 tetrahedron, and an outer C_4 tetrahedron. These two clusters interpenetrate one another, but importantly they both have the same set of point-group symmetries. Another way to think of this is as that of each methyl group pointing towards the centre of one face of the Li_4 tetrahedral cluster.

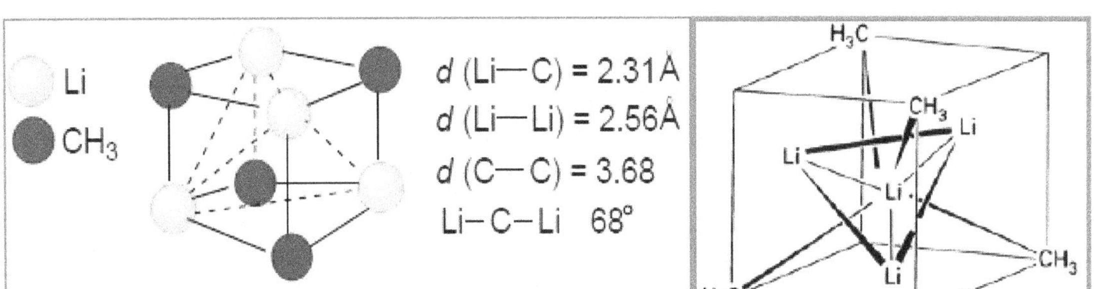

Schematic drawing of the unit $(LiCH_3)_4$

Solid methyllithium has cubic body-centered packing. The structure of $(LiCH_3)_4$ consists of Li_4-tetrahedron with methyl groups capping the triangular faces. It has been determined that in hydrocarbon solutions the dominant species is a hexamer when the alkyl groups are small. In the solid phase, the structure is body-centered cubic with the $(LiCH_3)_4$ units at each lattice site. Each unit is a tetramer in which the four lithium atoms reside at the corners of a tetrahedron and the methyl groups are located above the centers of the triangular faces. The carbon atoms of the alkyl groups are bound to the three lithium atoms at the corners of the triangle.

Bonding

The axial Li (sp^3) orbital interacts with the methyl groups of neighbouring $(LiCH_3)_4$ units. These intermolecular forces result in the low volatility and insolubility of $(LiCH_3)_4$ in nonsolvating media. Bonding of the methyl group to three lithium atoms involves an sp^3 orbital on the methyl group simultaneously overlapping three orbitals (they may be $2s$ or hybrids of $2s$ and $2p$ orbitals) on the Li atoms. Three-center two-electron bonds exist in numerous types of compounds (such as diborane), but it may be in this case that there is also some weak interaction between the lithium atoms.

Reactivity of Group I Organometallics- organolithium compounds

The degree of ionic character increases with increasing atomic number of alkali metal. For example, percentage ionic character of metal – carbon bond of metal alkyls increases in the order: Li – C(~43) < Na – C(~47) < K – C(~52) < Rb – C(~52) < Cs – C(~57)

As a consequence of this increase in ionic character, their reactivity (as nucleophile or carbanionic reagent) and solubility in organic solvents follows the same pattern:

LiR > NaR > KR ≈ RbR > CsR.

In fact, due to high reactivity of organometallics of sodium and its heavier congeners in polar solvents, their organometallic chemistry has been explored to a lesser degree compared to the current extensive organometallic chemistry of lithium.

I. Organolithium reagent as nucleophile

a) Carbolithiation reactions

As nucleophiles, organolithium reagents undergo carbolithiation reactions, whereby the carbon-lithium bond adds across a carbon-carbon double or triple bond, forming new organolithium species. This reaction is the most widely employed reaction of organolithium compounds. Carbolithiation is key in anionic polymerization processes, and *n*-butyllithium is used as a catalyst to initiate the polymerization of styrene, butadiene, or isoprene or mixtures thereof.

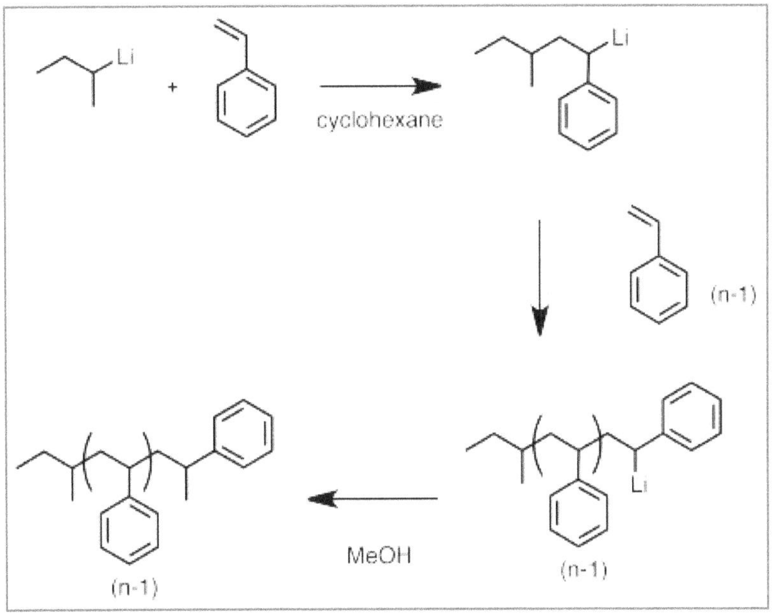

b) Addition to carbonyl compounds

Nucleophilic organolithium reagents can add to electrophilic carbonyl double bonds to form carbon-carbon bonds. They can react with aldehydes and ketones to produce alcohols. The addition proceeds mainly via polar addition, in which the nucleophilic organolithium species attacks from the equatorial direction, and produces the axial alcohol. Addition of lithium salts such as $LiClO_4$ can improve the stereoselectivity of the reaction.

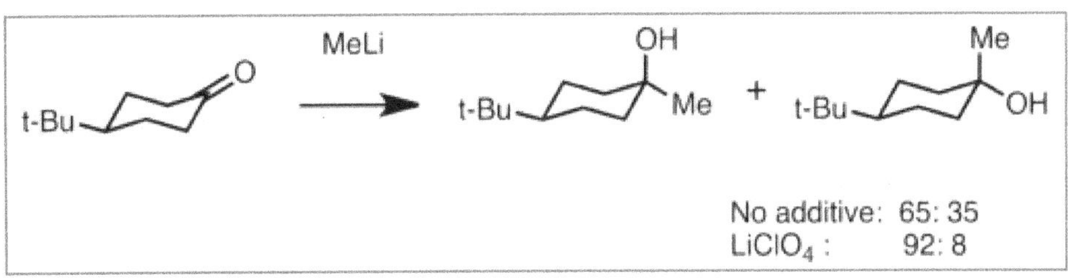

II. Organolithium reagent as base

Organolithium reagents provide a wide range of basicity. *tert*-Butyllithium, with three weakly electron donating alkyl groups, is the strongest base commercially available (pKa = 53). As a result, the acidic protons on -OH, -NH and -SH are often protected in the presence of organolithium reagents. Some commonly used lithium bases are alkyllithium species such as *n*-butyllithium and lithium dialkylamides ($LiNR_2$). Reagents with bulky R groups such as lithium diisopropylamide (LDA) and lithium bis(trimethylsilyl)amide (LiHMDS) are often sterically hindered for nucleophilic addition, and are thus more selective toward deprotonation. Lithium dialkylamides ($LiNR_2$) are widely used in enolate formation and aldol reaction.

III. Metalation

Metalation with organolithium reagents, also known as lithiation or lithium-hydrogen exchange, is achieved when a basic organolithium species, most commonly an alkyllithium, abstracts a proton and forms a new organolithium species.

$$R\text{-}H + R'Li \longrightarrow RLi + R'H$$

Common metalation reagents are often butyllithiums, *tert*-Butyllithium and *sec*-butyllithium are generally more reactive and have better selectivity than *n*-butyllithium, however, they are also more expensive and difficult to handle. Metalation is a common way of preparing versatile organolithium reagents. The position of metalation is mostly controlled by the acidity of the proton. Metalation often occurs at the α position of electron withdrawing groups, since they are good at stabilizing the electron-density of the anion.

IV. Transmetalation

Organolithium reagents are often used to prepare other organometallic compounds via transmetalation. Organocopper, organotin, organosilicon, organoboron, organophosphorus, organocerium and organosulfur compounds are frequently prepared by reacting organolithium reagents with appropriate electrophiles.

$$R\text{-}M + n\text{-}BuLi \longrightarrow R\text{-}Li + n\text{-}BuM$$

Organolithium can also be used in to prepare organozinc compounds

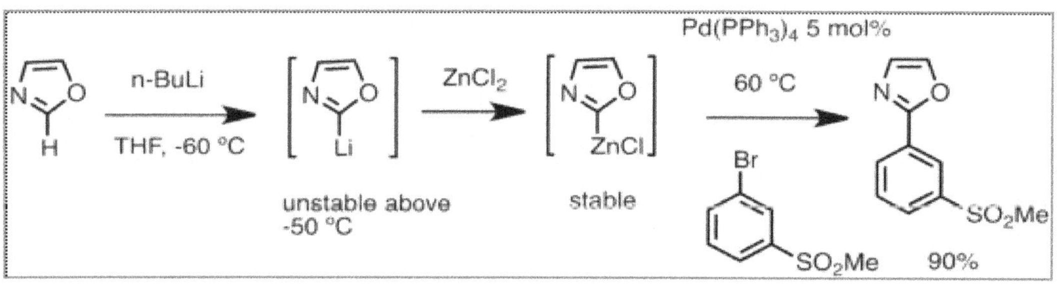

3.3 ORGANOMETALLICS OF GROUP IIA ELEMENTS: Organomagnesium compounds

Organomagnesium compounds are widespread. They are commonly found as Grignard reagents. Examples of Grignards are phenylmagnesium bromide and ethylmagnesium bromide.

Relevant organomagnesium reagents outside the scope of Grignards are magnesium anthracene with magnesium forming a 1,4-bridge over the central hexagon used as a source of highly active magnesium and butadiene magnesium an adduct with butadiene and a source for the butadiene dianion.

Grignard reactions and reagents were discovered by and are named after the French chemist François Auguste Victor Grignard (University of Nancy, France), who was awarded the 1912 Nobel Prize in Chemistry for this work. Grignard reagents are similar to organolithium reagents because both are strong nucleophiles that can form new carbon–carbon bonds.
A Grignard reagent has a formula RMgX where X is a halogen, and R is an alkyl or aryl (based on a benzene ring) group. For the purposes of this page, we shall take R to be an alkyl group.

A typical Grignard reagent might be CH_3CH_2MgBr.

Preparation of Grignard reagent

Grignard reagents form via the reaction of an alkyl or aryl halide with magnesium metal.

$$R{-}X + Mg \longrightarrow RMgX$$

same for Ar—X

Diethyl ether is most often used solvent. Tetrahydrofuran is also used.

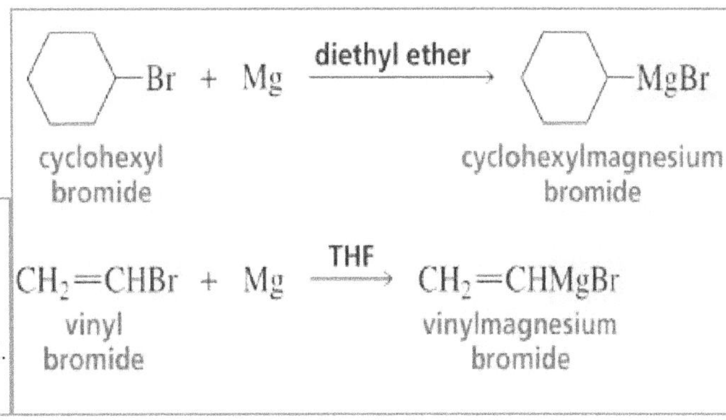

The reaction is conducted by adding the organic halide to a suspension of magnesium in an etherial solvent, which provides ligands required to stabilize the organomagnesium compound. Empirical evidence suggests that the reaction takes place on the surface of the metal. The reaction proceeds through single electron transfer: In the Grignard formation reaction, radicals may be converted into carbanions through a second electron transfer.

$$R{-}X + Mg \rightarrow R{-}X^{\bullet -} + Mg^{\bullet +}$$
$$R{-}X^{\bullet -} \rightarrow R^{\bullet} + X^{-}$$
$$R^{\bullet} + Mg^{\bullet +} \rightarrow RMg^{+}$$
$$RMg^{+} + X^{-} \rightarrow RMgX$$

However, even in solution " RMgX " exists as aggregates with the dimer being the dominant species. The equilibrium $2\ RMgX \rightleftharpoons (RMgX)_2$

can involve other species, and the composition depends on the nature of the alkyl group, which can function as an electron pair donor; the association is hindered by complex formation between the solvent and RMgX. Moreover, the situation is complicated by the equilibrium

$$2\ RMgX \rightleftharpoons MgR_2 + MgX_2$$

which is known as the Schlenk equilibrium. The dimer, $(RMgX)_2$, is believed to have a structure that can be shown as:

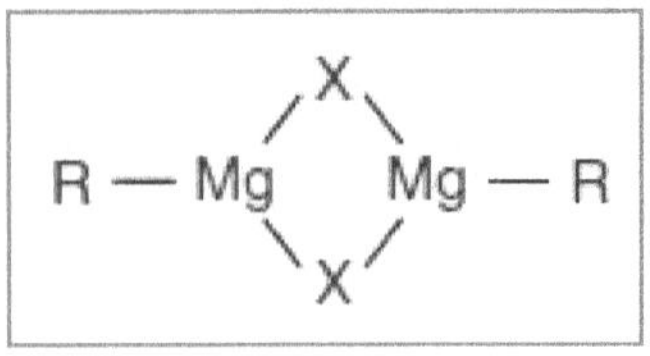

although the dimer can also be represented by the structure

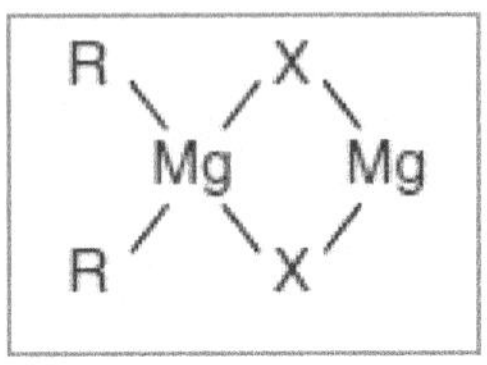

There may also be some ionization that can be represented as

$$2\ RMgX \rightleftharpoons RMg^+ + RMgX_2^-$$

The nature of the species present in a solution of a Grignard reagent is a complex issue with no single representation showing the true situation. Normally, the reactivity of Grignard reagents varies with the nature of the halogen as follows:

I >Br > Cl. It is also found that the alkyl compounds are more reactive than the aryl compounds.

Schlenk equilibrium

Grignard reagents exist not as a simple monomeric species in ether. The constitution of the Grignard reagent, therefore, has been a subject of intense scruitiny, but its main features are now almost clear. The solution of Grignard reagents, RMgX, is believed to contain many species, interconvertible via an extended form of the Schlenk equilibrium as shown below,

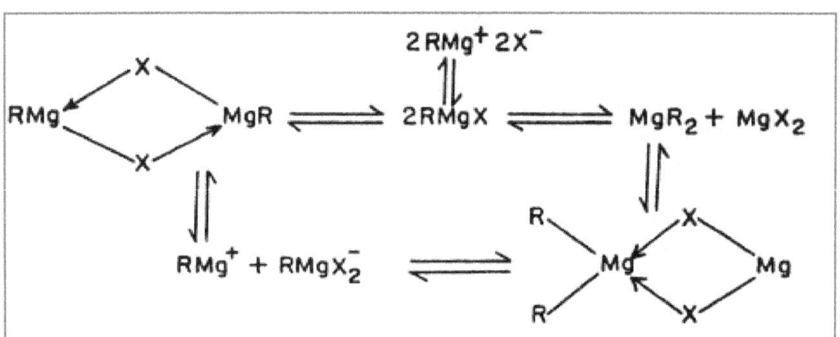

Schlenk equilibrium in MgRX system.
(All species are normally solvated; solvent molecules are not shown for convenience.)

The concentration of the various species shown in the scheme is largely dependent on a number of factors such as the nature of R and X, the solvent, temperature and concentration; e.g., the bulkier type of the alkyl group (R), and the larger size of the halide ions (X), as well as strongly donating solvents and lower concentrations favour the lesser associated or monomeric species.

Structures of organomagnesium compounds (Grignard Reagents-RMgX)

Organomagnesium compounds (Grignard reagents, RMgX) are generally tetrahedral in nature, but penta-coordinated trigonal bipyramidal species are also accessible.

Dialkylmagnesium compounds are associated with symmetrical bridging groups to afford linear polymeric molecules. In coordinating solvents, diorganomagnesium compounds depict similarity to those of Grignard reagents in forming structures with tetrahedral environment around the magnesium.

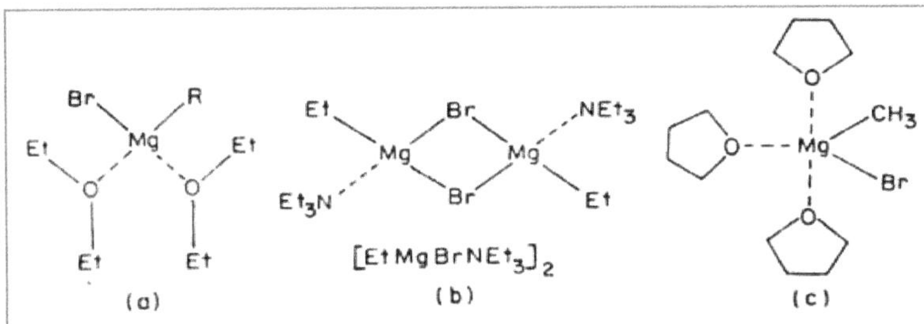

Structure of (a) $Mg(Br)R(Et_2O)_2$, (b) $[Mg(Br)Et(NEt_3)]_2$, and (c) $MgBr(Me)(THF)_3$.

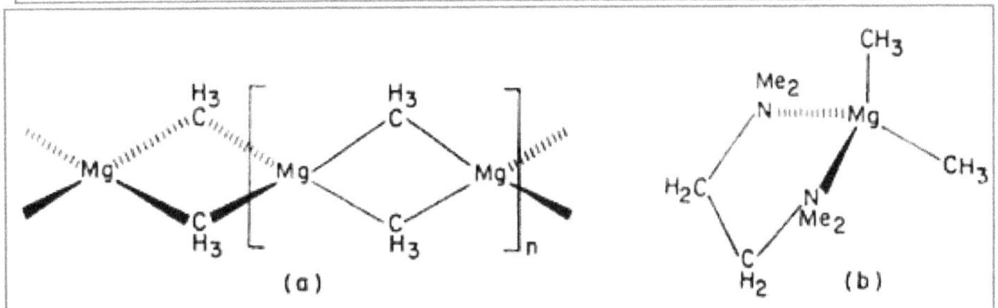

Structure of (a) $[Mg(CH_3)_2]_n$, and (b) $Mg(CH_3)_2(TMEDA)$.

Grignard reaction

The **Grignard reaction** is an organometallic chemical reaction in which alkyl- or aryl-magnesium halides (**Grignard reagents**) add to a carbonyl group in an aldehyde or ketone. This reaction is an important tool for the formation of carbon–carbon bonds. The reaction of an organic halide with magnesium is *not* a Grignard reaction, but provides a Grignard reagent.

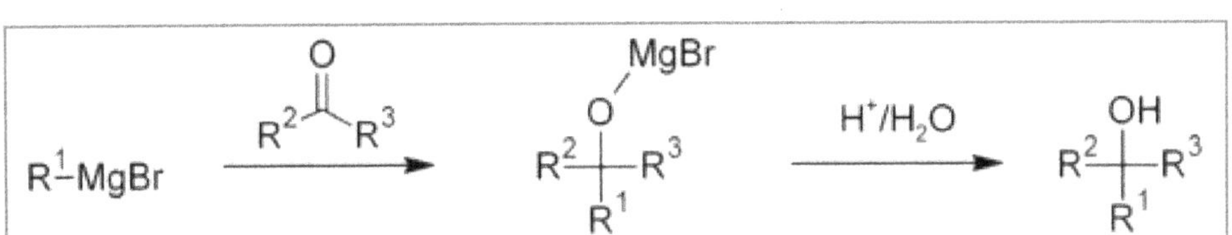

Reaction mechanism

The Grignard reagent functions as a nucleophile, attacking the electrophilic carbon atom that is present within the polar bond of a carbonyl group. The addition of the Grignard reagent to the carbonyl typically proceeds through a six-membered ring transition state.

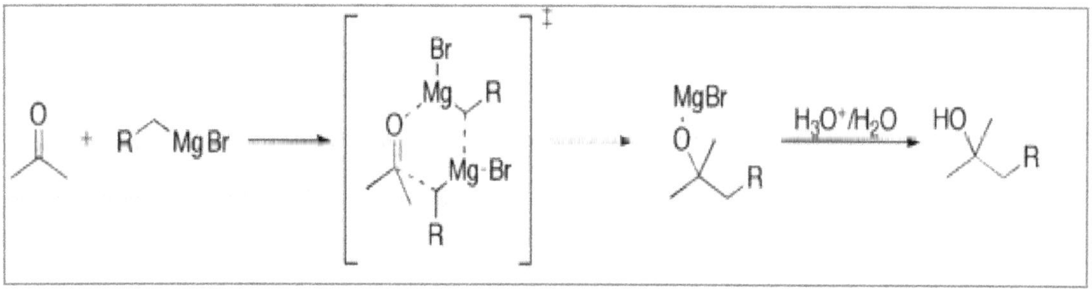

Reactivity and applications of Grignard reagents

I. Reactions with carbonyl compounds

Grignard reagents will react with a variety of carbonyl derivatives.

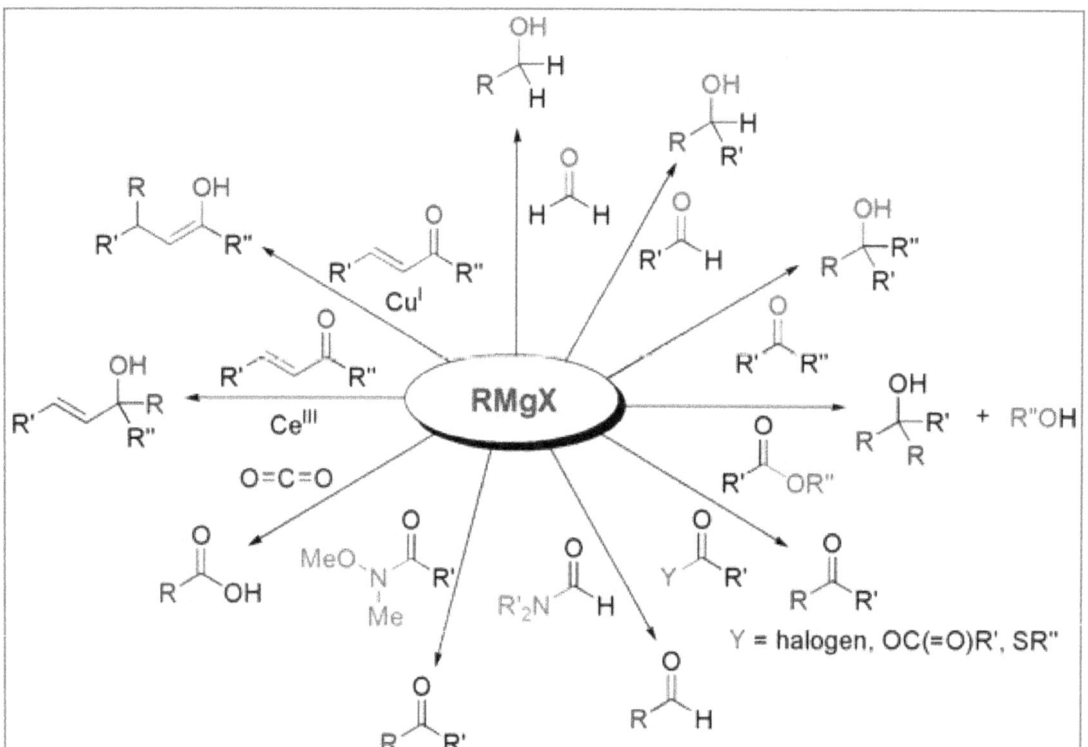

The most common application is for alkylation of aldehydes and ketones, as in this example:

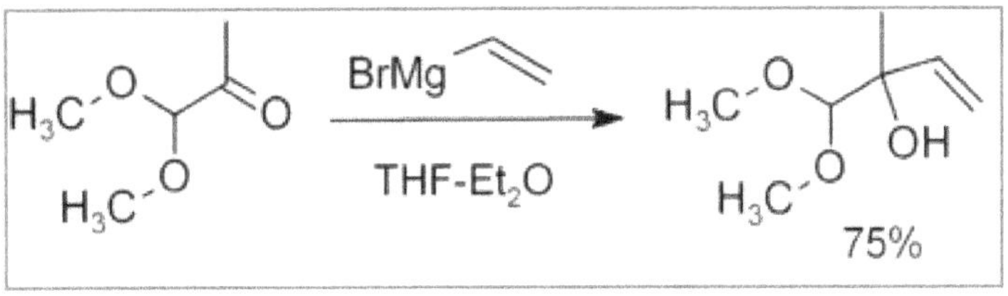

II. Reactions with other electrophiles

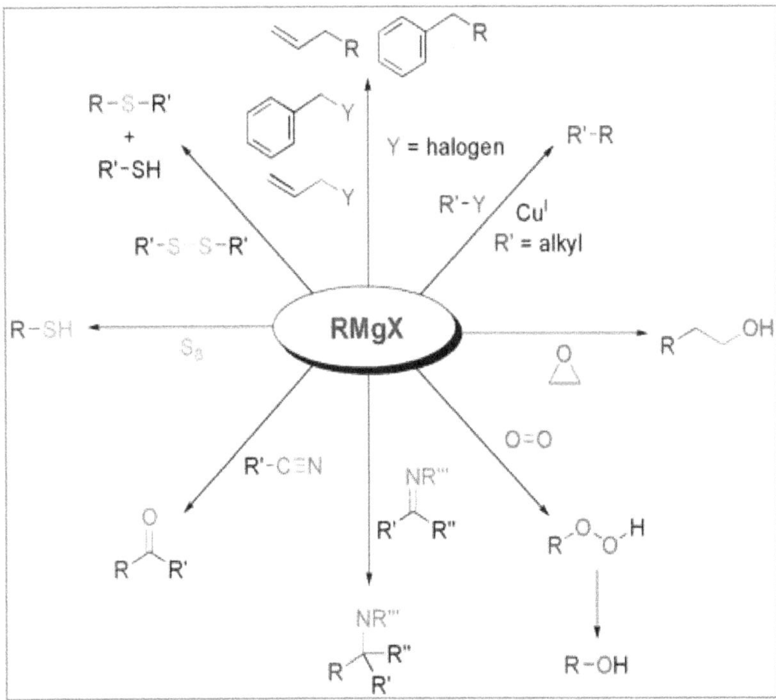

Grignard reagents will react with various other electrophiles, serving both as a nucleophile for many and as a base for protic substrates (this scheme does not show workup conditions, which typically includes water).Grignard reagents are basic and react with alcohols, phenols, etc. to give alkoxides (ROMgBr). The phenoxide derivative is susceptible to formylation paraformaldehyde to give salicylaldehyde.

III. Formation of bonds to B, Si, P, Sn

Like organolithium compounds, Grignard reagents are useful for forming carbon–heteroatom bonds.

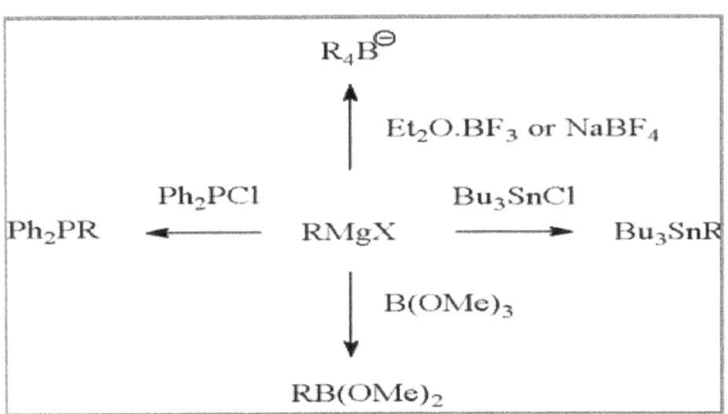

IV. Reaction with transition metal halides

Grignard reagents react with many metal-based electrophiles. For example, they undergo transmetallation with cadmium chloride ($CdCl_2$) to give dialkylcadmium:

$2\ RMgX + CdCl_2 \rightarrow R_2Cd + 2\ Mg(X)Cl$

Dialkylcadmium reagents are used for preparation of ketones from acyl halides:

$2\ R'C(O)Cl + R_2Cd \rightarrow 2\ R'C(O)R + CdCl_2$

V. Carbon–carbon coupling reactions

A Grignard reagent can also participate in coupling reactions. For example, nonylmagnesium bromide reacts with methyl *p*-chlorobenzoate to give *p*-nonylbenzoic acid, in the presence of Tris(acetylacetonato)iron(III) ($Fe(acac)_3$), after workup with NaOH to hydrolyze the ester, shown as follows. Without the $Fe(acac)_3$, the Grignard reagent would attack the ester group over the aryl halide.

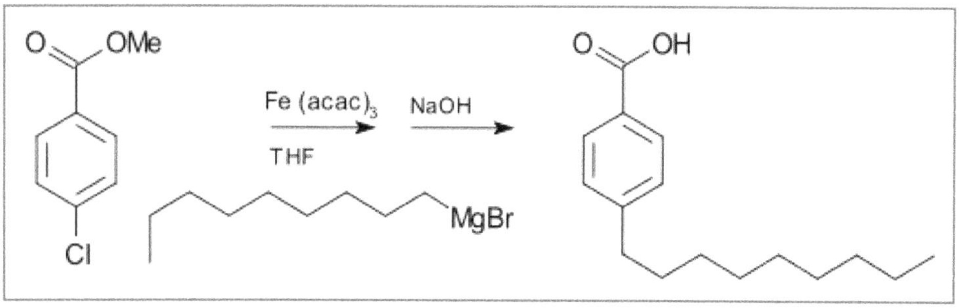

For the coupling of aryl halides with aryl Grignards, nickel chloride in tetrahydrofuran (THF) is also a good catalyst.

VI. Oxidation

Treatment of a Grignard reagent with oxygen gives the magnesium organoperoxide. Hydrolysis of this material yields hydroperoxides or alcohol. These reactions involve radical intermediates.

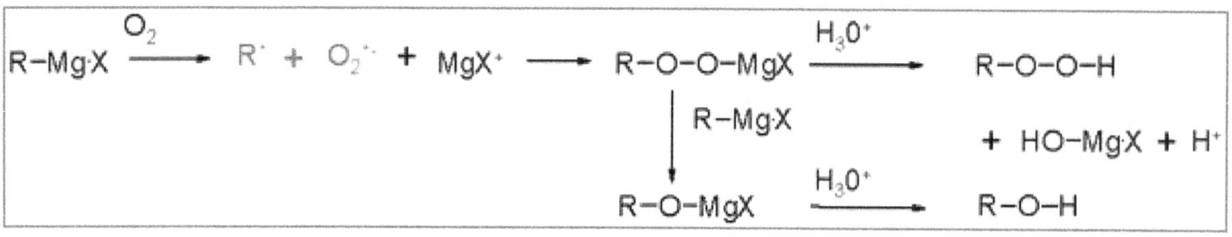

The simple oxidation of Grignard reagents to give alcohols is of little practical import as yields are generally poor. In contrast, two-step sequence via a borane that is subsequently oxidized to the alcohol with hydrogen peroxide is of synthetic utility.

3.4 ORGANOMETALLICS OF GROUP IIIA ELEMENTS:Organoaluminium compounds

The chemistry of the organometallics of the main group III elements (B→Al) particularly that of boron and aluminium has been the subject of intensive research during the past 3 to 4 decades.

In view of their ns^2 np^1 electron configuration in the valence shell, all these elements are predominantly trivalent and favour the formation of three normal σ (two-electron two-centre) bonds in typical MR_3 compounds. However, the availability of only three electrons to contribute to covalent bonding involving the four orbitals- s, p_x, p_y and p_z in organometallics of the type $(MR_3)_2$ renders their compounds 'electron deficient'. These, therefore, show a marked tendency to increase their covalency from three to four and thus, attain the effective atomic number of the nearest noble gas. Their strong electron pair acceptor behaviour (Lewis acidity) and high affinity for oxygen and nitrogen (hard bases) and even for carbanions lead to formation of a large number of compounds (including adducts) in different geometries.

The formation of $[MR_2]^+$ cations, isoelectronic with main group II, can be expected for main group III elements. Such cations are particularly stable in the case of Tl. The tendency to form $[MR_2]^+$ cationd and their stability increase from B→Tl. Although aluminium in purely inorganic compounds can utilize 3d-orbital in achieving higher coordination numbers (e.g., AlH_3, $[AlF_6]^{3-}$, $Al(acac)_3$) yet coordination numbers higher than four are extremely rare for aluminium. Two examples of pentacoordinated aluminium are shown below,

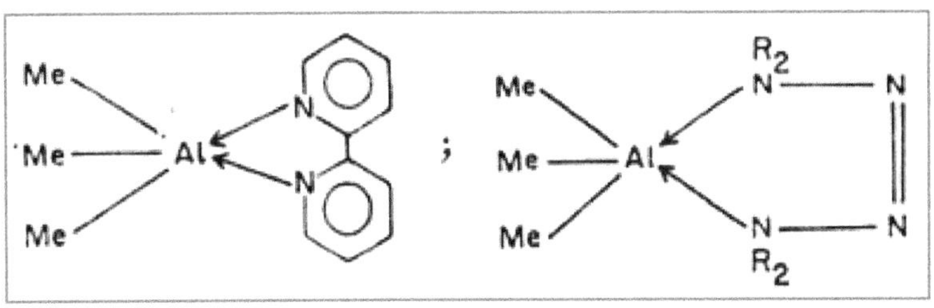

The element-carbon bond distance increases from boron to thallium and consequently element-carbon bond energies in these trialkyls fall with increasing atomic numbers. The same sequence, B > Al > Ga > In > Tl is followed in the thermal stability of the trialkyls. The electron deficient character of these trialkyls could be expected to lead to their dimerisation, but the reasons for this being observed only in the case of some aluminium trialkyls.

Organoaluminium compounds find unique applications in organic as well as organometallic synthesis and coordination catalysts for alkene polymerization and other reactions. The properties of organoaluminium compounds that make them a versatile reagent are the dynamic nature of Al-C bond; their alkylating ability and tendency to reduce certain transition metal salts; their ready reactions with oxygen and carbonyl containing compounds; their tendency to form bridged complexes containing other metals; and various reactions they undergo with alkenes.

Preparation of organoaluminium compounds

Synthetic routes to other group III organometallics can be represented by the following typical reactions.

$$AlX_3 + 3Mg(R)X \xrightarrow[-3MgX_2]{Et_2O} AlR_3 \cdot Et_2O \xrightarrow[-Et_2O]{heat} AlR_3$$

$$AlCl_3 + 3LiR \xrightarrow{heptane} AlR_3 + 3LiCl$$

Structure and bonding of organoaluminium compounds

Trimethylaluminium is known to be dimeric in solid state whereas trimethylindium and trimethylthalium are tetrameric. In its dimerisation tendency, trimethylaluminium shows a similarity to aluminium trichloride and aluminium tertiary butoxide. However, in both the latter compounds lone pair of electrons are available on the halogen or the alkoxy oxygen atoms giving rise to donor-acceptor linkages of the type:

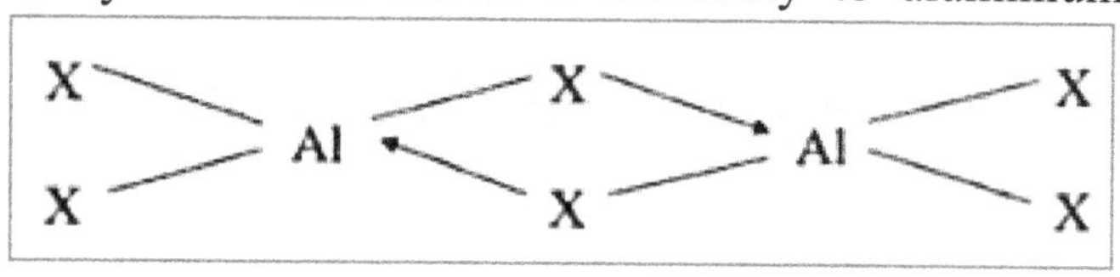

The bonding in $Al_2(CH_3)_2$ closely follows those of diborane, and suggestion has been put forward that linkages probably involve hydrogen bonding, or resonance as shown in the structures given below,

An alternative structure which may be referred to as a "methylated double bond" was by analogy with the as suggest protonated double bond description of diborane. The $[CH_3]^+$ groups are regarded as being embedded in the π-electron charge clouds of an Al-Al double bond.

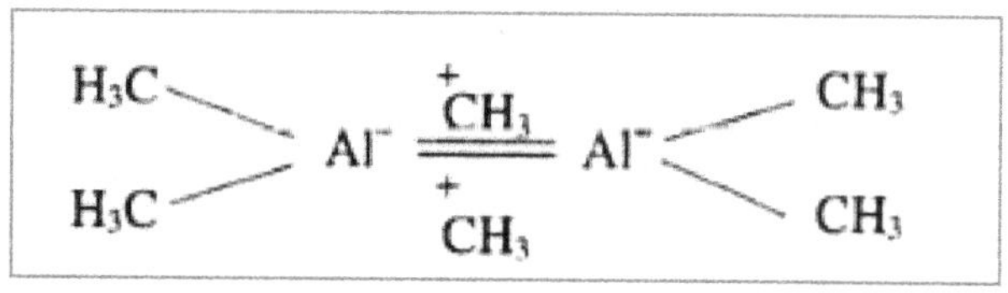

Probably the most convincing explanation for the bridge bond in the trimethylaluminium uses two-electron, three-centre bond description, as given below:

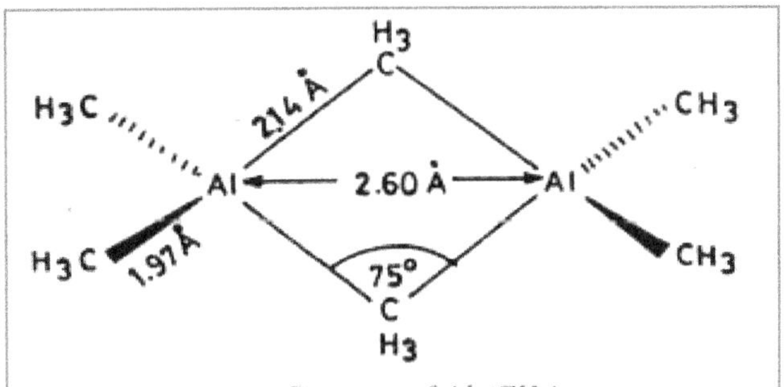

Structure of $Al_2(CH_3)_6$

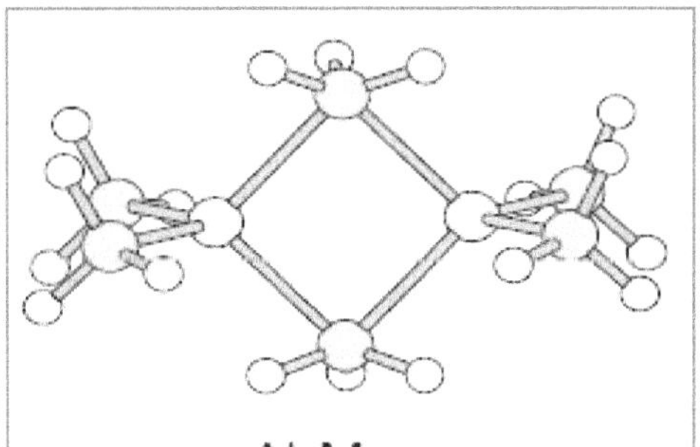

Al_2Me_6

The bonding situation in dimeric aluminium alkyls can be describes in two limiting ways,

Firstly, it is suggested that the 3s and 3p valence orbitals of aluminium are hybridized to produce four equivalent sp^3 hybrid orbitals, which could be represented schematically as,

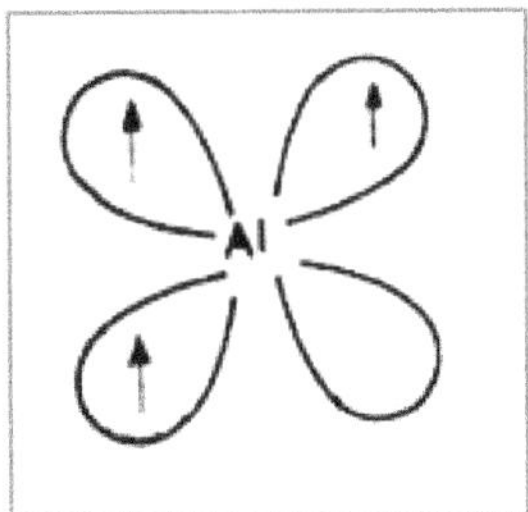

The formation of two terminal two-electron two-centre Al-CH_3 bonds in trimethylaluminium may be depicted as:

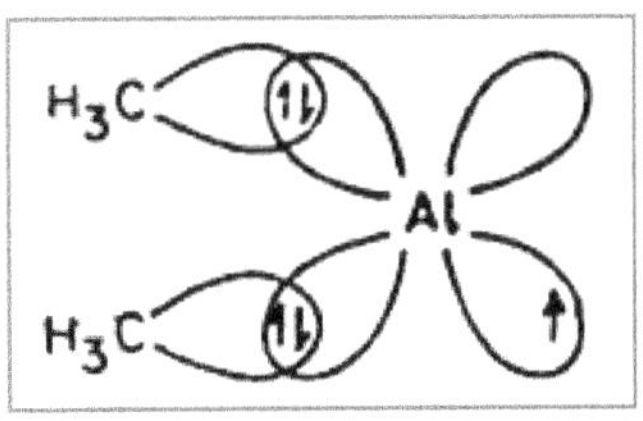

The bonding in the bridge involves bringing together two units of the above type, followed by inserting two methyl groups in between the units in such a manner as to give effective overlap between Al(sp^3), C(sp^3)and Al(sp^3) hybrid orbitals, i.e., the molecular orbital may be constructed from the sp^3 hybrid orbitals of the aluminium and carbon which will produce two bonding orbitals, two nonbonding orbitals, and two antibonding orbitals; each bonding molecular orbital encompasses two aluminium and one carbon atom, and is completely filled:

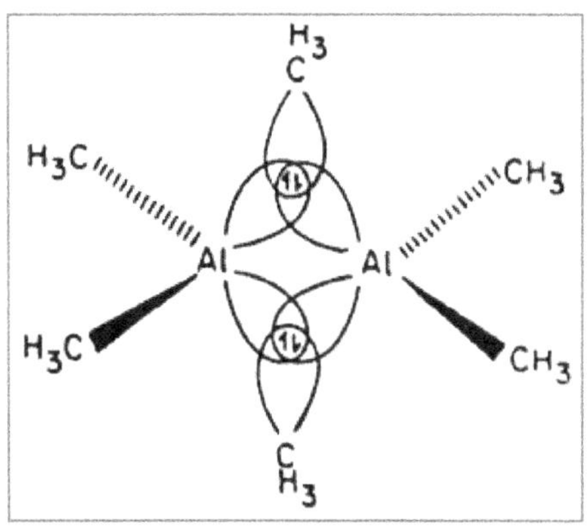

For maximum overlap of the carbon sp^3 hybrid orbital by two $Al(sp^3)$ hybrid orbitals, the Al-C-Al angle must be small.

In spite of a more general acceptance of the above mode of bonding, an alternative description has also been advanced to explain the short (260 pm) aluminium – aluminium distance.

In this, the bridge frame-work is constructed from two aluminium trigonal (sp^2) hybrid orbitals, two aluminium p atomic orbitals, and two carbon sp^3 hybrid orbitals, generating two molecular orbitals each of bonding, non-bonding and antibonding character. The terminal Al-CH_3 bonds are normal two-electron two-centre bonds resulting from the overlap of a sp^2 hybrid orbital of aluminium and a sp^3 hybrid orbital of carbon.

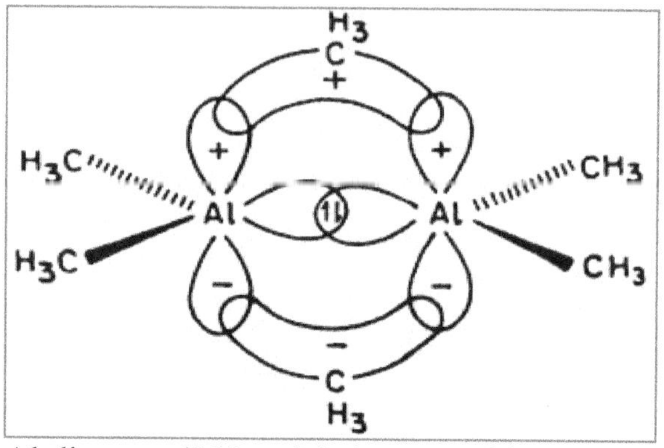

The advantage of this model is that the relatively short Al-Al distance (260 pm) is interpretable as Al-Al σ – bond and the very acute Al-C(bridge)-Al bond angle (75°) may also imply the existence of an Al-Al bond.

The acute bridging angle is a characteristic of organobridged electron deficient compounds of aluminium, beryllium and magnesium, and the relatively wide external C-Al-C andgle (123°) is a consequence of hybridisation of aluminium.

Reactivity of organoaluminium compounds

Organoaluminium compounds find unique applications in organic as well as organometallic synthesis and coordination catalysts for alkene polymerization and other reactions. The properties of organoaluminium compounds that make them a versatile reagent are the dynamic nature of Al-C bond; their alkylating ability and tendency to reduce certain transition metal salts; their ready reactions with oxygen and carbonyl containing compounds; their tendency to form bridged complexes containing other metals; and various reactions they undergo with alkenes.

The most important reaction of aluminum alkyls is in the polymerization of alkenes by the Ziegler-Natta process. In that process, $Al(C_2H_5)_3$ reacts as alkylating agent with $TiCl_4$, which then undergoes a reaction, in which an alkene molecule is inserted in the bond between Ti and the ethyl group, thus lengthening the chain. Ziegler and Natta shared the 1963 Nobel Prize for their enormously important work.

The mechanism involves the replacement of a chloride ion by an ethyl group or an ethyl group attaching at a vacant site as the initial step. After a molecule of ethylene is coordinated to the titanium, it undergoes migration and insertion in the bond between Ti and C_2H_5. Another ethylene molecule becomes coordinated at the vacant site in the coordination sphere of the Ti, and another insertion reaction occurs, which lengthens the chain, and so on. The reaction scheme shown below illustrates this process. The crucial step in the process is the insertion reaction by which the hydrocarbon chain is lengthened.

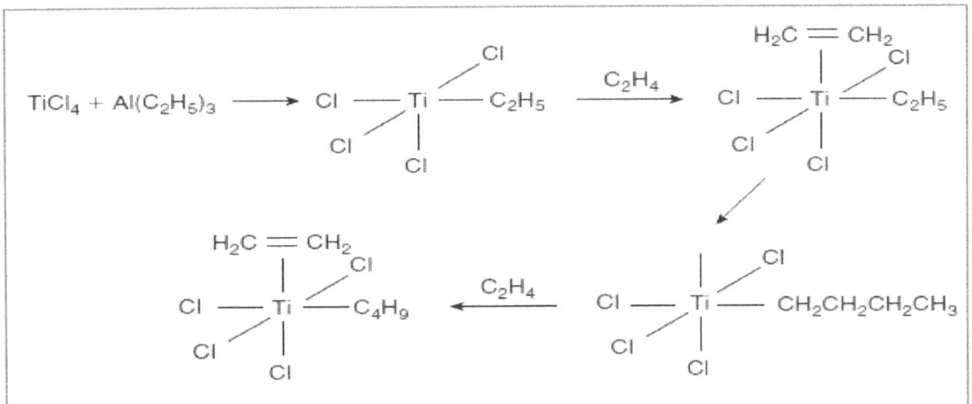

The catalytic scheme for polymerization by the Ziegler-Natta process.

Some of the general reactions of trialkyls of aluminium are summarized as follows,

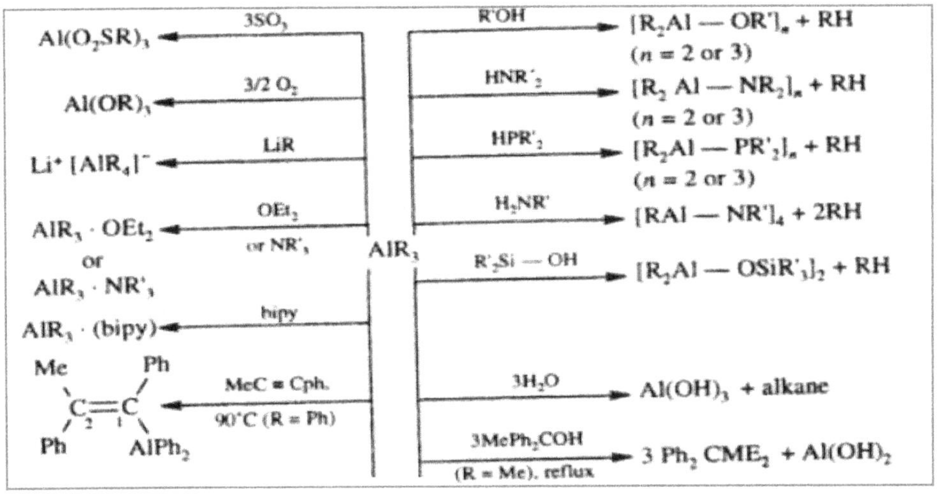

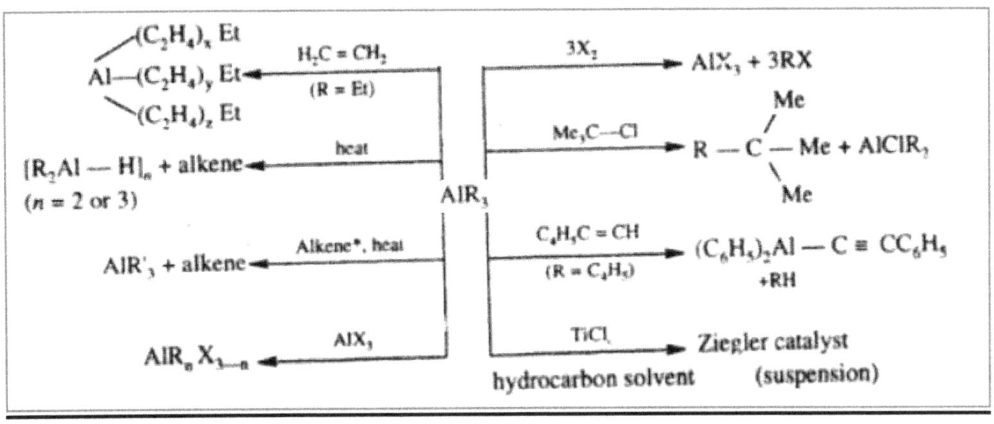

3.5 ORGANOMETALLICS OF GROUP IVA ELEMENTS: Organotin compounds

The synthetic reactions employed for organotin compounds are also of the same pattern as those described in the previous section for elements of group II and III,

$$SnCl_4 + 4MR \longrightarrow SnR_4 + 4MCl$$

(M = Li, Na, K, MgX, Zn, or Al, R = alkyl, aryl, C_5H_5 or C_5H_4Me)

$$3SnCl_4 + 4AlR_3 + 4NaCl \longrightarrow 3SnR_4 + 4AlCl_3 \cdot NaCl$$

Structure, bonding and reactivity of organotin compounds

Group IV elements have ns^2 np^2 electronic configuration in their valence shell and therefore, two oxidation states, i.e., +4 and +2 (due to inert s-pair effect) are possible. The extraordinary stability of organoderivatives of these elements in tetravalent state is reflected in their vast applications. The tetra-alkyl and tetra-aryl compounds of main group IV elements differ from the corresponding derivatives of the elements in neighbouring groups because of their low reactivity.

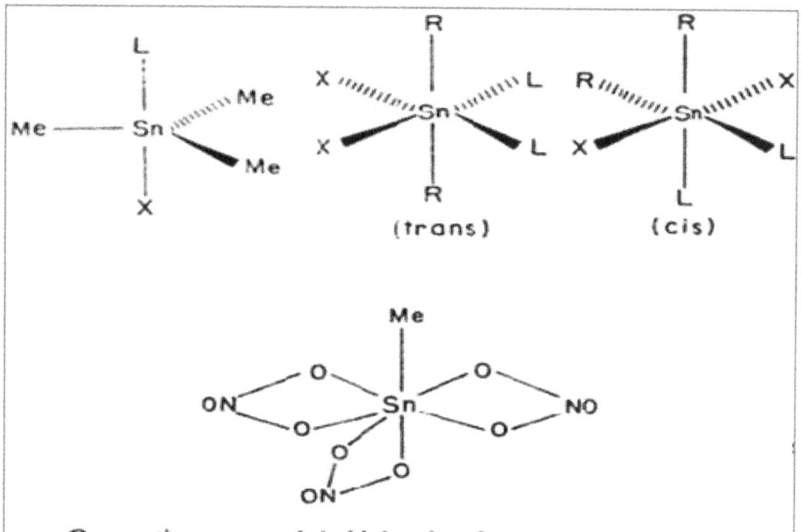

Organo-tin compounds in higher than four coordination numbers.

Within main group IV, the reactivity of the M-C bond in tetra-alkyls or tetra-aryls (MR_4) increases progressively from silicon to lead as the,

i) bond energy decreases in the sequence

$Si(C_2H_5)_4$ $Ge(C_2H_5)_4$ $Sn(C_2H_5)_4$ $Pb(C_2H_5)_4$

decreasing M—C bond energy →

decreasing thermal stability →

ii) As expansion of the coordination number of the metal (M) becomes easier with increasing atomic size and decreasing difference between np and nd orbitals.

General reactions of organo-tin compounds

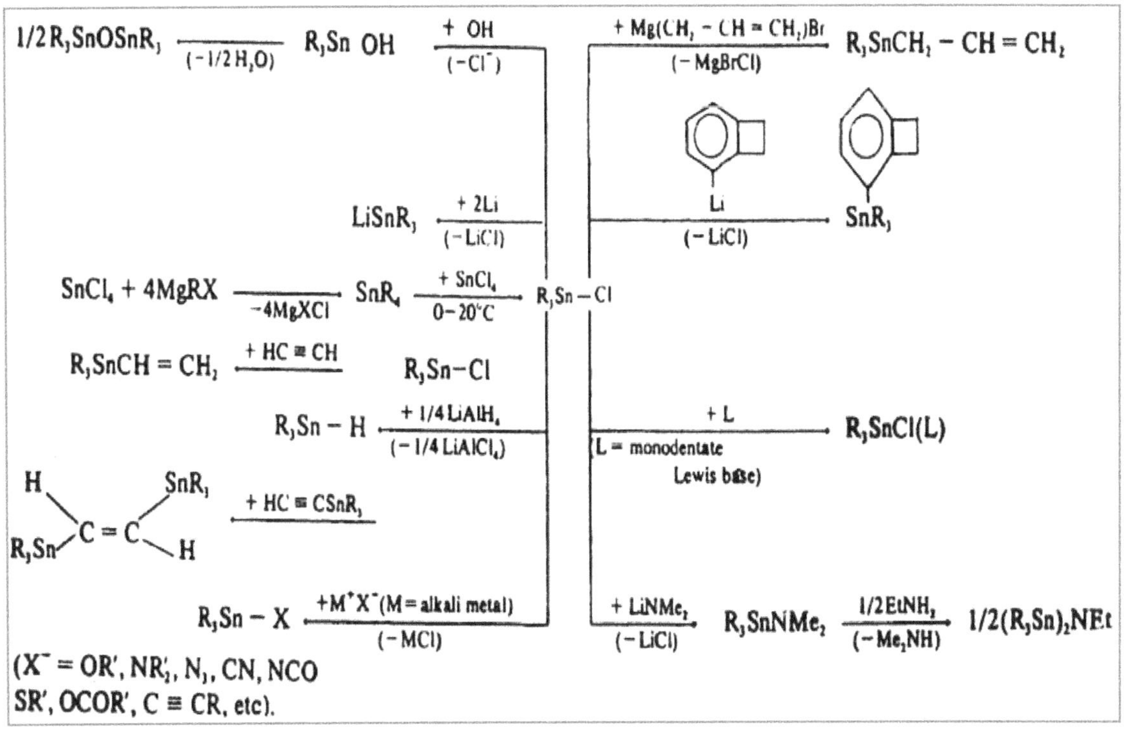

3.6 TRANSITION METAL ORGANOMETALLICS

Organometallic compounds, with their metal–carbon bonds (e.g., WMe_6), lie at the interface between classical organic and inorganic chemistry in dealing with the interaction between inorganic metal species and organic molecules. In the related metal–organic compound area, in contrast, the organic fragment is bound only by metal–heteroatom bonds [e.g., $Ti(OMe)_4$].

The organometallic field has provided a series of important conceptual insights, surprising structures, and useful catalysts both for industrial processes and for organic synthesis. The field is beginning to make links with biochemistry with the discovery of enzymes that carry out organometallic catalysis (e.g., acetyl CoA synthase). Ideas drawn from organometallic chemistry have helped interpret the chemistry of metal and metal oxide surfaces, both key actors in heterogeneous catalysis. The field is also creating links with the chemistry of materials because organometallic and metal–organic compounds are increasingly preferred as the precursors for depositing materials on various substrates via thermal decomposition of the metal compound. Nanoscience and nanotechnology are also benefiting with the use of such compounds as the most common precursors for nanoparticles. These small particles of a metal or alloy, with properties quite unlike the bulk material, are finding more and more useful applications in electronic, magnetic, or optical devices or in sensors.

The presence of d electrons in their valence shell distinguishes the organometallic chemistry of the elements of groups 3–12 of the periodic table, the transition elements, from that of groups 1–2 and 12–18, the main-group elements. Group 12, and to some extent also group 3, often show greater resemblance to the main-group elements.

Transition metal ions can bind *ligands* (L) to give a coordination compound, or *complex* MLn, as in the familiar aqua ions $[M(OH_2)_6]^{2+}$ (M = V, Cr,Mn, Fe,Co, or Ni). Organometallic chemistry is a subfield of coordination chemistry in which the complex contains an M–CorM–H bond [e.g., $Mo(CO)_6$]. Organometallic species tend to be more covalent, and the metal is often more reduced, than in other coordination compounds. Typical ligands that usually bind to metals in their lower oxidation states are CO, alkenes, and arenes, for example, $Mo(CO)_6$, $(C_6H_6)Cr(CO)_3$,or $Pt(C_2H_4)_3$.

The full variety of organometallic chemistry presents itself in the realm of transition metals. Contributing factors include:

- For transition metals, the *(n-1)d, ns* and *np* orbitals must all be regarded as regular valence orbitals.
- Partial occupation of these orbitals gives transition metals both electron-donor and electron-acceptor properties (σ -donor/ π -acceptor synergism).
- Various forms of metal-metal multiple bonding with or without additional bridging ligands
- The ability to change the coordination number together with the lability of metal-carbon σ -bonds offers possibilities for organometallic catalysis.

Organometallic chemistry is concerned with the metal–carbon bond, of which the simplest is the M–C single bond of metal alkyls. As σ-bonding ligands, alkyls are closely related to the ligands found in coordination compounds, such as Cl, H_2O, and NH_3. A larger class of organometallic ligands (CO, C_2H_4) are soft and can π bond. The structures of some typical organometallic compounds show many examples of such π-bonding ligands as butadiene, benzene, cyclopentadienyl (C_5H_5 or Cp), and allyl. There are several differences between complexes of these ligands and coordination compounds containing Cl^-, H_2O, and NH_3. The metals are more electron rich, in the sense that the metal bears a greater negative charge in the organometallic complex.

The M–L bonds are much more covalent and often have a substantial π component. The metal d orbitals are higher in energy and by back donation perturb the electronic structure of the ligands much more than are the case for coordination compounds. The organometallic ligands can be polarized and therefore activated toward chemical reactions, σ and π bonds in the ligands can be weakened or broken, and chemical bonds can be made or broken within and between different ligands. This rich pattern of reactions is characteristic of organometallic chemistry.

3.7 TYPES OF LIGANDS

Ligands, Bonding Types, Charges, and Electron Donor Numbers

Ligand Name	Bonding Type	Formal Charge	Electrons Donated
Molecular Hydrogen: H_2		0	2
Hydride: H^-	M-H	-1	2
Hydride: H^-	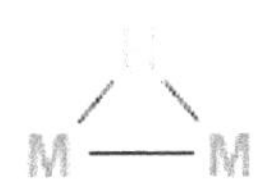	-1	2
Halide: X^-	M-X	-1	2
Halide: X^- μ bridging		-1	4 (2 to each M)
Halide: X^- μ_3 bridging	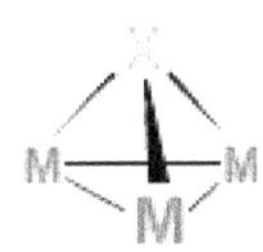	-1	6 (2 to each M)
Alkoxide, thiolate: OR^-, SR^-	M- OR	-1	2 or 4
Alkoxide, thiolate: OR^-, SR^-	M- OR	-1	2 or 4
Alkoxide, thiolate: OR^-, SR^- μ bridging	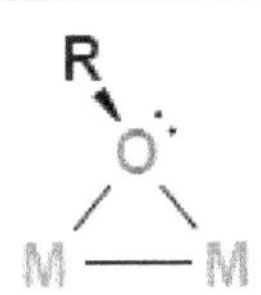	-1	4 (2 to each M)
Alkoxide, thiolate: OR^-, SR^- μ_3 bridging	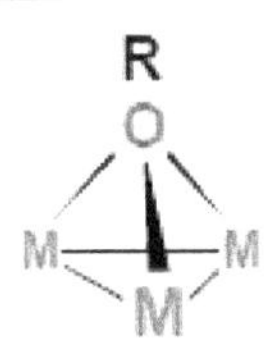	-1	6 (2 to each M)
Peroxide: O_2^{2-} η^2	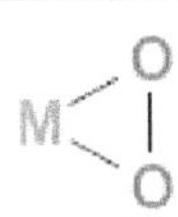	-2	4
Peroxide: O_2^{2-} μ_2 bridging	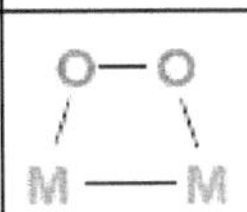 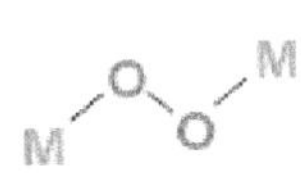	-2	4 (2 to each M)

Oxide, sulfide: O^{2-}, S^{2-}	M=O: M≡O:	-2	4 6
Oxide, sulfide: O^{2-}, S^{2-} μ bridging	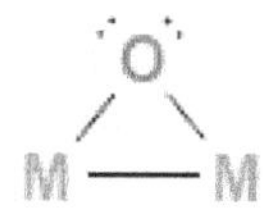	-2	4 (2 to each M)
Oxide, sulfide: O^{2-}, S^{2-} μ_3 bridging	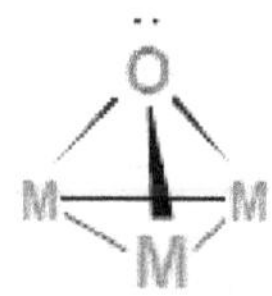	-2	6 (2 to each M)
Amine, phosphine, arsine: NR_3, PR_3, AsR_3	M-NR_3 M-PR_3	0	2
Amide, phosphide: NR_2^-, PR_2^-	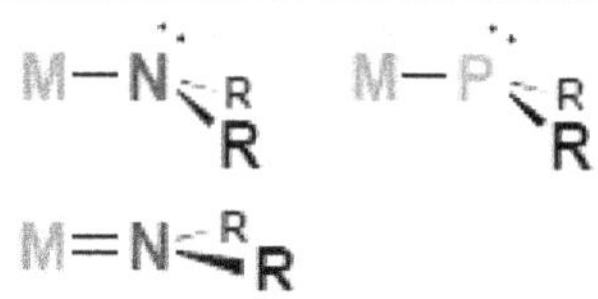	-1	2 4
Amide, phosphide: NR_2^-, PR_2^- μ bridging:	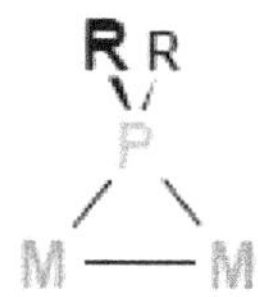	-1	4 (2 to each M)
Imido: NR^{2-}	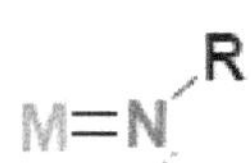 M≡N-R	-2	4 (bent) 6 (linear)
Imido: NR^{2-} μ bridging	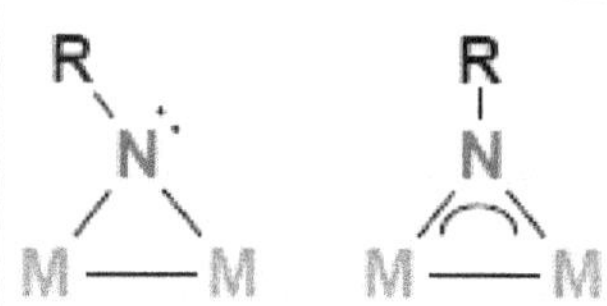	-2	4 or 6 (2 or 3 to each M)
Imido: NR^{2-} μ_3 bridging	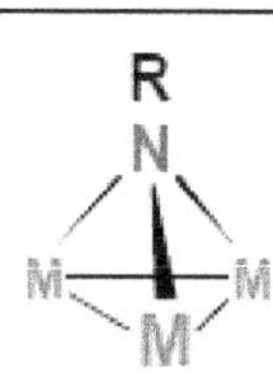	-2	6 (2 to each M)
Nitride, phosphide: N^{3-}, P^{3-}	M≡N:	-3	6

Nitride, phosphide: N^{3-}, P^{3-}, μ bridging	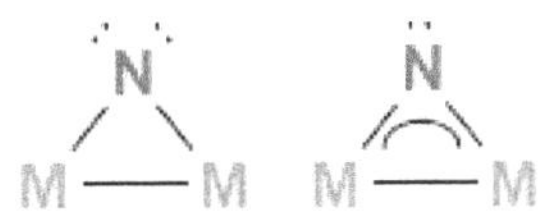	-3	4 or 6 (2 or 3 to each M)
Nitride, phosphide: N^{3-}, P^{3-}, μ_3 bridging		-3	6
Nitrosyl: $N{\equiv}O^{+}$ *linear form*	M−N≡O	+1	2
Nitrosyl: $N{\equiv}O^{-}$ *bent form*	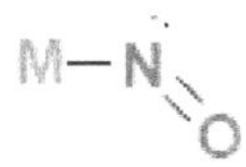	-1	2
Nitrosyl: $N{\equiv}O^{+}$ μ bridging	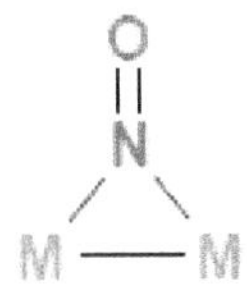	+1	1 to each metal
Nitrosyl: $N{\equiv}O^{+}$ μ_3 bridging	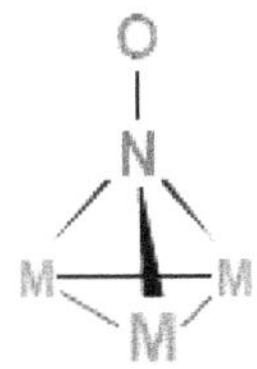	+1	1 to each metal
Carbonyl: C≡O	M−C≡O	0	2
Carbonyl: C≡O μ,η^2 mode	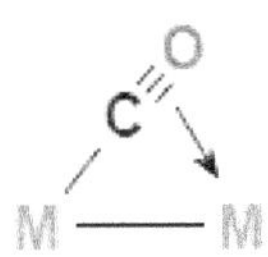	0	2 to each metal
Carbonyl: C≡O μ bridging	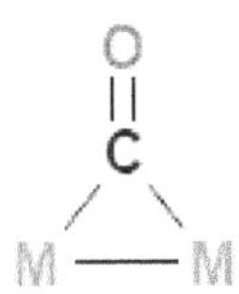	0	1 to each metal
Carbonyl: C≡O μ_3 bridging	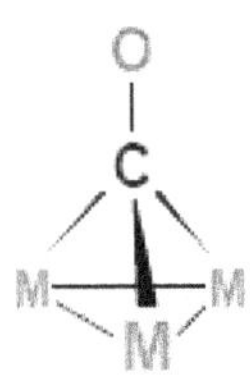	0	1 to each metal

Isocyanide: C≡N-R	M−C≡N−R	0	2
Isocyanide: C≡N-R μ bridging	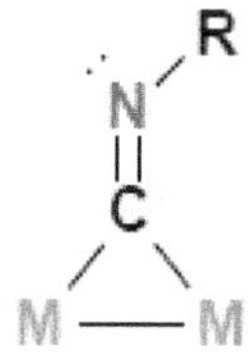	0	1 to each metal
Alkene: $R_2C{=}CR_2$ η^2 bonding mode	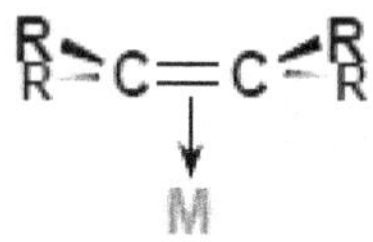	0	2
Alkyne: RC≡CR η^2 bonding mode	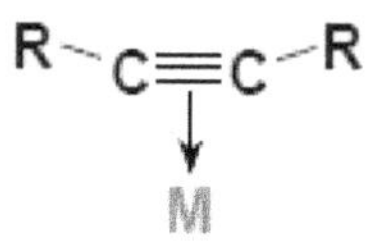	0	2 or 4
Alkyne: RC≡CR μ bridging	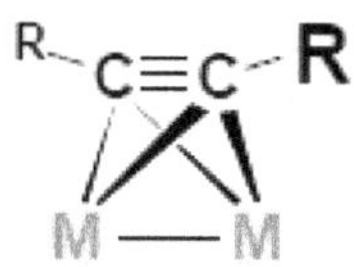	0	2 to each metal
Dienes: $R_2C{=}CH{-}CH{=}CR_2$ η^4 bonding mode		0	4
Dienes: $R_2C{=}CH{-}CH{=}CR_2$ μ bridging	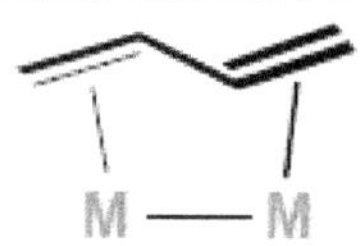	0	2 to each metal
Benzene: C_6H_6 η^6 bonding mode	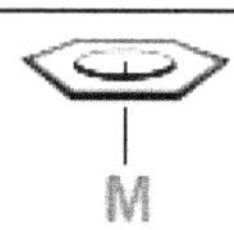	0	6
Ketones, aldehydes: η^2 bonding mode	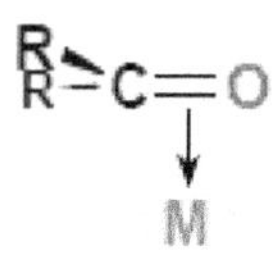	0	2
Alkyl: CH_3^-, CH_2R^- η^1 bonding mode	$M{-}CR_3$	-1	2

Alkyl: CH_3^-, CH_2R^-	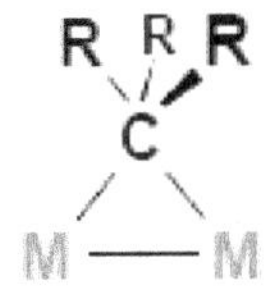	-1	2 (1 to each M) *assuming no agostic M-H interactions*
Aryl: $C_6H_5^-$ η^1 terminal	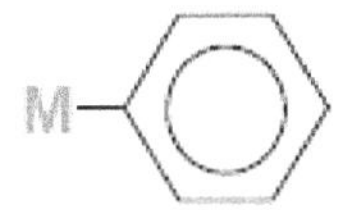	-1	2
Alkenyl (vinyl), alkynyl: $CH=CH_2^-$, $C\equiv CH^-$ η^1 terminal	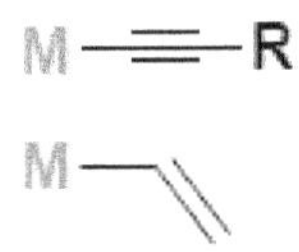	-1	2
Alkenyl (vinyl): $CH=CH_2^-$ η^2 terminal		-1	4
Alkenyl (vinyl), alkynyl: $CH=CH_2^-$, $C\equiv CH^-$ μ,η^2 terminal	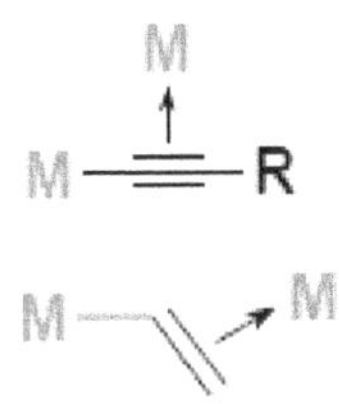	-1	4 (2 to each M)
Cyclopropenyl: $C_3H_3^-$ η^3 terminal	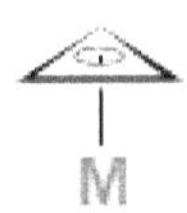	-1	4
Allyl: $C_3H_5^-$ η^3 terminal		-1	4
Allyl: $C_3H_5^-$ η^1 terminal	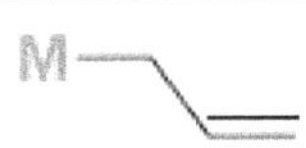	-1	2
Pentadienyl: $C_5H_7^-$ η^5 terminal		-1	6
Pentadienyl: $C_5H_7^-$ η^3 terminal		-1	4

Pentadienyl: $C_5H_7^-$ **η^1 terminal**		-1	2
Cyclopentadienyl: $C_5H_5^-$ **η^5**	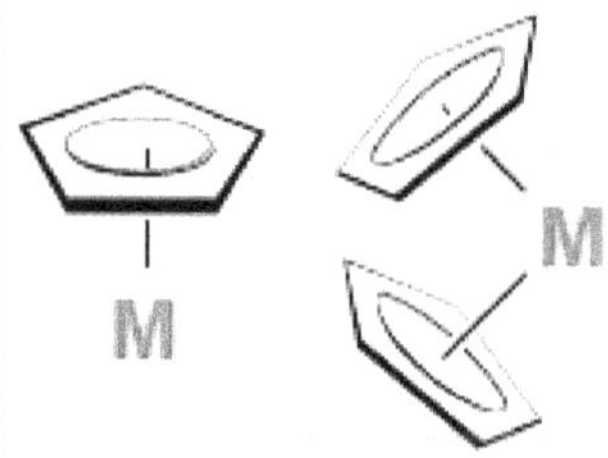	-1	6
Cyclopentadienyl: $C_5H_5^-$ **η^3**	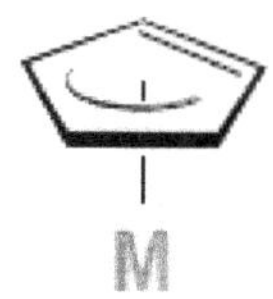	-1	4
Carbene: =CYR *where Y is a substituent capable of π interaction with the carbene carbon atom*	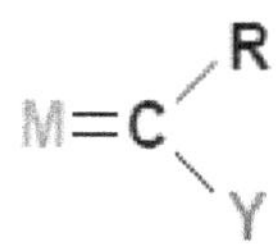	0	2
Alkylidene: $=CR_2^{2-}$ *where no substituents capable of π bonding to the carbene carbon atom are present*	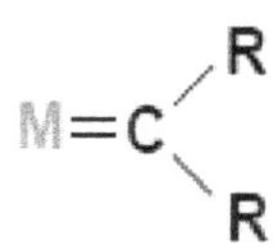	-2	4
Carbene: =CYR or $=CR_2^{2-}$ **μ bridging**	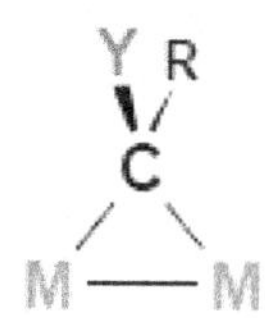	-2	2 to each metal
Carbyne or Alkyidyne: **$\equiv CR^{3-}$**	$M\equiv C-R$	-3	6
Carbyne or Alkyidyne: **$\equiv CR^{3-}$ μ_3 bridging**	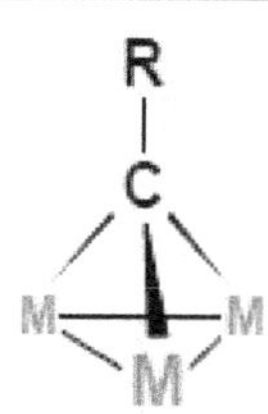	-3	2 to each metal

3.8 18 ELECTRON RULE (18e rule)

The 18 electron rule is a way to help us decide whether a given d-block transition metal organometallic complex is likely to be stable. Stable compounds, such as CH_4, have the noble gas octet, and so carbon can be thought of as following an 8e rule. This corresponds to carbon using its s and three p orbitals to form four filled bonding orbitals and four unfilled antibonding orbitals. On the covalent model, we can consider that of the eight electrons required to fill the bonding orbitals, four come from carbon and one each comes from the four H substituents. We can therefore think of each H atom as being a 1e ligand to carbon.

To assign a formal oxidation state to carbon in an organic molecule, we impose an ionic model by artificially dissecting it into ions. Each electron pair in any bond is assigned to the most electronegative of the two atoms or groups that constitute the bond. For methane, this dissection gives C^{4-}+$4H^+$, with carbon as the more electronegative element. This makes methane an 8e compound with an oxidation state of −4, usually written C (-IV). Note that the net electron count always remains the same, whether we adopt the covalent (4e {Catom}+ 4 × 1e {4H atoms}=8e) or ionic (8e{C^{4-}ion}+ 4 × 0e{$4H^+$ions}=8e) model. The 18e rule, which applies to many low-valent transition metal complexes, follows a similar line of reasoning. The metal now has one s,and three p orbitals, as before, but now also five d orbitals. We need 18e to fill all nine orbitals; some come from the metal, the rest from the ligands. Only a limited number of combinations of metal and ligand give an 18e count.

The 18e rule: characterizing and predicting the stability of metal complexes.

The rule predicts the stability of organometallic compounds and simply states that stability is attained or gained when all of the bonding and non-bonding orbitals are occupied. Or

The sum of the metal d electrons plus the electrons of conventionally regarded as being supplied by the ligands equals 18.

Valence shells of a transition metal can accommodate **18** electrons: 2 in each of the five ***d*** orbitals (10 in total); 2 in each of the three ***p*** orbitals (6 in total); and 2 in the ***s*** orbital.

Electron counting is the process of determining the number of valence electrons about a metal center in a given transition metal complex.

To determine the electron count for a metal complex:

1) determine the oxidation state of the transition metal center(s) and the metal centers resulting *d*-electron count. To do this one must:

a) note any overall charge on the metal complex

b) know the charges of the ligands bound to the metal center (ionic ligand method)

c) know the number of electrons being donated to the metal center from each ligand (ionic ligand method)

2) add up the electron counts for the metal center and ligands

Complexes with 18 e- counts are referred to as *saturated*, because there are no empty low-lying orbitals to which another incoming ligand can coordinate. Complexes with counts lower than 18e- are called *unsaturated* and can electronically bind additional ligands. The vast majority of stable diamagnetic organometallic compounds have 16 or 18 valence electrons due to the presence of the five *d* orbitals which can hold 10 more electrons relative to C, O, N, etc.

Cationic 2e- donor:	**NO^+ (nitrosyl)**
Neutral 2e- donors:	**PR_3 (phosphines), CO (carbonyl), $R_2C{=}CR_2$ (alkenes), $RC{\equiv}CR$ (alkynes, can also donate 4 e-), $N{\equiv}CR$ (nitriles)**
Anionic 2e- donors:	**Cl^- (chloride), Br^- (bromide), I^- (iodide), CH_3^- (methyl), CR_3^- (alkyl), Ph^- (phenyl), H^- (hydride)** *The following can also donate 4 e- if needed, but initially count them as 2e- donors (unless they are acting as bridging ligands):* **OR^- (alkoxide), SR^- (thiolate), NR_2^- (inorganic amide), PR_2^- (phosphide)**
Anionic 4e- donors:	**$C_3H_5^-$ (allyl), O^{2-} (oxide), S^{2-} (sulfide), NR^{2-} (imido), CR_2^{2-} (alkylidene)** *and from the previous list:* **OR^- (alkoxide), SR^- (thiolate), NR_2^- (inorganic amide), PR_2^-**
Anionic 6e- donors:	**Cp^- (cyclopentadienyl), N^{3-} (nitride)**

Examples:

- **$Fe(CO)_5$** *$3s^23p^63d^64s^2$*

Neutral counting: Fe 8e, CO 2e, 8 + 2(5) = 18 valence electrons

Conclusion: This is a special case, all fragments being neutral. Since this is an 18-electron complex, $Fe(CO)_5$ is a stable compound.

- **$Fe(C_5H_5)_2$, $FeCp_2$**

Neutral counting: Fe 8e, C_5H_5 5e: 8 + 2(5) = 18 electrons

Ionic counting: Fe^{2+} 6e, $C_5H_5^-$ 6e: 6 + 2(6) = 18 electrons

Conclusion: Ferrocene is expected to be a stable compound.

○ **$[CpFe(CO)_2]_2$**

Neutral counting:

Cp 5 + Fe 8 + 2 CO 4 + Fe-Fe 1 = **18**

Counting electrons for just one iron center can be done by considering the other iron as contributing 1 electron to the count.

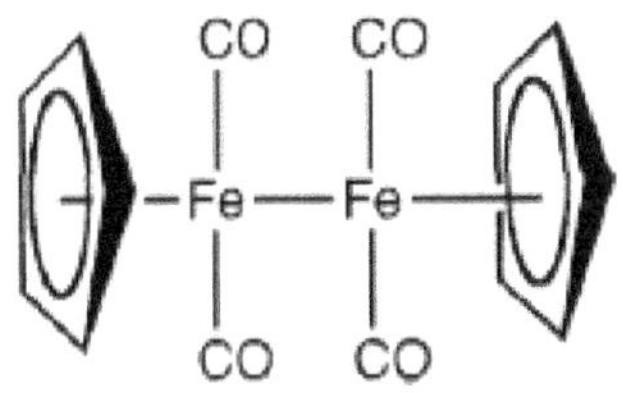

○ **$CpFe(CH_3)(CO)_2$**

Neutral counting:

Cp 5 + Fe 8 + CH_3 1 + 2 CO 4 = **18**

○ **$Fe_2(CO)_9$**

Neutral counting:

3 CO 6 + 3 μ-CO 3 + Fe 8 + Fe-Fe 1 = **18**

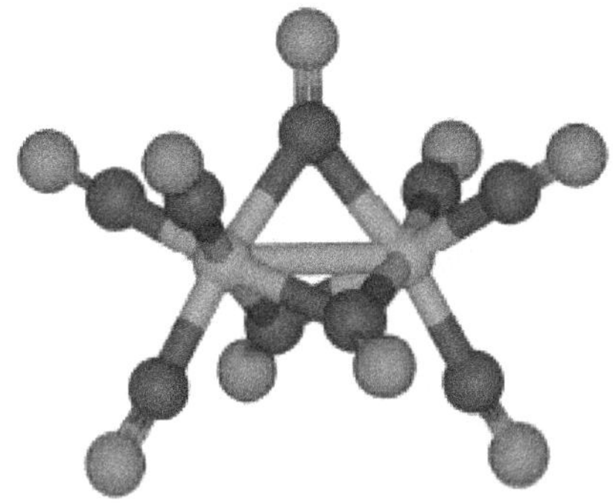

tri-μ-carbonyl-bis(tricarbonyliron)(*Fe-Fe*)

Carbonyl complexes and the 18-electron rule

To obey the 18-electron rule, many carbonyl complexes are anions or cations, as in:

$[V(CO)_6]^-$	$[Mn(CO)_6]^+$	$[Fe(CO)_4]^{2-}$
V(0) = d^5	Mn(0) = d^7	Fe(0) = d^8
6 CO = 12*e*	6 CO = 12*e*	4 CO = 8*e*
1- = +1*e*	1+ = -1*e*	2- = 2*e*
= 18e	= 18 e	= 18e
Formal oxidation state = V(-I)	Formal oxidation state = Mn(I)	Formal oxidation state = Fe(-II)

[NOTE: In applying the 18-electron rule, metal ions are always considered to be zero-valent, not the formal oxidn. state]

Exceptions to the 18e "Rule"

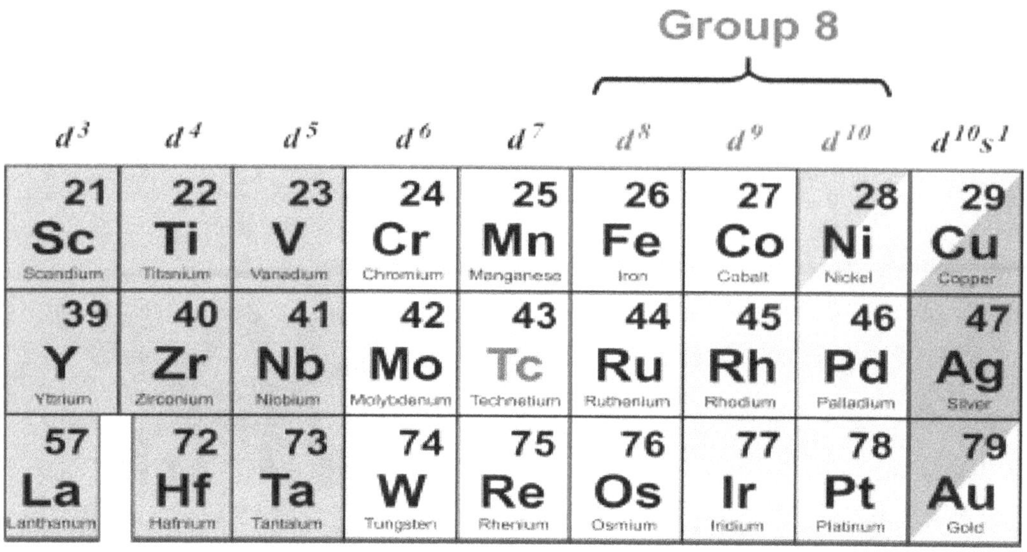

Early Transition Metals	Middle Transition Metals	Late Transition Metals
16e- and sub-16e- configurations are common	18e- configurations are common	16e- and sub-16e- configurations are common
Coordination geometries higher than 6 relatively common	Coordination geometries of 6 are common	Coordination geometries of 5 and lower are common: d^8 = square planar

Carbonyl complexes and the 18-electron rule

$[Ni(CO)_4]$	$[Fe(CO)_5]$	$[Cr(CO)_6]$
Ni(0) = d^{10}	Fe(0) = d^8	Cr(0) = d^6
4 x CO = 8	5 x CO 10	6 x CO = 12
18 e	18e	18e

possessed by the metal plus those donated by the ligands (2 per C≡O) must total eighteen:

A species such as $[Mn(CO)_5]$ would have only 17e. The 18e rule can be obeyed by two such entities forming a Mn-Mn bond, where each Mn contributes one electron to the valence shell of the other Mn, giving the metal-metal bonded species $[(CO)_5Mn\text{-}Mn(CO)_5]$. To check on the 18e rule, we look at one metal at a time:

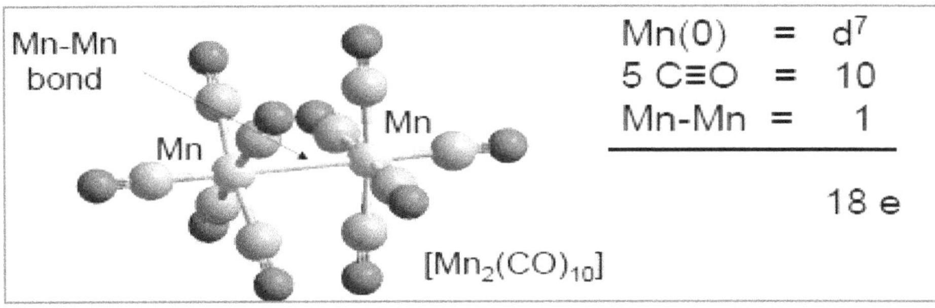

Here we see a Rhodium tetramer where each Rh forms three Rh-Rh bonds to the other Rh atoms. 18-electron rule: Focus on one Rh atom:

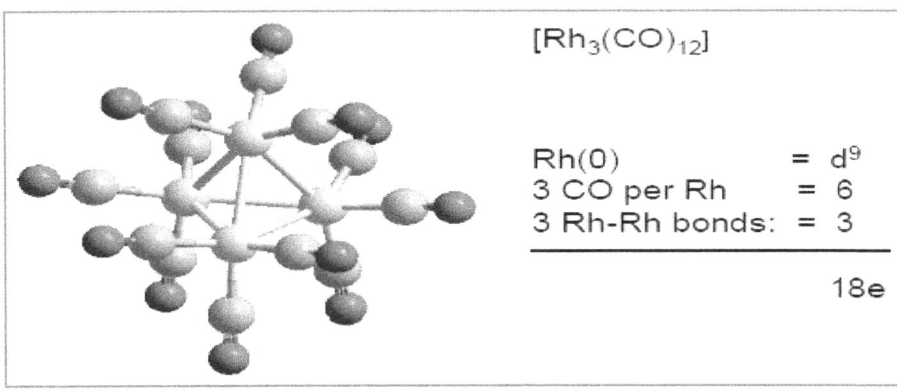

Carbonyls may form bridges between two metals, where they donate one electron to each metal in working out the 18 electron rule. In [Co2(CO)8] at left each Co has three terminal CO's, two bridging CO's, and a Co-Co bond:

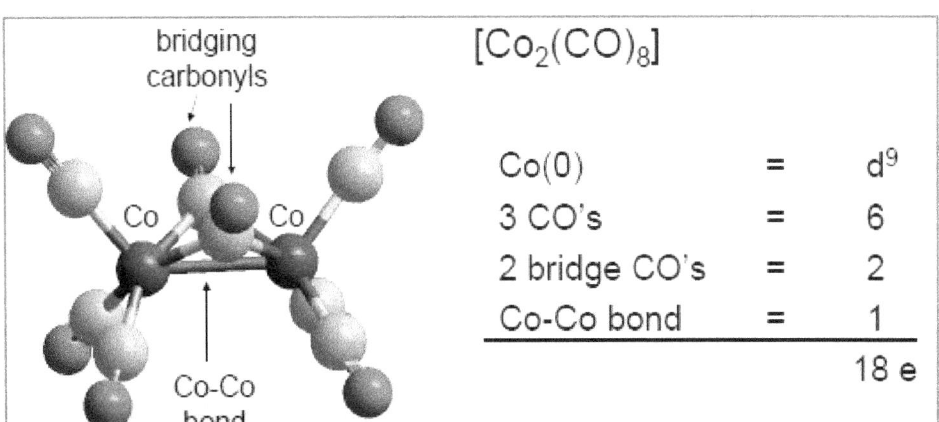

Fe2(CO)9 has each Fe with three terminal CO's, three bridging CO's, and an Fe-Fe bond. The 18 electron rule holds for each Fe atom as:

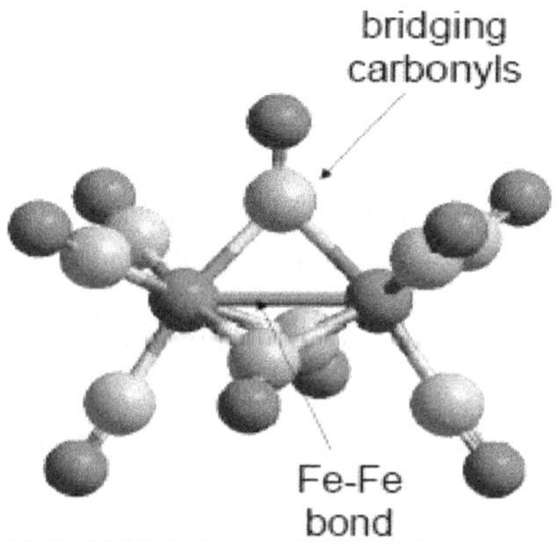

$[Fe_2(CO)_9]$

Fe(0)	=	d^8
3 CO's	=	6
3 bridge CO's	=	3
Fe-Fe bond	=	1
		18 e

Charged ligands and 18 electron rule

The formally charged ligands that are important in organometallic chemistry are mainly soft ligands such as Cl-, Br-, and I-, with CN- also occurring. Hydride (H-) is also very important. These mono-anionic ligands all contribute one electron for the 18-electron rule, as in the following examples:

$[Mn(CO)_5Cl]$			$[H_2Fe(CO)_4]$			$[HCo(CO)_4]$		
Mn(0)	=	d^7	Fe(0)	=	d^8	Co(0)	=	d^9
5 CO's	=	10	4 CO's	=	8	4 CO's	=	8
1 Cl	=	1	2 H	=	2	1 H	=	1
		18 e			18 e			18 e

In the following complex, since there is no overall charge on the complex (it is neutral), and since we have one anionic ligand present, the Re metal atom must have a +1 charge to compensate for the one negatively charged ligand. The +1 charge on the metal is also its oxidation state.

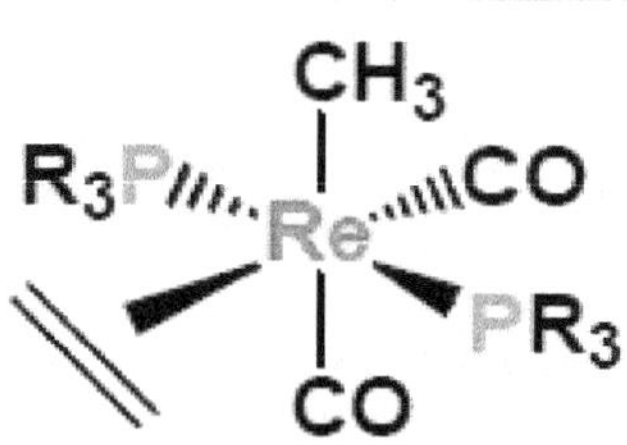

Re(+1)	d^6
2 PR_3	4e-
2 CO	4e-
CH_3^-	2e-
$CH_2=CH_2$	2e-
Total:	18e-

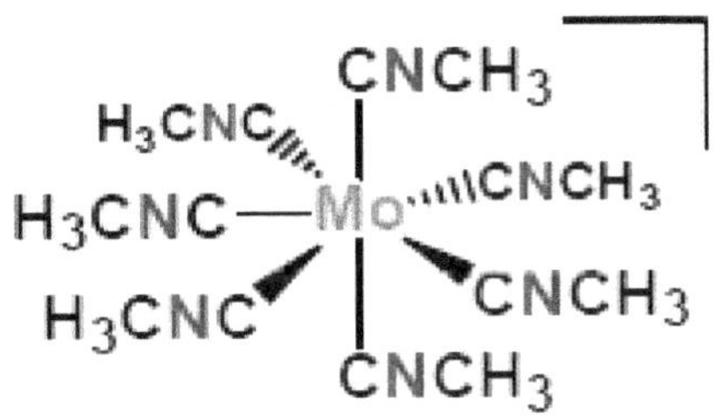

Mo(+2)	d^4
7 $CNCH_3$	14e-
Total:	18e-

Rule of Thumb: π-based carbon ligands that have *odd* number attachments to the metal center are usually *anionic* (except for CO and CNR [isocyanide] ligands) and donate one more electron than the number of carbons bonded to the metal center. So one ligand η^3-$C_3H_5^-$, allyl ligand is anionic and donates 4e- to the metal center. The other C_5H_5Me ligand is NOT a simple anionic cyclopentadienyl ligand. It is a neutral ligand with two alkenes that each act as neutral 2e- donors to the metal:

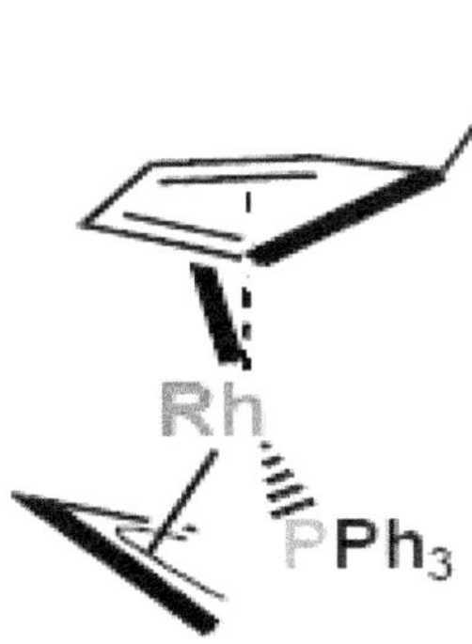

Rh(+1)	d^8
PR_3	2e-
η^4-C_5H_5Me	4e-
η^3-$C_3H_5^-$	4e-
Total:	18e-

Metal – Metal bonded examples

- Most M-M bonded complexes are symmetrical, which means that one only has to e- count one metal center. If the metals are different, however, one does need to figure out each metal center. The simple rule for M-M bonding is that if you have two metal atoms next to one another and each has an odd electron-count, you pair the odd electrons to make a M-M bond

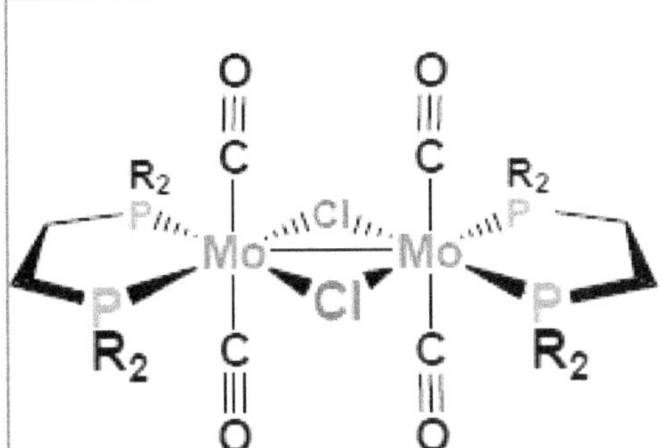

Mo(+1)	d^5
$2PR_3$	4e-
2CO	4e-
2μ-Cl^-	4e-
Sub-total:	*17e-*
Mo-Mo	1e-
TOTAL:	18e-

This is electron-counted as a 1e- donation from one metal to the other. This example also has μ-Cl ligands. Bridging ligands with at least 2 lone pairs almost always donate 2e- to each metal center.

Oxidation state determination: There are a total of two anionic ligands for two metal centers (overall complex is neutral). Thus each metal center needs to have a +1 oxidation state to balance the anionic ligands.

Note that each bridging chloride is donating 2e- to each Mo center. Although we only "allocate" one negative charge for figuring out the oxidation state of the Mo atom, one needs to include in the electron-counting the fact that both chlorides are donating electrons to each metal center. This structure is called an edge-sharing bioctahedron (i.e., two octahedral metal centers sharing a common edge).

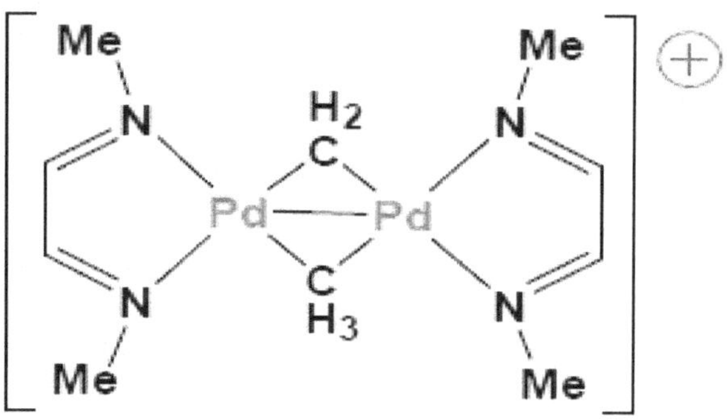

Ligand analysis: The chelating N ligand is a bis-imine, is neutral, with each N atom donating 2e-. There are two different bridging ligands – an anionic CH_3^- (methyl group) and a dianionic CH_2^{2-} (carbene or alkylidene). The CH_3^- only has one lone pair of electrons, so it has to split these between the two metals (1e- to each). The CH_2^{2-} alkylidene ligand, on the other hand, has 2 lone pairs so it can donate 2e- to each metal center. Oxidation state analysis: There are a total of 3 negative charges on the ligands (anionic methyl, dianionic alkylidene) and a positive charge on the complex. Therefore the two Pd centers must have a TOTAL of a +4 charge, or a +2 charge (oxidation state) on each. Note that Pd(+2) is generally happy with a 16e- configuration.

Pd(+2)	d^8
2 imines	**4e-**
μ-CH_3^-	1e-
μ-CH_2^{2-}	2e-
Sub-total:	***15e-***
Pd-Pd	1e-
TOTAL:	**16e-**

Problems in electron counting

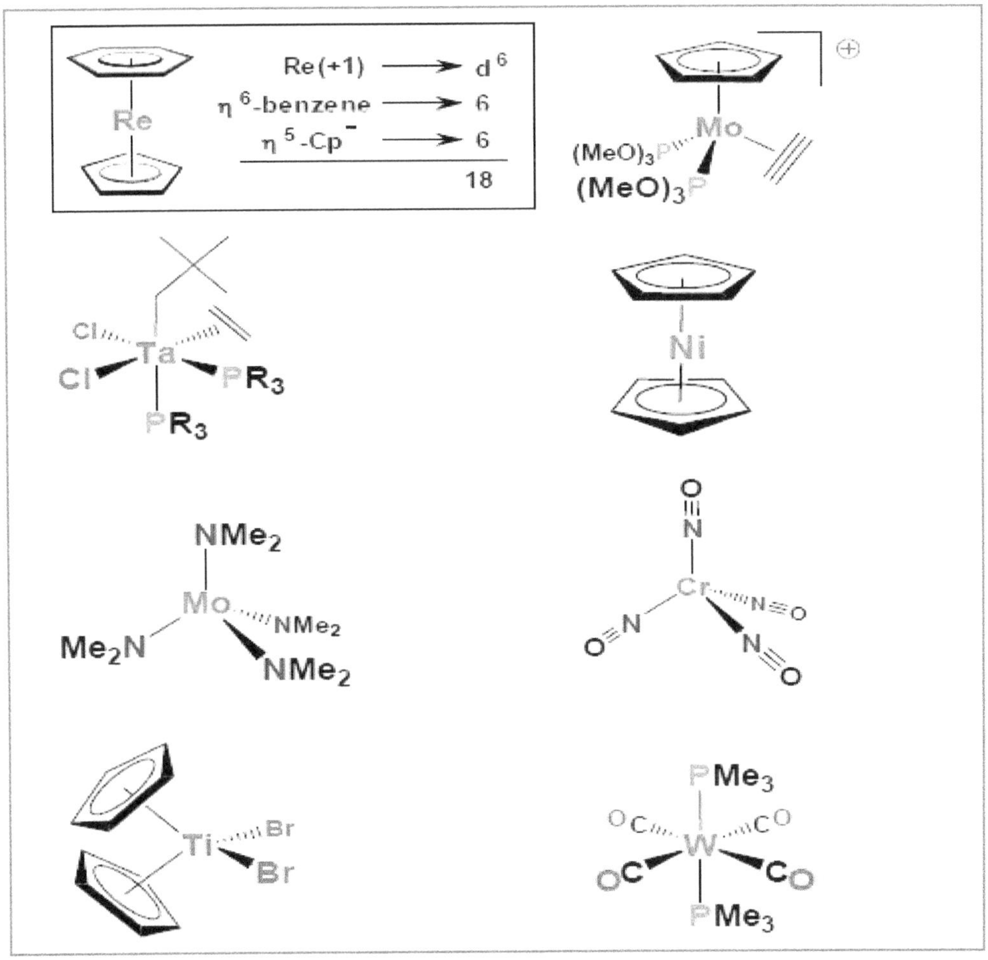

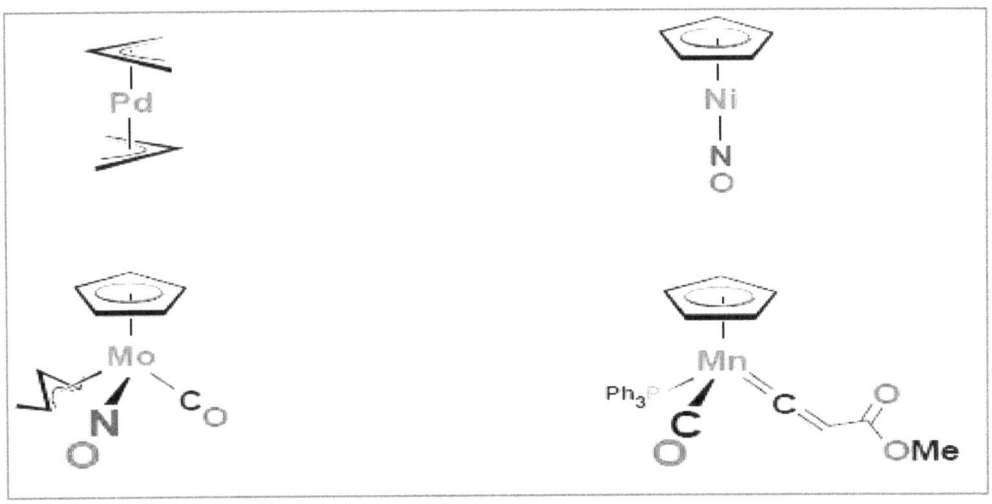

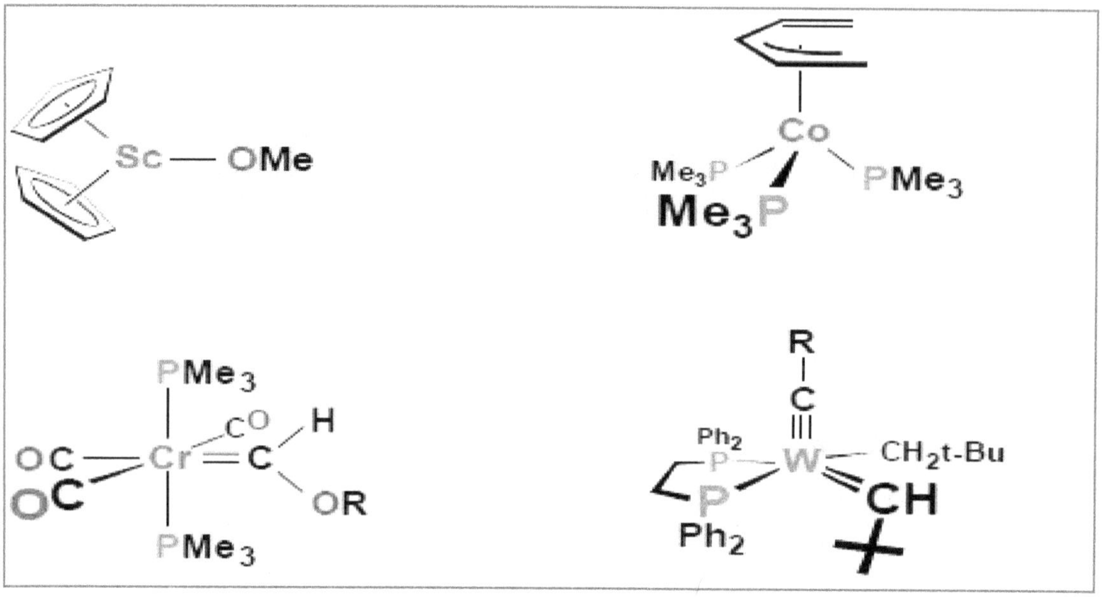
Sc
OMe
Co
Me_3P
PMe_3
Me_3P
R
C
Ph_2
P
W
CH_2t-Bu
P
CH
Ph_2
PMe_3
OC
Cr
CO
C
H
OC
OR
PMe_3

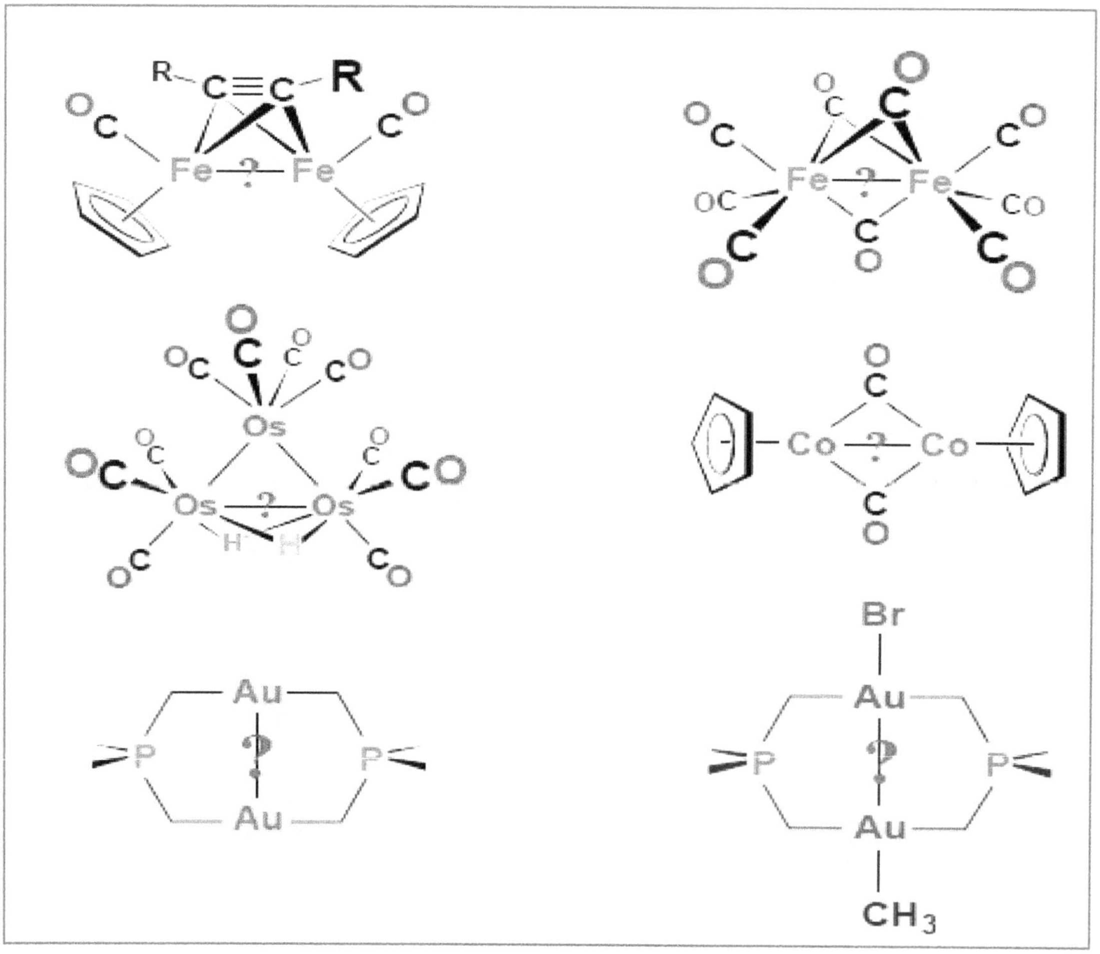
R
C≡C
R
OC
Fe
?
Fe
CO
O
C
C
O
OC
Fe
?
Fe
CO
OC
CO
C
O
C
O
C
O
O
C
C
O
OC
CO
Os
OC
Os
?
Os
CO
H
H
OC
CO
O
C
Co
?
Co
C
O
Au
P
?
P
Au
Br
Au
P
?
P
Au
CH_3

3.9 TYPES OF ORGANOMETALLIC REACTIONS

I. Oxidative addition/Reductive elimination

- **In oxidative addition**, a metal complex with vacant coordination sites and a relatively low oxidation state is oxidized by the insertion of the metal into a covalent bond (X−Y).
- Both the formal oxidation state of the metal and the electron count of the complex increase by two.
- Oxidative additions can occur with the insertion of a metal into many different covalent bonds, they are most commonly seen with H-H and C_{sp3}-halogen bonds.

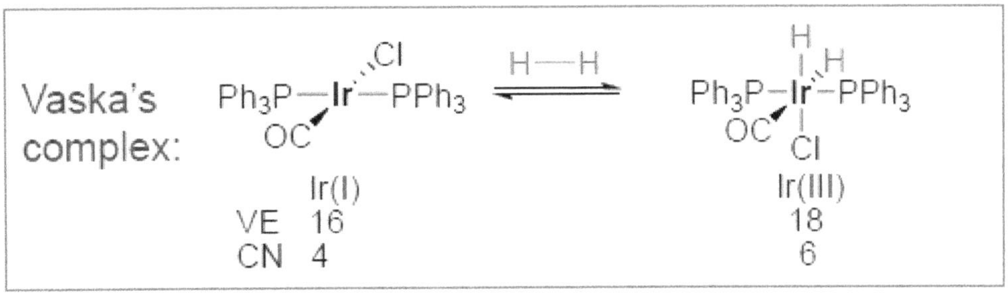

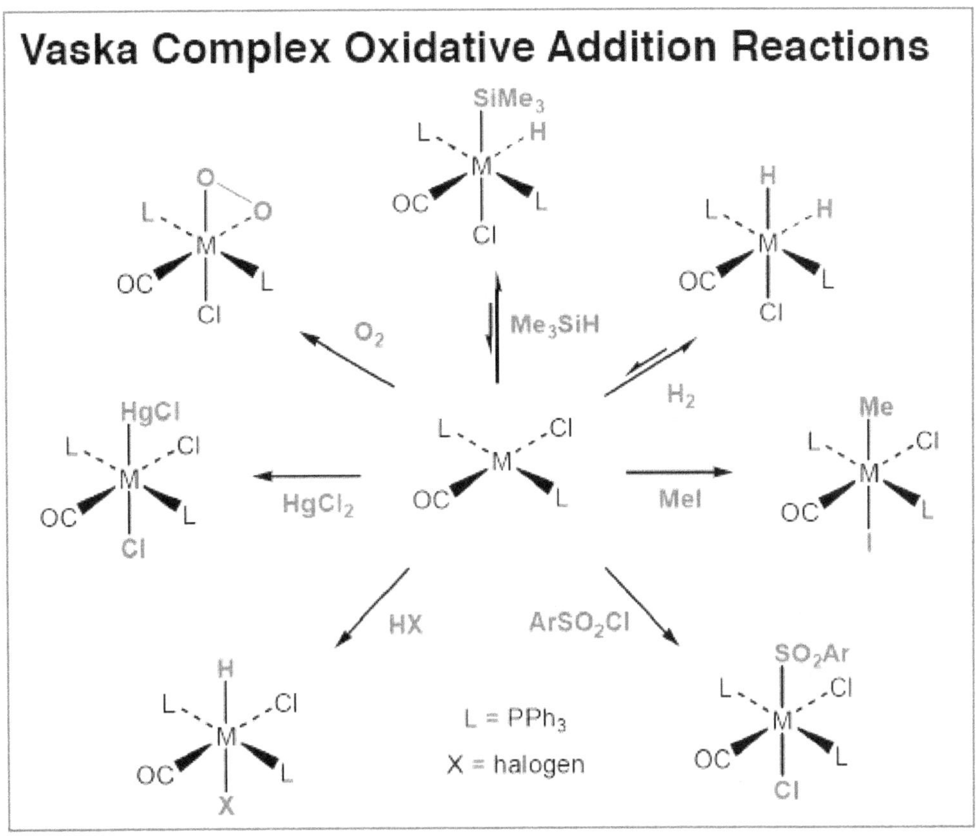

A **reductive elimination** involves the elimination or expulsion of a molecule from a transition metal complex. In the process of this elimination, the metal center is reduced by two electrons.

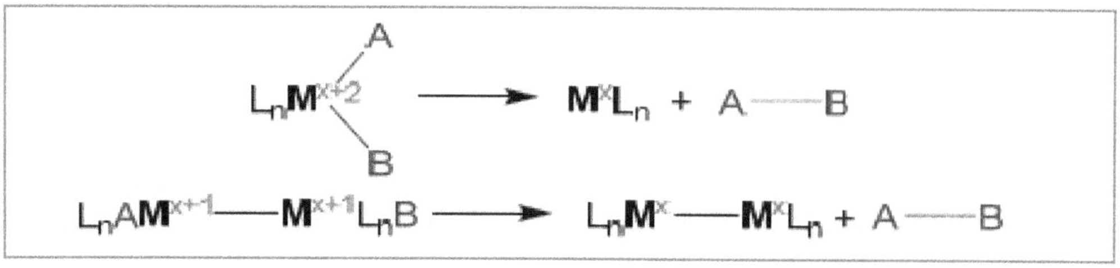

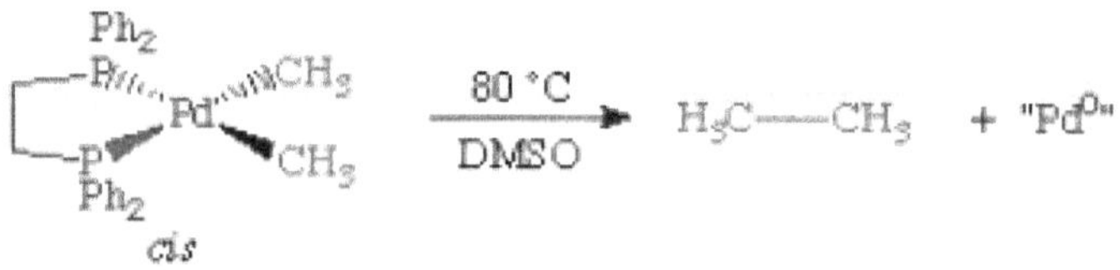

➢ The groups being eliminated must be in a mutually *cis* orientation.

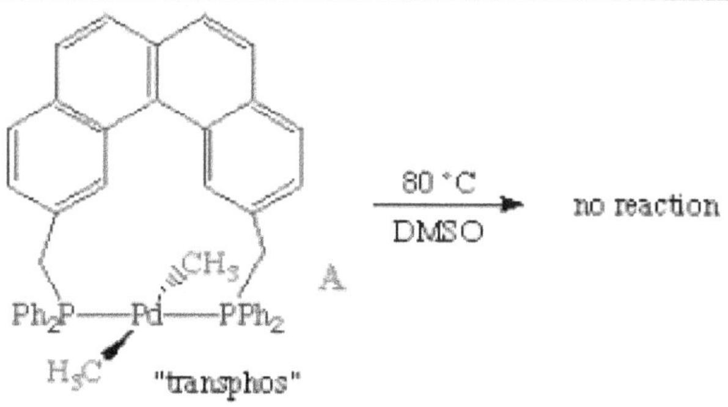

➢ A series of reactions involving an oxidative addition, a rearrangement and then a reductive elimination form the basis for a variety of industrially important catalytic cycles.

II. Transmetalation

➢ Transmetalation is a general chemical reaction type describing the exchange of ligands between two metal centers. The metal centers need not be the same. The ligands R and R' can be organic or inorganic.

$$M—R + M'—R' \longrightarrow M—R' + M'—R$$

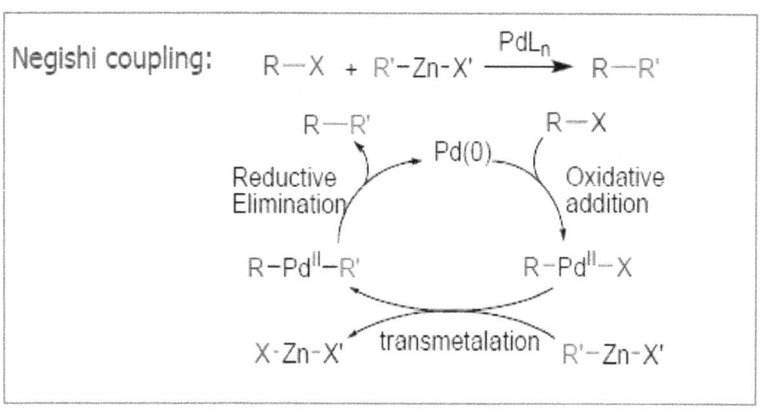

➢ Transmetalation is important in the synthesis of various organometallic compounds. This reaction type also appears frequently in the catalytic cycle of various metal catalyzed organic reactions.

III. Carbometalation

- Carbometalation is an reaction involving the nucleophilic addition to alkenes and alkynes of a diverse range of organometallic reagents such as organolithium compounds, organocopper compounds and Grignard reagents according to the following general alkyne scheme:

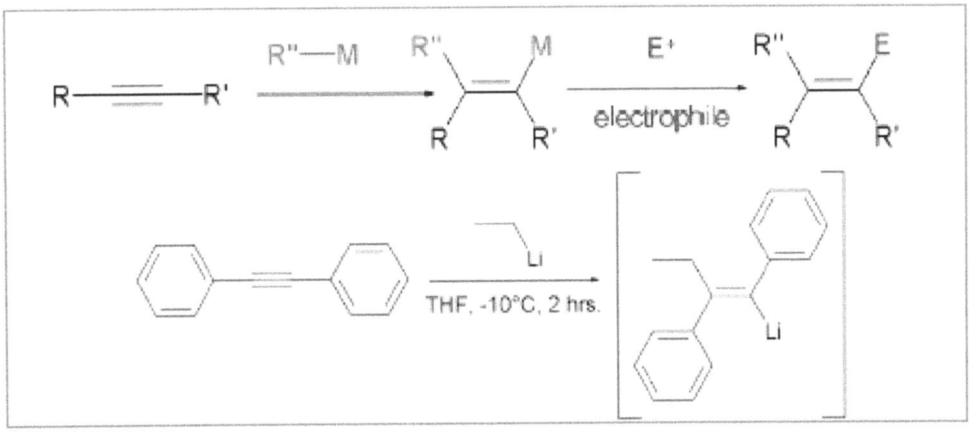

IV. Migratory insertion

- A migratory insertion reaction is when a cisoidal anionic and neutral ligand on a metal complex couple together to generate a new coordinated anionic ligand. This new anionic ligand is composed of the original neutral and anionic ligands now bonded to one another.

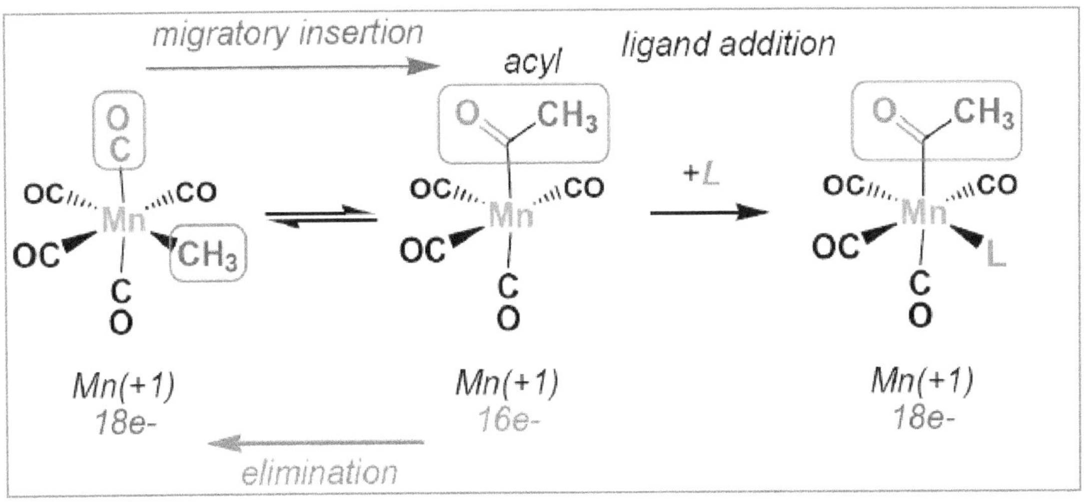

Anionic: **H$^-$, R$^-$ (alkyl)**, Ar-(aryl), acyl-, O^{2-} (oxo).

Neutral: **CO, alkenes**, alkynes, carbenes.

Alkene Migratory Insertions

Alkene and hydride/alkyl migratory insertion is the basis for almost all transition metal-based polymerization catalysts.

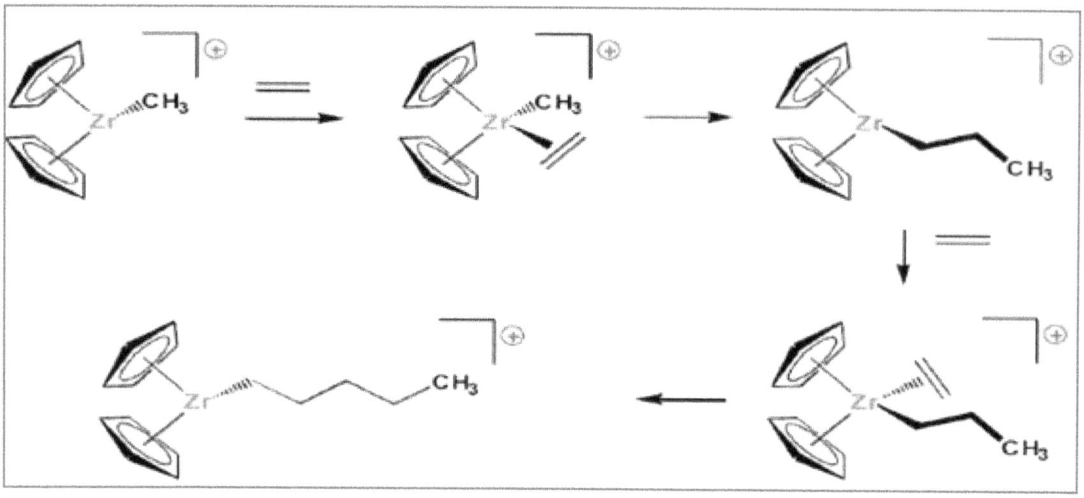

This is the basis for almost all transition metal-based polymerization catalysts. A polymerization reaction is just many, many migratory insertions of an alkene and alkyl (the growing polymer chain) interspaced with alkene ligand addition reactions. An alkene and a hydride usually react via a migration of the hydride to the coordinated alkene ligand:

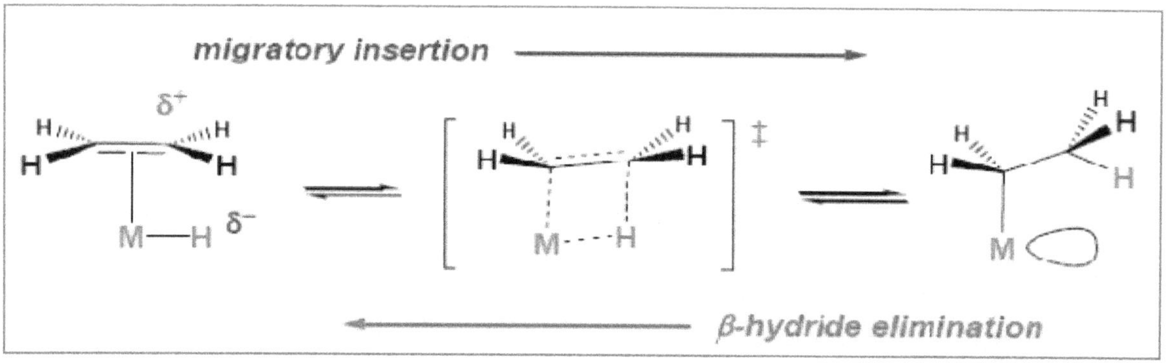

The backwards reaction, of course, is a β-hydride elimination and is usually quite favorable if there is an empty orbital *cis* to the alkyl ligand. Thus, the general importance of having a trapping ligand to coordinate to the empty orbital generated from the migratory insertion.

Alkynes can also do migratory insertions to produce vinyl groups:

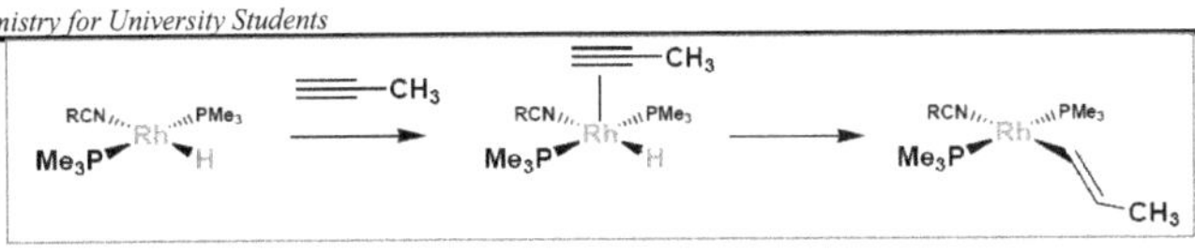

V. β -Hydride elimination

Elimination reactions are just the reverse of migratory insertionreactions.

- β -Hydride elimination is a reaction in which an alkyl group bonded to a metal centre is converted into the corresponding metal-bonded hydride and an alkene.

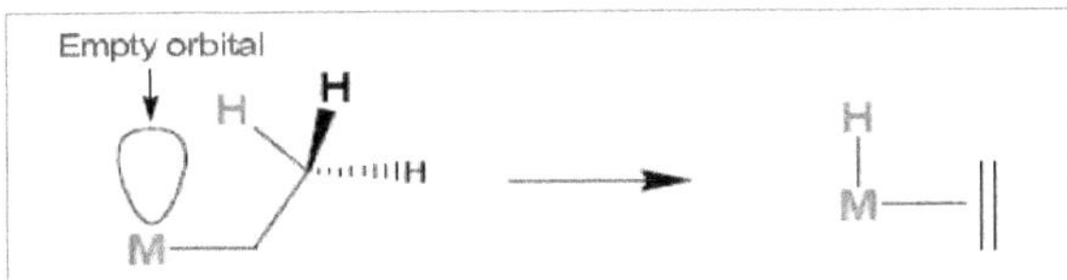

The key points to remember are:

i. No change in formal oxidation state (exception: alkylidenes).

ii. A vacant orbital is cisoidal to the group to do an elimination reaction on. Alternatively, a cisoidal labile ligand that can easily dissociate to open up an empty orbital.

Avoiding β-hydride elimination:

i. to employ an alkyl ligand that lacks a β-hydrogen (methyl or neopentyl).

ii. It is also inhibited when the reaction would produce a strained alkene.

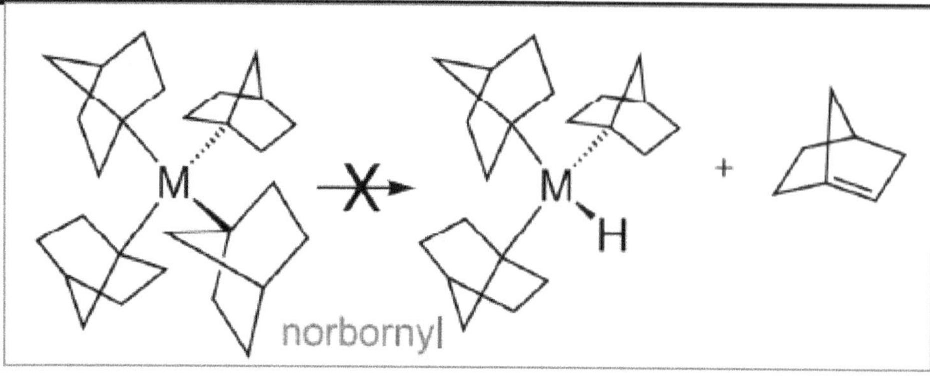

iii. The beta position may be blocked by non-hydrogen atoms.

iv. If the metal center does not have empty coordination sites, for example, by the complex already having 18 electron configuration, β- hydride elimination is not possible as well.

3.10 TRANSITION METAL ALKYL AND ARYL COMPLEXES

The story of metal alkyls starts with the main-group elements, particularly Li, Mg, Zn, Hg, As, and Al, and alkyls of some of these elements are still very widely useful. Indeed, it is scarcely possible to carry out an organic synthesis of any complexity without using an alkyllithium reagent. Organometallic chemistry began in a dramatic way. Working in a Parisian military pharmacy in 1757, Cadet made the appallingly evil-smelling cacodyl oxide (Greek: kakos = stinking), later shown by Bunsen to be $Me_2As–O–AsMe_2$, from As_2O_3 and $KOOCCH_3$. In 1848, an attempt by Edward Frankland to prepare free ethyl radicals by reaction of ethyl iodide with metallic zinc instead gave a colorless liquid that proved to be diethylzinc.

Because arsenic is a semimetal, $ZnEt_2$ was the first unambiguous case of a molecular compound with a metal–carbon bond and Frankland is often considered a founder of organometallic chemistry. Polar alkyls all react with traces of humidity to hydrolyze the M–C bond to form M–OH and release RH. Air oxidation also occurs very readily, and so polar organometallics must be protected from both air and water. Alkyls of the early transition metals, such as Ti or Zr, can also be very air and water sensitive, but as we move to less electropositive metals by moving to the right and down the periodic table, the compounds become much less reactive, until we reach Hg, where the Hg–C bond is so stable that $[Me–Hg]^+$ cation is indefinitely stable in aqueous sulfuric acid in air. As we go from the essentially ionic NaCH3, to the highly polar covalent Li and Mg species, to the essentially covalent late metal alkyls, the reactivity falls steadily along the series, showing the effect of changing metal.

The inherent stability of the R fragment plays a role, too. As an *sp*^{3} ion, CH_3^- is the most reactive. As we move to *sp*^{2} $C_6H_5^-$ and evenmore to *sp* $RC{\equiv}C^-$, where the lone pair of the anion is stabilized by being in an orbital with more s character, the intrinsic reactivity falls off. The same trend makes the acidity of the hydrocarbons increase as we go from CH_4 (pKa =~50) to C_6H_6 (pKa =~43) and to RC≡CH (pKa =~25), so the anion from the latter is the most stable and least reactive.

Following the successful syntheses of main-group alkyls, many attempts were made to form transition metal alkyls. Pope and Peachey's Me_3PtI (1909) was an early but isolated example of a d-block metal alkyl. Attempts during the 1920s through 1940s to make further examples of d-block alkyls all failed. This was especially puzzling because by then almost every nontransition element had been shown to form stable alkyls. These failures led to the view that transition metal–carbon bonds were unusually weak; for a long time after that, few serious attempts were made to look for them. In fact, we now know that such M–C bonds are strong (30–65 kcal/mol is typical). It is the existence of several easy decomposition pathways that makes many transition metal alkyls unstable. Kinetics, not thermodynamics, was to blame for the synthetic failures.

Stable Alkyls

To have a kinetically stable alkyl, we must block the β-elimination pathway for decomposition. This can happen for

1. Alkyls that have no β hydrogen:

WMe_6 $Ti(CH_2Ph)_4$ $W(CH_2SiMe_3)_6$ $TaCl_2(CH_2CMe_3)_3$

$C_2F_5Mn(CO)_5$ $LAuCF_2CF_2Me$ $Pt(C{\equiv}CCF_3)_2L_2$

$Pt(CH_2COMe)Cl(NH_3)_2$

2. Alkyls for which the β hydrogen is unable to approach the metal as a result of the geometry or bulk of the ligand:

$PtH(C{\equiv}CH)L_2$ $PdPh_2L_2$ $Cr(CMe_3)_4$ $Cr(CHMe_2)_4$

$CpL'_3MoCH{=}CHCMe_3$

3. Alkyls in which the M–C–C–H unit cannot become *syn*-coplanar:

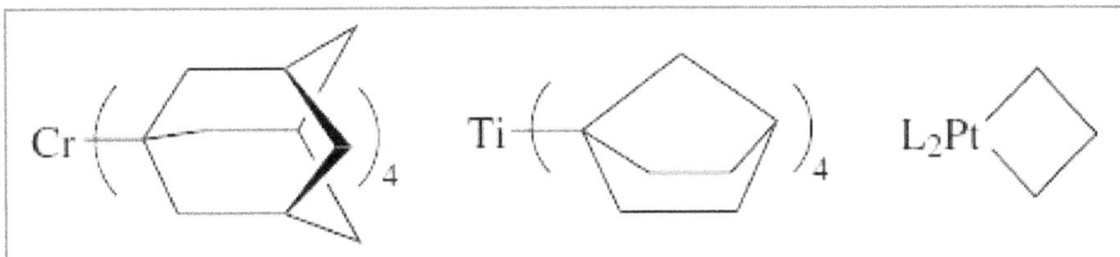

$[Cr(1\text{-adamantyl})_4]$ $Ti(6\text{-norbornyl})_4$ $L_2Pt(CH_2)_3$

The first two would give "forbidden" anti-Bredt olefins if they were to β-eliminate.

4. An 18-electron species with firmly bound ligands, which will not dissociate to generate a vacant site:

$Cp(CO)_2FeCH_2CH_3$, $Cp(CO)_3MoCH_2CH_3$,

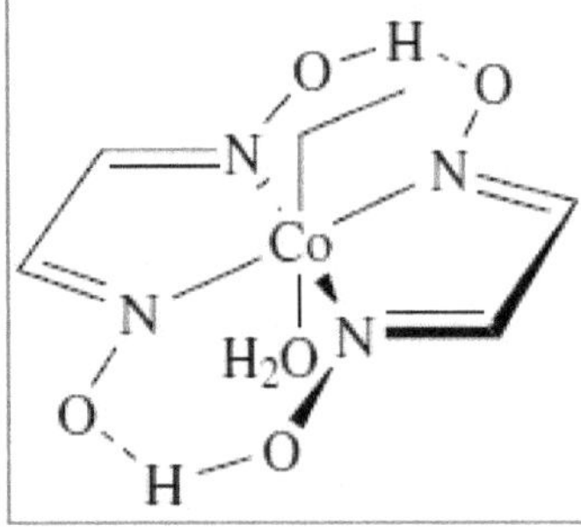

$Cp(CO)IrPrH$, $[Cr(H_2O)_5Et]^{2+}$

5. Some d^0 alkyls:

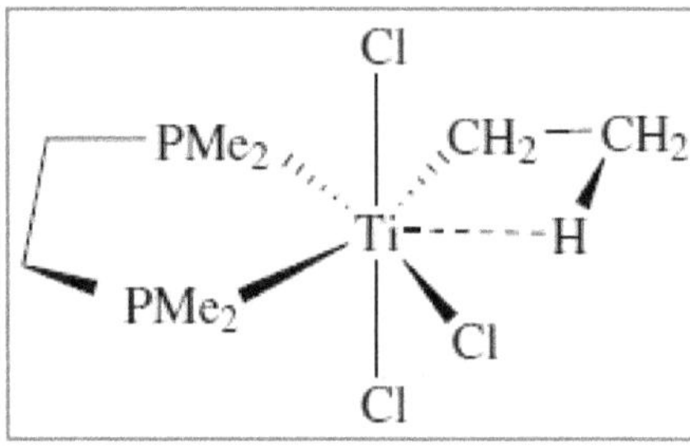

Preparation of metal alkyls

The chief methods for the synthesis of alkyls involve (1) an $\bar{R}$ reagent, (2) an $\overset{+}{R}$ reagent, (3) oxidative addition, and (4) insertion. Typical examples of these are shown below,

1. From an R^- reagent (nucleophilic attack on the metal):

$$WCl_6 \xrightarrow{LiMe} WMe_6 + LiCl$$

$$NbCl_5 \xrightarrow{ZnMe_2} NbMe_2Cl_3 + ZnCl_2$$

2. From an R^+ reagent (electrophilic attack on the metal):

$$Mn(CO)_5^- \xrightarrow{MeI} MeMn(CO)_5 + I^-$$

$$CpFe(CO)_2^- \xrightarrow{Ph_2I^+} Cp_2Fe(CO)_2Ph + PhI$$

$$[Mn(CO)_5]^- \xrightarrow{CF_3COCl} CF_3COMn(CO)_5 \xrightarrow{-CO} CF_3Mn(CO)_5$$

3. ***By oxidative addition:***

$$IrCl(CO)L_2 \xrightarrow{MeI} MeIrICl(CO)L_2$$

$$PtL_4 \xrightarrow{MeI} MePtIL_2 \quad (L = PPh_3)$$

$$Cr^{2+}(aq) \xrightarrow{MeI} CrMe(H_2O)_5{}^{2+} + CrI(H_2O)_5{}^{2+}$$

4. ***By insertion:***

$$PtHCl(PEt_3)_2 + C_2H_4 \longrightarrow PtEtCl(PEt_3)_2$$

$$Cp(CO)_3MoH \xrightarrow{CH_2N_2} Cp(CO)_3MoCH_3$$

ARYL (BENZENE) AND RELATED COMPOUNDS

Although we have considered compounds of $C_5H_5^-$ as though the ligand is a donor of six electrons, benzene is also donor of six electrons. Therefore, two benzene molecules each donating six electrons would raise the total to 36 if the metal initially has 24. Such is the case if the metal is Cr^0.

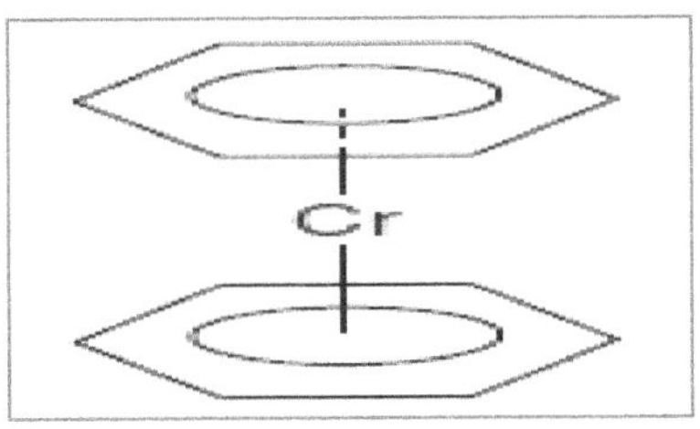

Thus, $Cr(\eta^6 - C_6H_6)_2$ obeys the 18-electron rule, and its structure is shown in the figure.

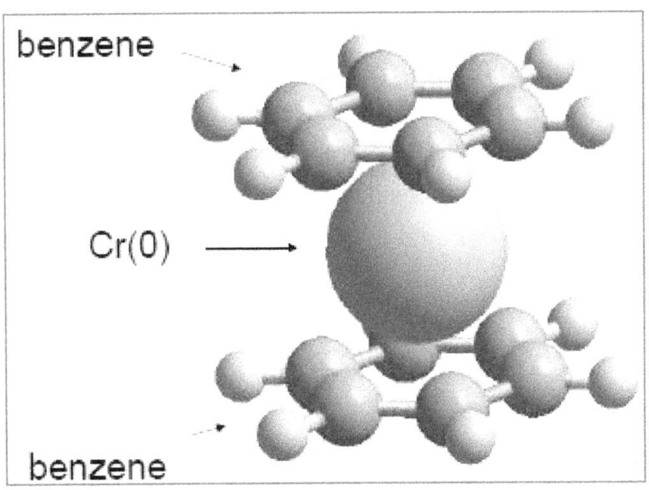

This compound, bis(benzene)chromium(0), has been prepared by several means including the following reactions:

$$3\ CrCl_3 + 2\ Al + AlCl_3 + 6\ C_6H_6 \rightarrow 3\ [Cr(\eta^6\text{-}C_6H_6)_2]AlCl_4$$

$$2\ [Cr(C_6H_6)_2]AlCl_4 + S_2O_4^{2-} + 4\ OH^- \rightarrow 2\ [Cr(\eta^6\text{-}C_6H_6)_2] + 2\ H_2O + 2\ SO_3^{2-} + 2\ AlCl_4^-$$

Bis(benzene)chromium(0) is easily oxidized, and mixed complexes are obtained by means of substitution reactions. For example,

$$Cr(CO)_6 + C_6H_6 \rightarrow C_6H_6Cr(CO)_3 + 3\ CO$$

The structure of benzenetricarbonylchromium(0) can be shown as:

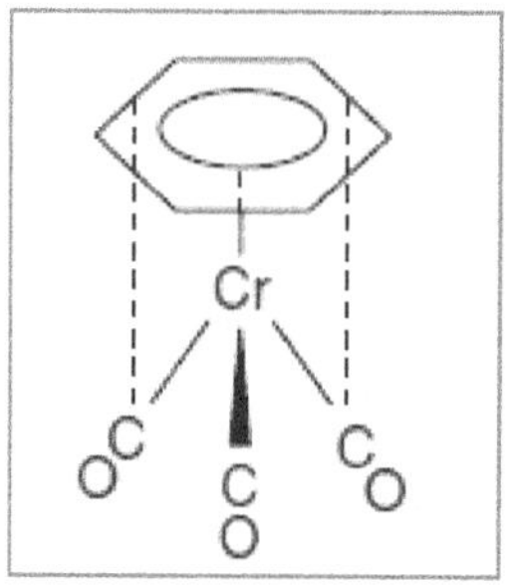

Similar compounds have also been prepared containing Mo and W. Methyl-substituted benzenes such as mesitylene (1,3,5-trimethylbenzene), hexamethylbenzene, and other aromatic molecules have been used to prepare complexes with several metals in the zero oxidation state. For example, $Mo(CO)_6$ will react with 1,3,5-C_6H_3 $(CH_3)_3$, 1,3,5- trimethylbenzene, which replaces three carbonyl groups.

CYCLOPENTADIENYL AND RELATED COMPOUNDS

In addition to complexes in which electron density is donated from localized σ bonds to metals, many complexes are known in which electrons in aromatic molecules are donated to metals. A brief survey of the chemistry of these interesting compounds will be presented here. However, the area is so vast that the interested reader will need to consult the references listed at the end of the chapter for further details.

Hydrogen atoms on the cyclopentadiene ring are very slightly acidic. Therefore, metallic sodium reacts with a solution of cyclopentadiene dissolved in tetrahydrofuran (THF) to liberate hydrogen, leaving sodium cyclopentadienide in solution.

$$2\ Na + 2\ C_5H_6 \rightarrow H_2 + 2\ Na^+C_5H_5^-$$

The reaction of this product with $FeCl_2$, which can be shown as

$$2\ NaC_5H_5 + FeCl_2 \rightarrow 2\ NaCl + Fe(C_5H_5)_2$$

yields an orange solid (m.p. 173–174°C) that is insoluble in water but soluble in organic solvents.

The product, known as ferrocene, $Fe(C_5H_5)_2$, can also be produced in other ways. For example, when a solution of $FeCl_2 . 4H_2O$ is added to a basic solution of cyclopentadiene, the following reaction occurs:

$$8\ KOH + 2\ C_5H_6 + FeCl_2{\cdot}4H_2O \rightarrow Fe(C_5H_5)_2 + 2\ KCl + 6\ KOH{\cdot}H_2O$$

As mentioned earlier, hydrogen atoms on the cyclopentadiene ring are very slightly acidic, and KOH facilitates the removal of H^+, leaving $C_5H_5^-$. Ferrocene was first produced in 1951, and it quickly became the focus of a great deal of organometallic chemistry.

Ferrocene is more correctly named as bis(η5 -cyclopentadienyl)iron(II), and the molecule has a structure that can be shown as follows:

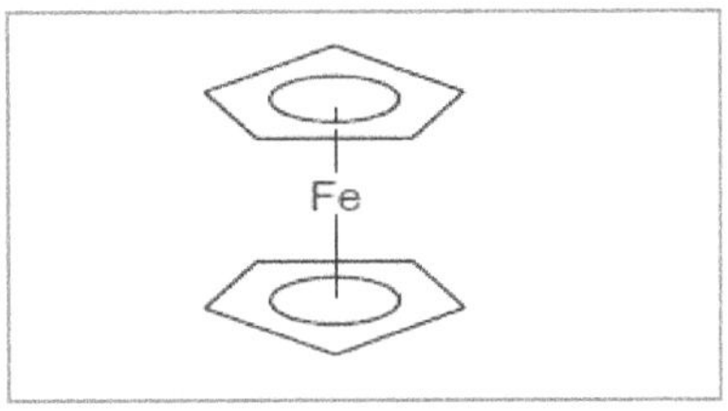

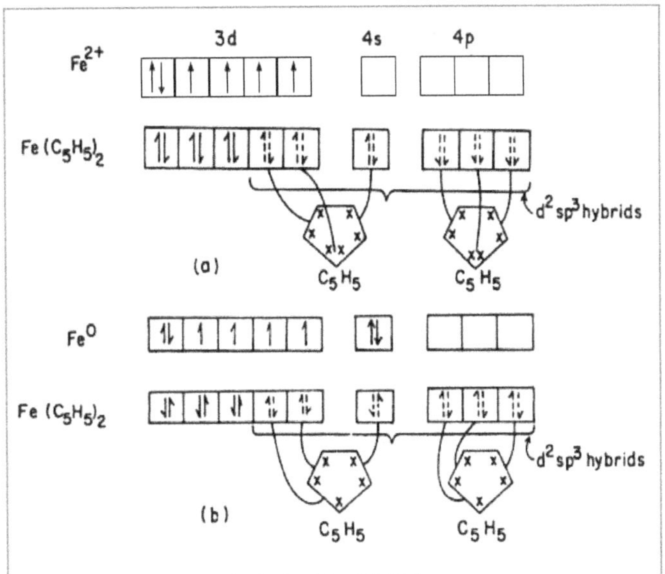

Bonding description (VBT) in ferrocene.

In ferrocene and similar complexes, the $C_5H_5^-$ ion behaves as an aromatic electron donor. In order for an organic ring structure to achieve aromaticity, there must be 2, 6, or 10 electrons to form the π electron system ($4\ n + 2$ where $n = 0$, 1, 2, …). Benzene is aromatic and forms complexes in which it behaves as if there were three double bonds that donate a total of six electrons. For cyclopentadiene, aromaticity is achieved when there is one additional electron as is the case in the anion, $C_5H_5^-$.

Formally, in ferrocene the ligands are considered as negative ions and the iron as Fe^{2+}. The Fe^{2+} has 24 electrons, and each ligand behaves as a donor of 6 electrons to Fe^{2+}, giving a total of 36 electrons around Fe in accord with the 18-electron rule.

If the iron were Fe^0 and each ligand were a donor of five electrons, the same total number of electrons around iron would result.

The simplest approach to illustrate the nature of bonding in ferrocene is the valence bond theory (VBT), with the help of which one can make prediction about the molecular geometry of a metal complex in terms of the number and types of metal orbitals available for bonding. For example, two alternative formulations of the bonding in ferrocene (Shown above) were suggested by Fisher (a) as well as by Dunitz and Orgel (b):

Reactions of ferrocene and other metallocenes

Ferrocene is a very stable compound that melts at 173°C and can be sublimed without disruption of the metal complex. Many reactions exhibited by ferrocene are essentially those of an aromatic organic compound. For example, sulfonation of ferrocene can be achieved as follows:

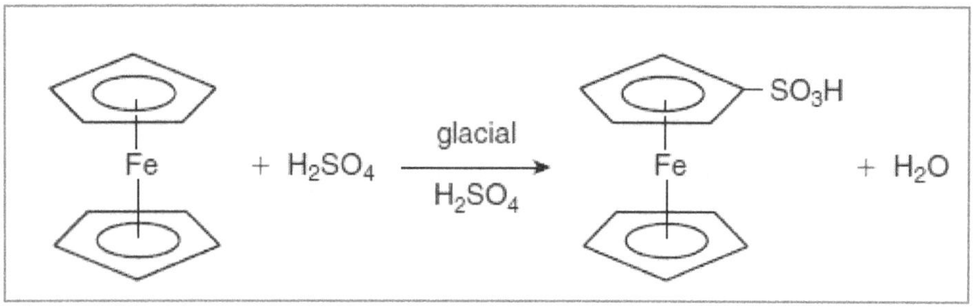

Acylation can be carried out by means of a Friedel-Crafts reaction,

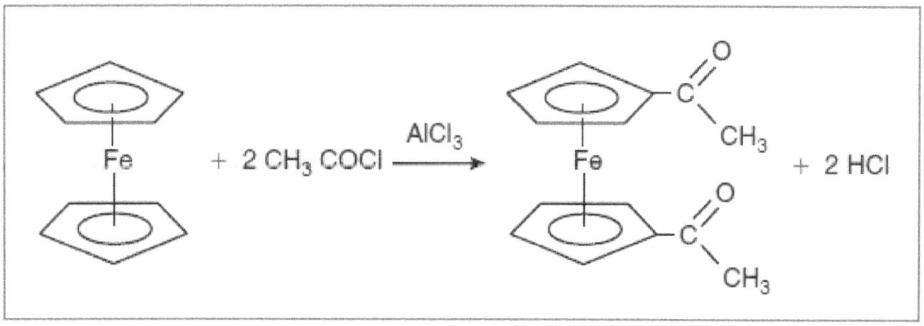

Although the disubstituted product is shown in the foregoing equation, some monosubstituted product is also obtained. In this process, $AlCl_3$ may react with CH_3COCl to generate the $CH_3C{=}O^+$ attacking species. It is also possible introduce the acetyl group by a reaction with acetic anhydride with phosphoric acid as the catalyst. In this reaction under less strenuous conditions, the dominant product is the monosubstituted derivative.

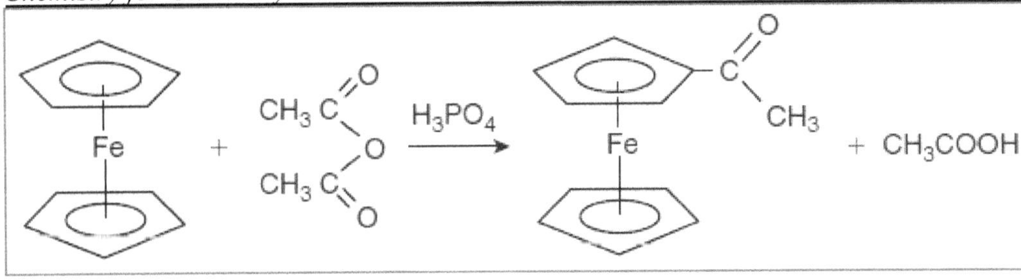

By a sequence of reactions, the two cyclopentadienyl rings can be connected. Using C_2H_5 $OCOCH_2COCl$ as the starting acid chloride, it is possible to attach the acyl end to one of the C_5H_5 rings. Hydrolysis to remove the ethyl group and conversion of that end to an acid chloride followed by another Friedel-Crafts reaction and reduction of the product leads to

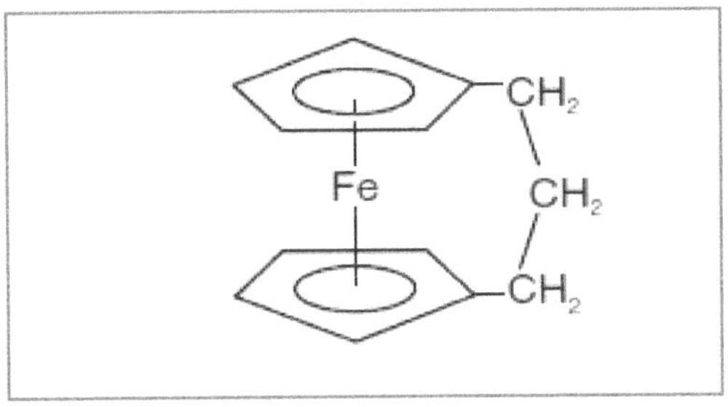

One of the most useful reactions of ferrocene with regard to synthesizing derivatives is *metallation*. In this reaction, the lithium derivative is first prepared by the reaction of ferrocene with butyllithium, a reaction showing the slight acidity of the hydrogen atoms.

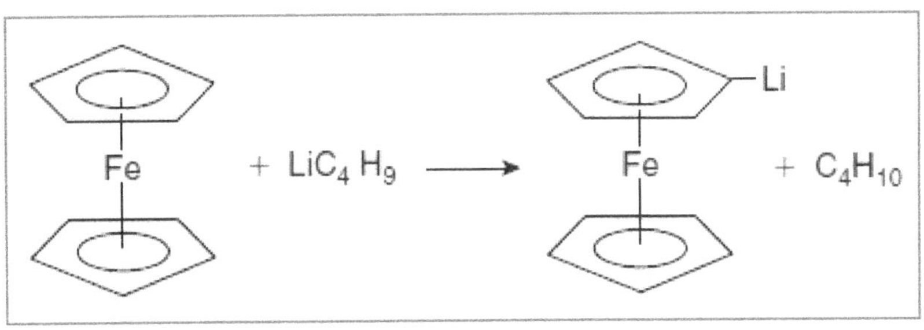

The derivative containing two lithium atoms can also be obtained by the subsequent reaction,

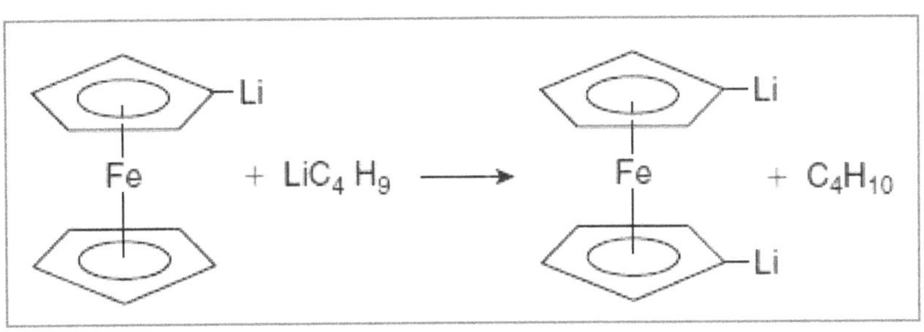

3.11 TRANSITION METAL CARBONYL COMPLEXES

Metal carbonyl compounds are organometallic in various aspects of their bonding, structure and reactions, and they are a good model system for understanding of the essence of transition metal organometallic chemistry. A remarkable characteristic of these compounds is that because the ligands are neutral molecules, the metals are present in the zero oxidation state. Depending on the net charge on the other ligands, the metal may or may not be in the zero oxidation state. In any event, the metals in metal carbonyls occur in low oxidation states because of the favorable interaction of the soft ligands (Lewis bases) with metals in low oxidation states (soft Lewis acids).

Generally, the metals that form stable carbonyl complexes are those in the first transition series from V to Ni, in the second row from Mo to Rh, and in the third row from W to Ir. There are several reasons for these being the metals most often found in carbonyl complexes. First, these metals have one or more *d* orbitals that are not completely filled so they can accept electron pairs from σ electron donors. Second, the *d* orbitals contain some electrons that can be involved in back donation to the π^* orbitals on the CO ligands. Third, the metals are generally in the zero oxidation state or at least a low oxidation state so they behave as soft Lewis acids, making it favorable to bond to a soft Lewis base like CO. The first metal carbonyl prepared was $Ni(CO)_4$, which was obtained by Mond in 1890. This extremely toxic compound was prepared by first reducing nickel oxide with hydrogen,

$$NiO + H_2 \xrightarrow{400\,°C} Ni + H_2O$$

then treating the Ni with CO,

$$Ni + 4\ CO \xrightarrow{100\,°C} Ni(CO)_4$$

Mononuclear metal carbonyl compounds take highly symmetric polyhedral coordination structures. Hexa-coordinate chromium, molybdenum, and tungsten hexacarbonyl, $M(CO)_6$, assume a regular octahedral, penta-coordinate pentacarbonyliron, $Fe(CO)_5$, a triangular bipyramid, and tetracarbonylnickel, $Ni(CO)_4$, a regular tetrahedron coordination. The carbon atoms of carbonyl ligands coordinate to the metal, and the CO moieties are oriented along the direction of the metal-carbon axis. Binuclear metal carbonyl $Mn_2(CO)_{10}$ has an Mn-Mn bond joining two square pyramidal $Mn(CO)_5$ parts. In $Fe_2(CO)_9$, two $Fe(CO)_3$ sub-units are bridged by three CO ligands, and in $Co_2(CO)_8$, two $Co(CO)_3$ sub-units are connected by both three CO bridges and a Co-Co bond. There are a number of cluster metal carbonyl compounds with metal-metal bonds joining three or more metals, and terminal CO, μ-CO (a bridge between two metals), and μ_3-CO (a bridge capping three metals) are coordinated to the metal frames. Many cluster carbonyls are formed by a pyrolysis reaction of mononuclear or binuclear carbonyl compounds. Typical metal carbonyls and their properties are shown below:

Stable metal carbonyl compounds						
	5	6	7	8	9	10
4	$V(CO)_6$ Black solid d.70	$Cr(CO)_6$ White solid d.130	$Mn_2(CO)_{10}$ Yellow solid mp 154	$Fe(CO)_5$ Yellow liquid bp 103	$Co_2(CO)_8$ Red solid mp 51	$Ni(CO)_4$ Colorless liquid bp 42.1
5		$Mo(CO)_6$ White solid sublime	$Tc_2(CO)_{10}$ White solid mp 160	$Ru_3(CO)_{12}$ Orange solid d.150	$Rh_6(CO)_{16}$ Black solid d.220	
6		$W(CO)_6$ White solid sublime	$Re_2(CO)_{10}$ White solid mp 177	$Os_3(CO)_{12}$ Orange solid mp 224	$Ir_4(CO)_{12}$ Yellow solid d.220	

STRUCTURES OF METAL CARBONYLS

The mononuclear metal carbonyls contain only one metal atom, and they have comparatively simple structures. For example, nickel tetracarbonyl is tetrahedral. The pentacarbonyls of iron, ruthenium, and osmium are trigonal bipyramidal, whereas the hexacarbonyls of vanadium, chromium, molybdenum, and tungsten are octahedral. These structures are shown below,

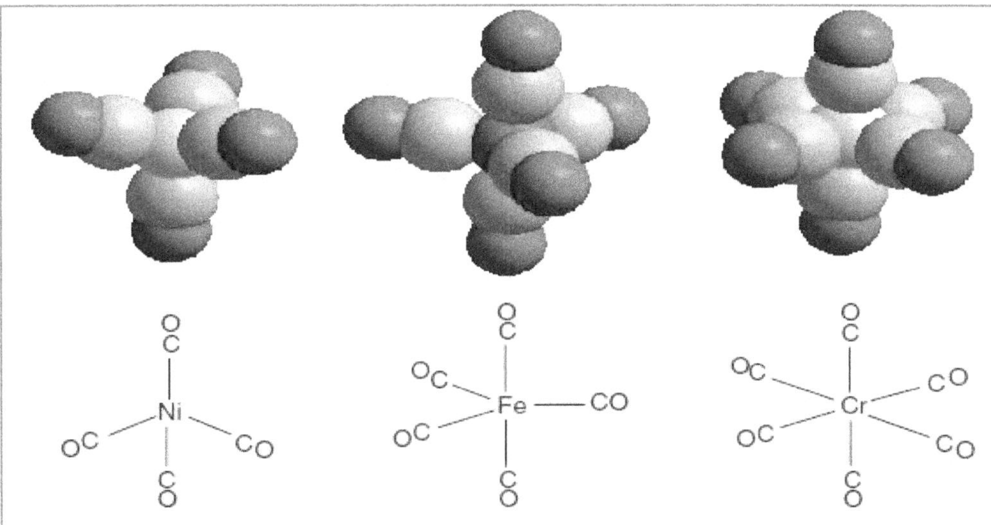

The structures of the mononuclear carbonyls of nickel, iron, and chromium.

The structures of the dinuclear metal carbonyls (containing two metal atoms) involve either metalmetal bonds or bridging CO groups, or both. For example, the structure of $Fe_2(CO)_9$, diiron nonacarbonyl, contains three CO ligands that form bridges between the iron atoms, and each iron atom also has three other CO groups attached only to that atom. Carbonyl groups that are attached to two metal atoms simultaneously are called bridging carbonyls, whereas those attached to only one metal atom are referred to as terminal carbonyl groups. The structures of $Mn_2(CO)_{10}$, $Tc_2(CO)_{10}$, and $Re_2(CO)_{10}$ actually involve only a metal-metal bond, so the formulas are more correctly written as $(CO)_5M$– $M(CO)_5$. Two isomers are known for $Co_2(CO)_8$. One has a metal-metal bond between the cobalt atoms; the other has two bridging CO ligands and a metalmetal bond.

The structure for $Mn_2(CO)_{10}$ is sometimes drawn to appear as if four CO ligands form a square plane around the Mn atom. In fact, the four CO groups do not lie in the same plane as the Mn atom. Rather, they lie about 12 pm from the plane containing the Mn on the side opposite the metal-metal bond. Structures of the dinuclear, trinuclear, tetranuclear, and higher compounds are best thought of as metal clusters containing either metal-metal bonds or bridging carbonyl groups.

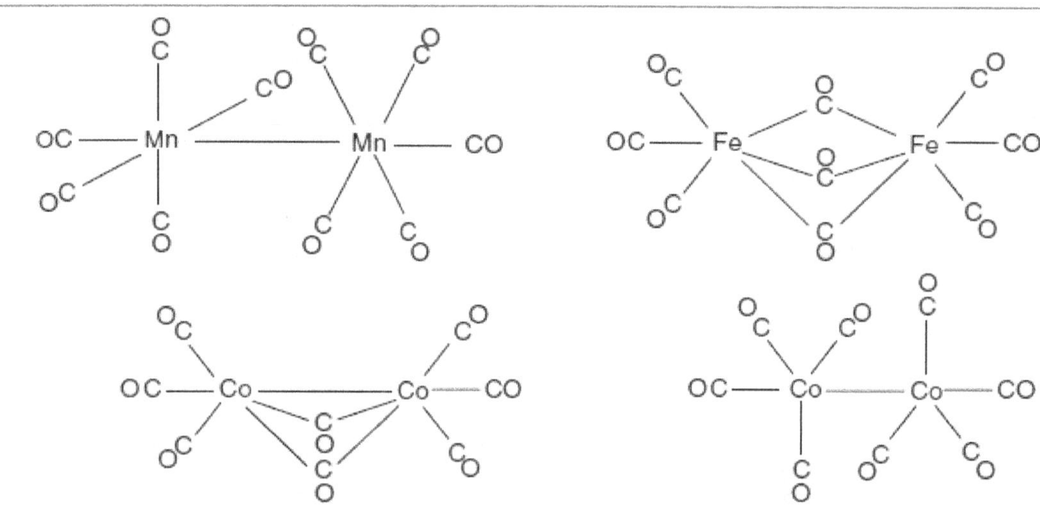

The structures of the binary metal carbonyls.

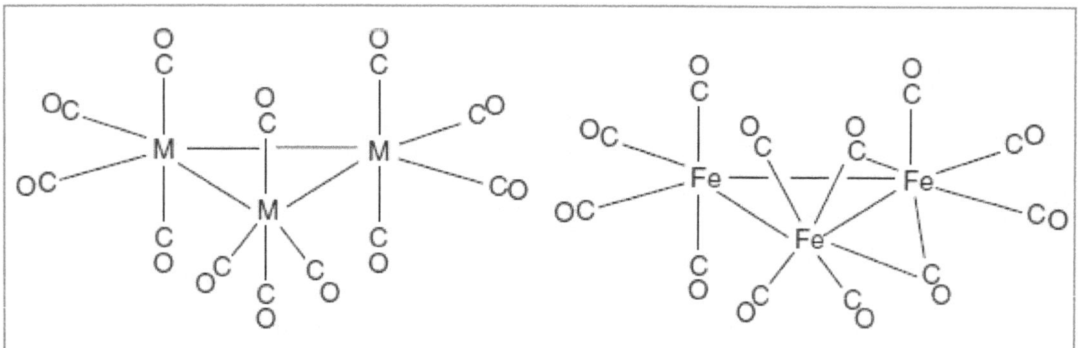

Structures of trinuclear metal carbonyls. In the structure on the left, M represents Ru or Os.

Carbonyls containing four metal atoms in the cluster are formed by Co, Rh, and Ir in which the metal atoms are arranged in a tetrahedron with 12 CO ligands, nine in terminal positions and three that are bridging. These structures are shown below:

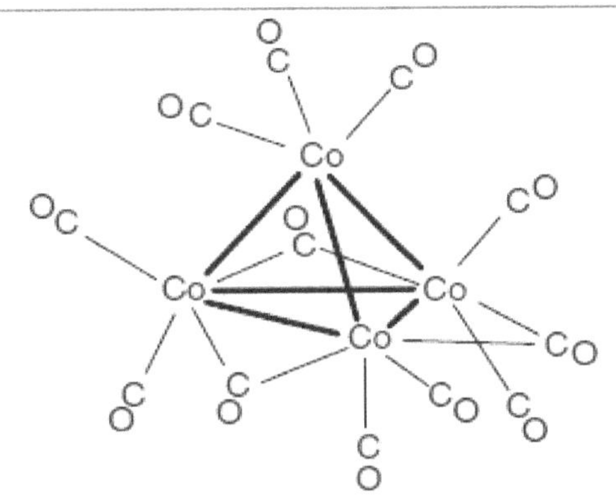

The structure of $[Co_4(CO)_{12}]$, which is also the structure of the analogous rhodium and iridium compounds.

BONDING OF CARBON MONOXIDE TO METALS

Carbon monoxide has the valence bond structure shown as with a triple bond between C and O. The formal charge on the oxygen atom is +1, whereas that on the carbon atom is -1. Although the electronegativity of oxygen is considerably higher than that of carbon, these formal charges are consistent with the dipole moment being small (0.12 D) with the carbon being at the negative end of the dipole. The carbon end of the CO molecule is thus a softer electron donor, and it is the carbon atom that is bound to the metal.

For CO, bond order = 3, a triple bond. For gaseous CO in which the triplebond exists, the C – O stretching band is observed at 2143 cm^{-1}. However, typically, in metal carbonyls the C – O stretching band is seen at 1850 to 2100 cm^{-1} for terminal CO groups.

The carbonyl ligand is a π-acid. This is an acid in the Lewis sense, where it receives electrons from the metal ion, and it is a π–acid because this involves π- bonding. The π-bonding involves overlap of the π* orbitals of the CO with d orbitals from the t_{2g} set of the metal, and so is dπ-pπ bonding.

What stabilizes CO complexes is M→C π–bonding. The lower the formal charge on the metal ion, the more willing it is to donate electrons to the π–orbitals of the CO. Thus, metal ions with higher formal charges, e.g. Fe(II) form CO complexes with much greater difficulty than do zero-valent metal ions such as Cr(O) and Ni(O), or negatively charged metal ions such as V(-I). The π–overlap is envisaged as involving dorbitals of the metal and the π*–orbitals of the CO:

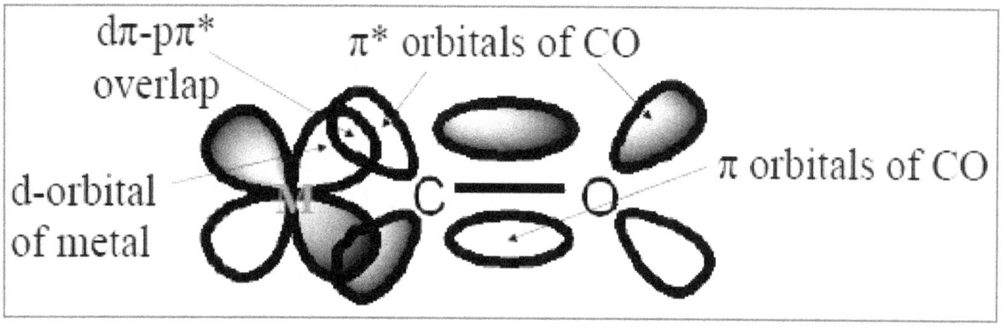

CO acts as a sigma donor and a π acceptor capable of stabilizing low oxidation states

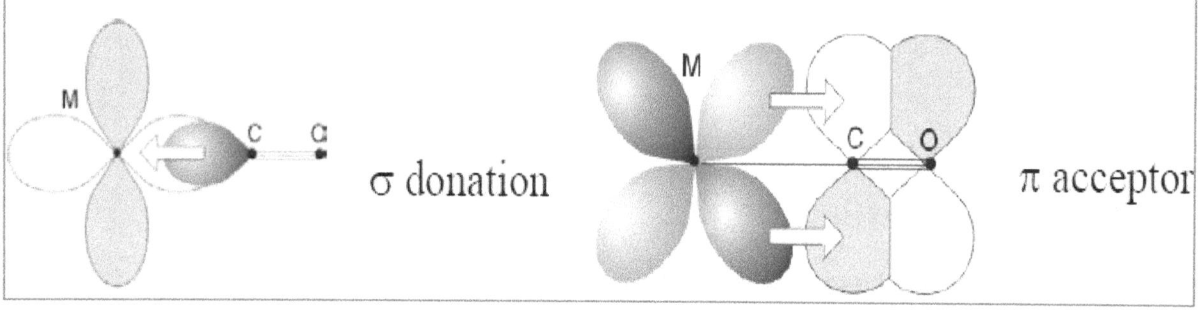

This bonding is synergistic. The more sigma donation by the carbonyl (or other sigmadonors on the metal center), the stronger the π -backbonding interaction. In $Fe(CO)_5$ the formal charge on iron is -5, and in $Cr(CO)_6$ the formal charge on chromium is -6. We should expect that the back donation would be more extensive in either of these compounds than it is in the case of $Ni(CO)_4$. Because the greater back donation results in a greater reduction in C – O bond order, the infrared spectra of these compounds should show this effect. The positions of the CO stretching bands for these compounds are as shown:

$Ni(CO)_4$	$2057\ cm^{-1}$
$Fe(CO)_5$	$2034\ cm^{-1}$
$Cr(CO)_6$	$1981\ cm^{-1}$

PREPARATION OF METAL CARBONYLS

As we have described previously, $Ni(CO)_4$ can be prepared directly by the reaction of nickel with carbon monoxide. However, most of the binary metal carbonyls cannot be obtained by this type of reaction. A number of preparative techniques have been used to prepare metal carbonyls, and a few general ones are given below,

1. **Reaction of a Metal with Carbon Monoxide**

The reactions of carbon monoxide with Ni and Fe proceed rapidly at low temperature and pressure.

$$Ni + 4\ CO \rightarrow Ni(CO)_4$$
$$Fe + 5\ CO \rightarrow Fe(CO)_5$$

For most other metals, high temperatures and pressures are required to produce the metal carbonyl. By this direct reaction, $Co_2(CO)_8$, $Mo(CO)_6$, $Ru(CO)_5$, and $W(CO)_6$ have been prepared when suitable conditions are used.

2. **Reductive Carbonylation**

This type of reaction involves reducing a metal compound in the presence of CO. The reducing agents may include a variety of materials depending on the particular synthesis being carried out. For example, in the synthesis of $Co_2(CO)_8$, hydrogen is used as the reducing agent.

$$2\,CoCO_3 + 2\,H_2 + 8\,CO \rightarrow Co_2(CO)_8 + 2\,H_2O + 2\,CO_2$$

Lithium aluminum hydride, $LiAlH_4$, has been used as the reducing agent in the preparation of $Cr(CO)_6$ from $CrCl_3$. Reduction by metals such as Na, Mg, or Al has also been used, as shown by the preparationof the $V(CO)_6{}^-$ ion:

$$VCl_3 + 4\,Na + 6\,CO \xrightarrow[\text{high pressure}]{\text{Diglyme, }100\,^\circ C} [(diglyme)_2Na][V(CO)_6] + 3\,NaCl$$

3. **Displacement Reactions**

Certain metal carbonyls have been prepared by the reaction of metal compounds directly with CO owing to the fact that CO is a reducing agent. For example,

$$2\,IrCl_3 + 11\,CO \rightarrow Ir_2(CO)_8 + 3\,COCl_2$$

$$Re_2O_7 + 17\,CO \rightarrow Re_2(CO)_{10} + 7\,CO_2$$

REACTIONS OF METAL CARBONYLS

Metal carbonyls undergo reactions with a great many compounds to produce mixed carbonyl complexes. A large number of these reactions involve the replacement of one or more carbonyl groups by a substitution reaction. Such reactions have also been studied kinetically in some cases.

1. Substitution Reactions

Many substitution reactions occur between metal carbonyls and other potential ligands. For example:

$$Cr(CO)_6 + 2\,py \rightarrow Cr(CO)_4(py)_2 + 2\,CO$$

$$Ni(CO)_4 + 4\,PF_3 \rightarrow Ni(PF_3)_4 + 4\,CO$$

$$Mo(CO)_6 + 3\,py \rightarrow Mo(CO)_3(py)_3 + 3\,CO$$

2. Reactions with Halogens

Reactions of metal carbonyls with halogens lead to the formation of carbonyl halide complexes by substitution reactions or breaking metal-metal bonds. The reaction given below, involves the rupture of the Mn – Mn bond, and one Br is added to each Mn.

$$[Mn(CO)_5]_2 + Br_2 \rightarrow 2\,Mn(CO)_5Br$$

In the reaction, $$Fe(CO)_5 + I_2 \rightarrow Fe(CO)_4I_2 + CO$$

one CO is replaced on the iron by two iodine atoms so that the coordination number of the iron is increased to 6. The formulas for these carbonyl halides obey the EAN rule.

3. Reactions with NO

The nitric oxide molecule has one unpaired electron residing in an antibonding π^* molecular orbital. When that electron is removed, the bond order increases from 2.5 to 3, so in coordinating to metals, NO usually behaves as though it donates three electrons. The result is formally the same as if one electron were lost to the metal,

$$NO \rightarrow NO^+ + e^-$$

followed by coordination of NO^+, which is iso-electronic with CO and CN^-. Because NO^+ is the nitrosyls ion, the products containing nitric oxide and carbon monoxide are called *carbonyl nitrosyls*.

The following reactions are typical of those producing this type of compound:

$Co_2(CO)_8 + 2\ NO \rightarrow 2\ Co(CO)_3NO + 2\ CO$
$Fe_2(CO)_9 + 4\ NO \rightarrow 2\ Fe(CO)_2(NO)_2 + 5\ CO$
$[Mn(CO)_5]_2 + 2\ NO \rightarrow 2\ Mn(CO)_4NO + 2\ CO$

4. Disproportionation

A number of metal carbonyls undergo disproportionation reactions in the presence of other coordinatingligands. For example, in the presence of amines, $Fe(CO)_5$ reacts as follows:

$$2\ Fe(CO)_5 + 6\ Amine \rightarrow [Fe(Amine)_6]^{2+}[Fe(CO)_4]^{2-} + 6\ CO$$

This reaction takes place because of the ease of formation of the carbonylate ions and the favorable coordination of the Fe^{2+} produced. The reaction of $Co_2(CO)_8$ with NH_3 is similar.

$$Co_2(CO)_8 + 6\ NH_3 \rightarrow [Co(NH_3)_6][Co(CO)_4]_2$$

These disproportionation reactions are quite useful in the preparation of a variety of carbonylate complexes. For example, the $[Ni_2(CO)_6]^{2-}$ ion can be prepared by the reaction,

$$3\,Ni(CO)_4 + 3\,phen \rightarrow [Ni(phen)_3][Ni_2(CO)_6] + 6\,CO$$

5. Carbonylate Anions

Several carbonylate anions such as $Co(CO)_4^-$, $Mn(CO)_5^-$, $V(CO)_6^-$, and $Fe(CO)_4^{2-}$ obey the EAN rule. One type of synthesis of these ions is that of reacting the metal carbonyl with a reagent that loses electrons readily, a strong reducing agent. Active metals are strong reducing agents, so the reactions of metal carbonyls with alkali metals should produce carbonylate ions. The reaction of $Co_2(CO)_8$ with Na carried out in liquid ammonia at -75°C is one such reaction.

$$Co_2(CO)_8 + 2\,Na \rightarrow 2\,Na[Co(CO)_4]$$

$$Mn_2(CO)_{10} + 2\,Li \xrightarrow{THF} 2\,Li[Mn(CO)_5]$$

Although $Co(CO)_4$ and $Mn(CO)_5$ do not obey the 18-electron rule, the anions $Co(CO)_4^-$ and $Mn(CO)_5^-$ do.

A second type of reaction leading to the formation of carbonylate anions is the reaction of metal carbonyls with strong bases. For example,

$$Fe(CO)_5 + 3\,NaOH \rightarrow Na[HFe(CO)_4] + Na_2CO_3 + H_2O$$

$$Cr(CO)_6 + 3\,KOH \rightarrow K[HCr(CO)_5] + K_2CO_3 + H_2O$$

With $Fe_2(CO)_9$, the reaction is

$$Fe_2(CO)_9 + 4\,OH^- \rightarrow Fe_2(CO)_8^{2-} + CO_3^{2-} + 2\,H_2O$$

6. Carbonyl Hydrides

Some reactions of carbonyl hydrides will be illustrated in Chapter 22. Such species are involved in catalytic processes in which metal carbonyls function as hydrogenation catalysts. Generally, carbonyl hydrides are obtained by acidifying solutions containing the corresponding carbonylate anion or by the reactions of metal carbonyls with hydrogen. The following reactions illustrate these processes:

$$Co(CO)_4^- + H^+(aq) \rightarrow HCo(CO)_4$$

$$[Mn(CO)_5]_2 + H_2 \rightarrow 2\,HMn(CO)_5$$

3.12 EXERCISE QUESTIONS

TYPE 1 QUESTIONS

1. Explain briefly the thermal stability of tetramethyllead and tetramethyltin compounds.
2. Explain briefly the alkyl group transfer reaction method of preparation of organometallic compounds with suitable examples.
3. Explain the metal-halogen exchange method of preparation of organolithium compounds.
4. Explain oxidative-additive reaction method of preparation of organolithium compounds.
5. What is meant by Transmetallation? Explain with an example.
6. The reactivity of the organometallic compounds of group I metals in organic solvents decreases down the group. Justify.
7. Explain lithiation reaction with an example.
8. Give the polymeric structure of $(BeMe_2)_n$.
9. What is Grignard reagent? Give few examples.
10. Explain schlenk equilibrium with all inter-convertable species.
11. Write a very short note on Grignard reaction.
12. Explain carbon-carbon coupling reactions with Grignard reagents.
13. Give the general method of preparation of organo-aluminium compounds with suitable reaction.
14. Write the structure of $Al_2(CH_3)_6$. Mention bond lengths and bond angles on the structure.
15. The general reactivity pattern and Lewis acid strength of main group III trialkyls vary in the sequence;
    $BR_3 < AlR_3 < GaR_3 > InR_3 > TlR_3$. Justify.
16. Discuss the structure and bonding in methyl lithium.
17. Explain the applications of organo-lithium compounds as nucleophile in the following cases;
    i. Carbolithiation reactions
    ii. SN^2 reactions
18. Explain a method of preparation of organoberyllium compounds
19. Explain a preparative method for organomagnesium compounds with reactions.
20. Explain the reaction mechanism involved in Grignard reaction.
21. Explain the use of Grignard reagents in the industrial production of Tamoxifen.
22. Discuss briefly the bonding and structure involved in triethylaluminium.

23. Explain the application of organoborane in hydroboration-oxidation reaction.
24. Discuss briefly the application of triethylaluminium compound in the polymeristaion of ethylene with the mechanism.

TYPE 2 QUESTIONS

1. Explain very briefly 18 electron rule with suitable examples. Give electron count for the following complexes and comment on their stabilities based on the concept of 18 electron rule:

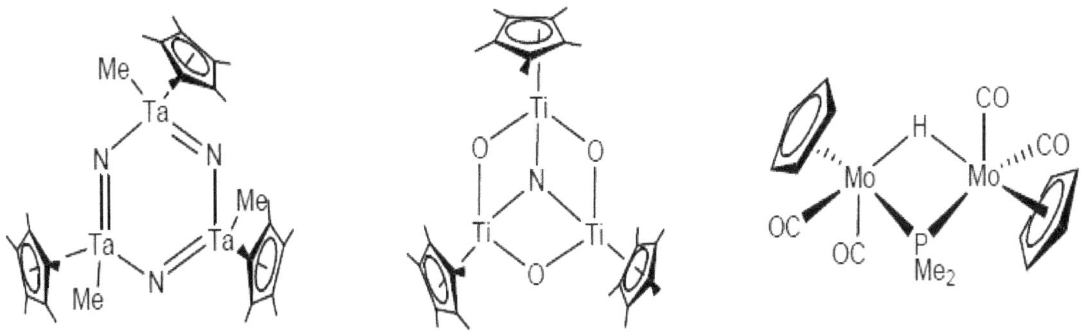

2. One μ_2- bridging ligand is missing in each of the given dimeric rhodium cyclopentadienyl complexes and all of them obey the 18 electron rule. Find out and insert the most suitable from the list of ligands provided.[Cl^-, H^-, CO, CH_2^{2-}]

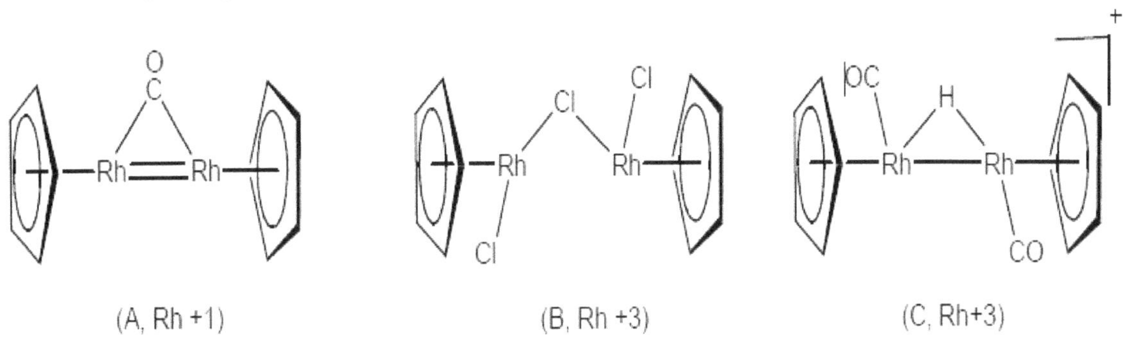

3. Although the 17 electron species V(CO)6 has not been found to dimerize to give V2(CO)12, the latter has been found to form along with V(CO)6 and remain stable at extremely low temperatures when a V/CO mixture in the ratio 1:102 was condensed into a pure CO matrix at 6-12 K. Infrared spectral analysis of V2(CO)12 showed three bands at 2014, 2050 and 1850 cm-1. Given that this dimer obeys the 18 electron rule and vanadium has a coordination number of eight, propose a structure for the same.
4. Give electron count for the following complexes:

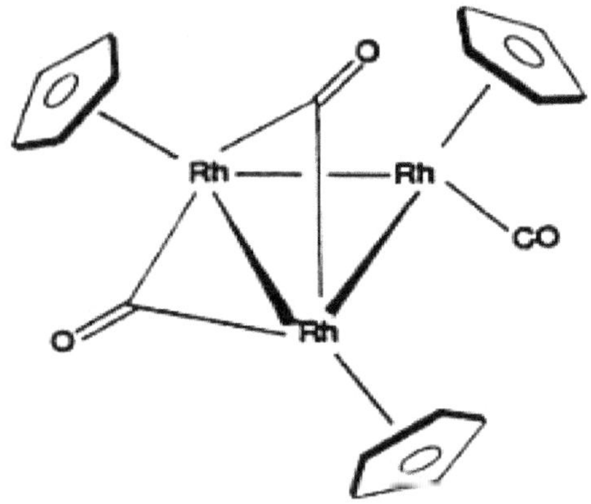

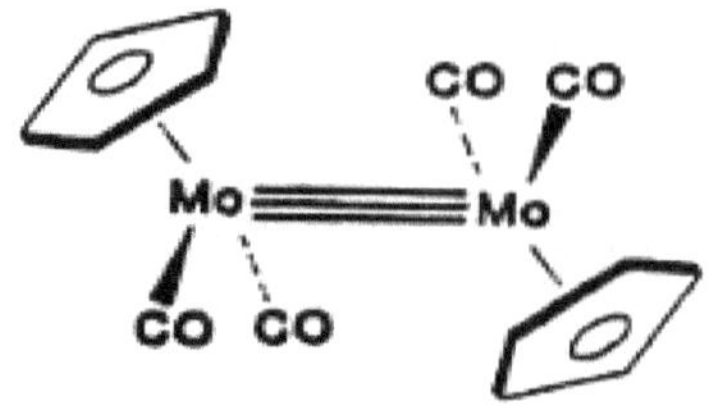

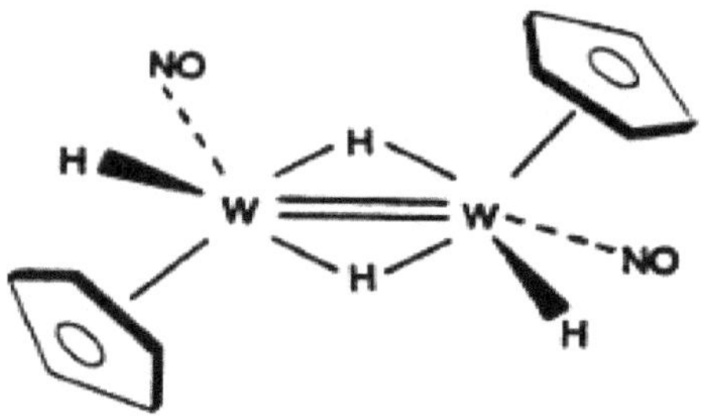

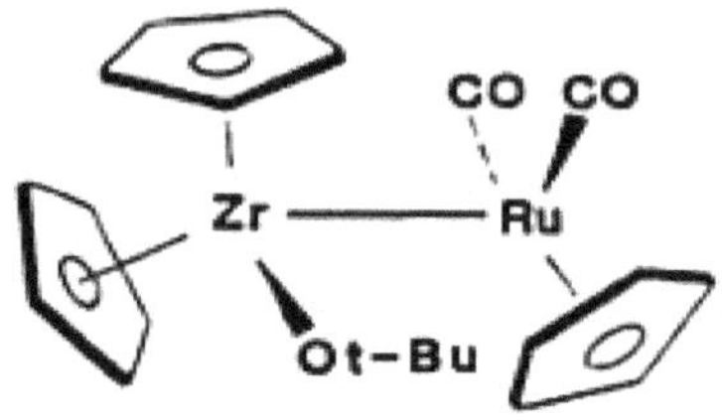

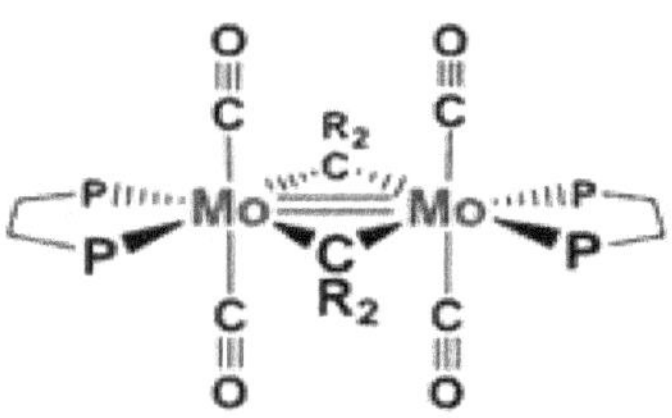

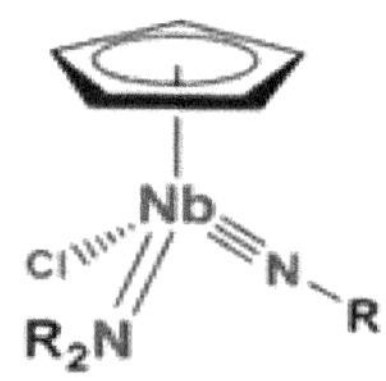

5. Explain migratory insertion reaction with a suitable organometallic compound.
6. Write a short note on Gilman reagent and Lipshutz reagents.
7. Give the reactions of ferrocene with the following reagents;
    i. H_2SO_4
    ii. LiC_4H_9
8. Explain the applications of organozinc compounds in the following reactions:
    i. Reformatsky reaction
    ii. Simmons-Smith reaction.
9. Predict the number of unpaired electrons in the following:
    i. Cp_2V
    ii. Cp_2Cr
    iii. $[Cp_2Ti]^+$
    iv. $[Cp_2Cr]^+$

10. Write the structures of the following:
    i) $(\eta^6\text{-}C_6H_6)_2\ (\eta^1\text{-}C_5H_5)_2\ Cr$ ii) $(CO)_3\ Mn(\eta^6\text{-}B_3N_3(CH_3)_6)$
11. For the following pairs of complexes, which one will have the highest average CO infrared stretching frequency? Briefly explain your reasoning.
    i. $Ni(CO)_4$ or $[V(CO)_6]^-$
    ii. $Cr(CO)_3(PMe_3)_3$ or $MnCl(CO)_5$
12. For the following pairs of complexes, which one will have the weakest M-CO bonding? Briefly explain your reasoning.
    i. $Tc(CH_3)(CO)_5$ or $Re(F)(CO)_5$
    ii. $HCo(CO)_4$ or $HMn(CO)_4(PMe_3)$

CURRICULUM VITAE OF THE AUTHOR

Dr. H. C. Ananda Murthy M.Sc., M.Phil., Ph.D
Lecturer, Department of Chemistry,
School of Physical Sciences,
College of Natural and Mathematical Sciences,
The University of Dodoma, P.O. Box. 259,
Dodoma, Tanzania.

1. Personal Information

Name	:Dr. H.C.Ananda Murthy
Date of Birth	: 06-07-1973 **Age**: 43 years Gender: Male
E-Mail Id&Contact Nos.	: anandkps350@gmail.com. anandkps7@yahoo.com & +255 768 406 138
Nationality	: Indian
Marital Status	: Married Number of Children: 2

2A. Academic Information		**(Completed studies at Government Universities in India)**		
Qualification	**Discipline / Subject**	**School / College / University**	**Year of passing**	**Class with Percentage**
B.Sc.	PCM	Bangalore University	1995	**First Class 60.16 %**
M.Sc.	Chemistry (Inorganic Chemistry)	Bangalore University	1997	**First Class 66.60 %**
M.Phil.	Chemistry (Composite materials)	Bharathidasan University, Tirchy.	2006	**First Class 76.40 %**
Ph.D.	Chemistry (Electrochemistry of composite materials) Physical Chemistry	Visvesvaraya Technological University, Belgaum, Karnataka, India.	2012	

2B.Title of the Ph.D. Degree

A study on the effect of ceramic reinforcement on the corrosion behavior of Aluminium metal matrix composites.

2C. Major Courses Studied

Sl. No.	Degree	Courses Learned
1.	B.Sc.	Physics, Chemistry & Mathematics.
2.	M.Sc.	Coordination Chemistry, Bio-Inorganic Chemistry, Organometallic Chemistry & Catalysis, Molecular Spectroscopy, Solid State Chemistry, Industrial and Nuclear Chemistry, Analytical Chemistry.
3.	M.Phil.	Spectroscopy, Inorganic chemistry, Research Methodology, Corrosion Theory.
4.	Ph.D.	Corrosion Science , Material Science, Instrumental Methods of Analysis, Research Methodology.

3A. Teaching Experience: Total - 19 Years

Period	Institution	Designation	Total Years of teaching
Before Ph.D (1997- 2012)	Vivesvaraya Technological University, India.	Lecturer (9 years) Senior Lecturer (3 years) Asst. Prof. & HOD (3 years)	15 years
After Ph.D (2012- 2016)	Vivesvaraya Technological University, India. The University of Dodoma, Tanzania.	Asst. Prof. & HOD (1 year) Lecturer (3 Years)	4 years

3B. Major Courses Taught for U G Program (Taught for 19 Years)

1. Coordination and Organometallic Chemistry
2. Transition Elements Chemistry
3. Organometallic Chemistry
4. Electrochemistry and Corrosion Protection
5. General Chemistry
6. Engineering Chemistry

3C. Major Courses Taught for P G Program (Taught for 1 year)

1. Advanced Coordination Chemistry
2. Surface Chemistry and Corrosion Science
3. Advanced Spectroscopy
4. Advanced Environmental Chemistry
5. Theoretical and Applied Electrochemistry

4A. Scientific Research Publications: Total Number of Publications = 17

National and International Journal Papers = 13, Conference Papers = 4.

Sl. No	Title of the Published Papers	Cooperated authors	Journal (Volume, Issue, date)
1.	Adsorption of Mercury (II) from Aqueous Solution Using Gum Acacia-Silica Nanocomposite: Kinetics, Equilibrium and Thermodynamics Studies	Somit Kumar Singh Vandana Singh	Advanced Materials letters VBRI Press, Sweden – 2016. (Accepted for publication in March 2016)
2.	Electroanalytical study on the corrosion behaviour of TiO_2 particulate reinforced Al 6061 composites	Single author	Material Science Research India, Vol. 12(2), 112-126 (2015). doi.org/10.13005/msri/12205
3.	Influence of TiC particulate reinforcement on the corrosion behaviour of Al 6061 Metal matrix composites	Somit K Singh	Advanced Materials letters VBRI Press, Sweden – 2015, 6(7), 633-640.doi: 10.5185/amlett.2015.5654

4.	Effect of TiN particulate reinforcement on corrosion behaviour of Aluminium 6061 composites in chloride medium	V Bheemaraju C Shiva kumara	Bulletin of Material Science – Vol.36, No. 6, Nov 2013, pp. 1057. Indian Academy of Sciences.(Springer)
5.	Electrochemical investigation of corrosion inhibition of AA6063alloy in 1M hydrochloric acid using Schiff base compounds	V Bheemaraju P F Sanaulla H B Lokesh	IOSR Journal of Applied Chemistry (IOSR-JAC) ISSN: 2278-5736. Volume 2, Issue 5 (Nov. – Dec. 2012), PP 37-47
6.	Electrochemical Behaviour of AA6063 Alloy in Hydrochloric Acid using Schiff Base Compounds as Corrosion Inhibitors	V Bheemaraju P F Sanaulla H B Lokesh	International Journal of Engineering Research and Applications (IJERA) ISSN: 2248, Vol. 2, Issue 5, 2012.
7.	Corrosion inhibition of AA 6061 and AA6063 alloys in hydrochloric acid media by Schiff base compounds	V Bheemaraju P F Sanaulla	Journal of Chilean Chemical Society, 57, N° 4 (2012)
8.	Corrosion Characterization of TiO_2 Particulate Reinforced Al-6063 Composites in Chloride and Nitrate Media	V Bheemaraju P F Sanaulla	Asian Journal of Chemistry, ISSN 0970-7077, 2011, Volume 23, Issue 4, 1664 – 1668.
9.	Influence of chlorides, nitrate and sulphate media on corrosion behaviour of TiO_2 particulate reinforced Al-6061 composites	V Bheemaraju P F Sanaulla	PortugaliaeElectrochimicaActa, ISSN 1647-1571, *2010, 28(5), 309-320*
10.	Corrosion behaviour of TiO_2 particulate reinforced Al-6063 composites in sodium sulphate medium	V Bheemaraju P F Sanaulla	International Journal of Applied Chemistry, ISSN 0973-1792, Vol. 6, 2, (2010) p. 225

11.	A Study on the effect of TiO_2 particulate on the corrosion behaviour of Al-6063 based composites	V Bheemaraju H B Lokesh	Material Science, An Indian Journal, MSAIJ, 4(4), June 2008 [328-334]
12.	Estimation of copper in brass by iodo-potentiometric technique of analysis	O G Palanna	*Analytical Chemistry*, An Indian Journal, ACAIJ, 7(9), August 2008 [694-702]
13.	Estimation of copper in copper salts at different pH conditions and buffer media by iodo-potentiometric technique of analysis	O G Palanna	*Analytical Chemistry*, An Indian Journal, ACAIJ, 7(9), July 2008 [684-693]

4B. Paper Presentations in National / International Conferences

14 **Ananda Murthy H C, Kusuma, ShobhaBhaskara, BheemaRaju V**

Corrosion behaviour of 6061 Al and its TiB_2 particulate reinforced composites in chloride medium, **International Conference on Electrochemical Science and Technology(ICONEST-2014**), August 7-9, ECSI, Bengaluru, India.

15 **H.C. Ananda Murthy,** Suba Rameshand V. BheemaRaju

Evaluation of corrosion behaviour of TiO_2 particulate reinforced Al-6061 composites by polarisation techniques**, National Conference on Recent Advances in Chemical and Environmental Sciences (NCRACES- 2011)** held on 28 & 29th of Dec 2011.

16 **H.C. Ananda Murthy**, O.G.Palanna, T. Jeevananda, Subaramesh, RamanathPrabhu, N. Suma

A Novel iodo-Potentiometric method of estimation of percentage of MnO_2 in Pyrolusite, **National Conference on Recent Trends in Materials Chemistry and Engineering (RTMCE-2011)**, September 29-30, 2011, Department of Chemistry, RNS Institute of Technology, Bangalore 560098.

17 **H.C. Ananda Murthy** and V. BheemaRaju,
Corrosion behaviour of TiO_2 particulate reinforced Al-6061 in chloride, nitrate, sulphate and acidic media, International conference on Recent Advances in Industrial Electrochemical Science and Technology (ICRAIEST-2009), November 2009, Department of Chemistry, Mangalore University, Mangalagangothri, Mangalore.

5. Seminars and Training Programmes

Seminars / Course	Duration	Organisers
Workshop on "Research Methodologies and Report Writing	2 days	Visvesvaraya Technological University, Mysore on May 28th and 29th 2009
TEQUIP sponsored Short Term Training Programme	5 days	BMS college of Engineering And Dr. A I T, Bangalore from 26th 2006.
Workshop on "Industrial Corrosion Awareness & Prevention"	1 day	Mangalore University, Mangalagangothri, Mangalore on 25th February 2006.
AICTE-ISTE Sponsored short tern training programme	5 days	from 23rd to 27th August 2004 at Dr. Ambedkar Institute of Technology, Bangalore.

6. Written Books, Book Chapter and Compendia

Title (Written)	Publisher	Year
1. Corrosion Science for University Students	Lulu Press, Inc. 3101 Hillsborough St, Raleigh, NC 27607. ISBN: 978-1-329-97036-6	2016
2. Corrosion Behaviour of Reinforced Aluminium MMCs	LAMBERT Academic Publishing, Germany. ISBN: 978-3-659-56260-0	2014
3. Electrochemistry Chemistry and Corrosion protection (CH 314) for 3rd Year B.Sc. course.	The university of Dodoma, Tanzania.	2014
4. General Chemistry (CH 111) for 1st Year B.Sc. course.	The university of Dodoma, Tanzania.	2014
5. Coordination chemistry and Organo-metallic chemistry (CH 221) for 2nd Year B.Sc. course.	The university of Dodoma, Tanzania.	2014
6. Organometallic chemistry (CH 310) for 3rd year B.Sc. course.	The university of Dodoma, Tanzania.	2014
7. Engineering Chemistry	Wiley-India ISBN: 978-81-265-1988-0	2011

7. Details of Scholarship / Awards / Honours

1. Prathibha puraskar-2012 by Devanga Employees Association, Bangalore.
2. NagarikaSanman- 2012 by Dr AIT employees welfare association, Bangalore.

8. Passport Details

Passport Number: K8778065

Date of Issue: 17/09/2012 Place of Issue: BANGALORE

Date of Expiry: 16/09/2022 Passport Validity: 10 years.

9. DETAILS OF EXTRA-CURRICULAR ACTIVITIES

<u>The Innovation</u>: JAL-BHOOMI -THE AMPHIBIOUS CYCLE

I have successfully completed a project titled "Amphibious bicycle" sanctioned by National Innovation Foundation of India (NIF) with Three students of 7th semester Electronics and communications Engineering Branch in the year 2008 in RNS Institute of Technology. This Amphibious bicycle was displayed in VahanYatra held at Karnataka Trade Promotion Centre, Near ITPL, Bangalore in December 2008.

The amphibious bicycle comprises of a conventional bicycle modified with two extra attachments that enables it to run on both water and land.

I certify that the information given above is true, complete and correct to the best of my knowledge and belief.

Place: Dodoma, Tanzania.

REFERENCES

1. Housecraft, C.E. and Sharpe, A.G. Inorganic Chemistry, Prentice Hall, Second Edition 2004.
2. Douglas, B.E., Mcdaniels, D.H. and Alexander, J.J. Concepts and Models in Inorganic Chemistry, Wiley 1994.
3. Purcell, K.F. and Kotz, J.C., Inorganic Chemistry, Prentice Hall 1980.
4. Elschenbroich, C. and Salzer, A. Organometallics, A Concise Introduction, Wiley VCH Verlag, 2nd edition 1992.
5. House, J. Inorganic Chemistry, Elsevier Science and Technology 2008.
6. Elschenbroich, C. and Salzer, A. Organometallics, A Concise Introduction, Wiley VCH Verlag, 2nd edition 1992.
7. M. Gerloch, E.G. Constable, Transition Metal Chemistry, VCH Verlagsgesellschaft mbH, D-69451 Weinheim (Federal Republic of Germany), 1994.
8. S.F.A.Kettle, Physical Inorganic Chemistry, Spectrum, 1996.
9. R Gopalan and V. Ramalingam, Concise coordination chemistry, Vikas publishing house Pvt Ltd, 2014.
10. Electronic (Absorption) Spectra of 3d Transition Metal Complexe S. Lakshmi Reddy, Tamio Endo and G. Siva Reddy
11. ChandraShekara Reddy, Advanced Coordination Chemistry, Campus Books International, India, 2015.
12. P.S.Sindhu, Elements of Molecular Spectroscopy, New Age International Publishers, India, 2010.
13. R.C. Mehrotra, Organometallic chemistry, New Age International publishers, 2015.
14. Robert H Crabtree, The organometallic chemistry of transition elements, Wiley-Interscience, 2005.
15. Perez, N. Electrochemistry and Corrosion Science. Kluwer Academic Publishers, 2004.
16. Mars G Fontana, Corrosion Engineering Tata McGraw Hill, 2005
17. Engineering Chemistry, John Wiley International, 2010.
18. Koryta, J. Principles of Electrochemistry, 2nd ed. J. Wiley & Sons, 1993.

19. Plieth, W. Electrochemistry for Material Science. Elsevier B.V. 2008.

20. Comninellis, C. and Chen, G. Electrochemistry for the Environment. Springer, 2010.

21. H C Ananda Murthy, Corrosion Behaviour of Reinforced Al MMCs, Lambert Academic Publishing, Germany, 2014.

22. H C Ananda Murthy, Corrosion Science, Lulu Press Inc, USA, 2016.

23. Vitousek, P. M., Mooney, H. A., Lubchenco, J., and Melilli, J. M. Human domination of Earth's ecosystems. *Science*, **277**, July, 1997, 494--99.)

24. Various internet sources.

www.ingramcontent.com/pod-product-compliance
Lightning Source LLC
LaVergne TN
LVHW060034210926
848481LV00013B/3864

9789987070541